Coastal Research Library

Volume 12

Series editor
Charles W. Finkl
Department of Geosciences
Florida Atlantic University
Boca Raton, FL 33431
USA

The aim of this book series is to disseminate information to the coastal research community. The Series covers all aspects of coastal research including but not limited to relevant aspects of geological sciences, biology (incl. ecology and coastal marine ecosystems), geomorphology (physical geography), climate, littoral oceanography, coastal hydraulics, environmental (resource) management, engineering, and remote sensing. Policy, coastal law, and relevant issues such as conflict resolution and risk management would also be covered by the Series. The scope of the Series is broad and with a unique cross-disciplinary nature. The Series would tend to focus on topics that are of current interest and which carry some import as opposed to traditional titles that are esoteric and non-controversial. Monographs as well as contributed volumes are welcomed.

More information about this series at http://www.springer.com/series/8795

Giovanni Randazzo • Derek W.T. Jackson
J. Andrew G. Cooper
Editors

Sand and Gravel Spits

Springer

Editors
Giovanni Randazzo
Dipartimento di Fisica e di Scienze della Terra
University of Messina
Messina, Italy

Derek W.T. Jackson
Environmental Sciences Research Institute
University of Ulster
Coleraine, UK

J. Andrew G. Cooper
Environmental Sciences Research Institute
University of Ulster
Coleraine, UK

ISSN 2211-0577 ISSN 2211-0585 (electronic)
Coastal Research Library
ISBN 978-3-319-13715-5 ISBN 978-3-319-13716-2 (eBook)
DOI 10.1007/978-3-319-13716-2

Library of Congress Control Number: 2015936450

Springer Cham Heidelberg New York Dordrecht London

Printed on acid-free paper

Springer International Publishing AG Switzerland is part of Springer Science+Business Media (www.springer.com)

Contents

Introduction

Among the most dynamic of coastal landforms, spits have been studied for many years. Early studies described their physical form with little attention to their formational processes. Among the first geomorphological studies of spits were those of Douglas Johnston (1919) in the USA. Subsequently Evans (1942) presented a landmark paper on evolutionary aspects of spit development, particularly focused on recurve development. Zenkovich (1959) developed an important model of cuspate spit development based on his work on the Caspian Sea. Schwartz (1972), motivated by the intensification of coastal occupation for industrial, residential and recreational purposes, edited a volume on spits and bars, involving a collection of previously published papers that set out the then state of the art. Since then many studies of individual spits have been undertaken and understanding of the formative processes has been much improved by advances in instrumentation to measure wave and current dynamics, higher frequency of aerial observation and topographic survey and application of new technologies such as ground-penetrating radar. New approaches to conceptual modelling of spit evolution (e.g. Ashton et al. 2001) have shown how small irregularities can create the nucleus for spit formation and, importantly, that such irregularities grow to create previously enigmatic rhythmic shoreline features (including spits) over long timescales.

Spits typically form from longshore transport of material from an up-drift source and accumulation in or across a coastal re-entrant. In several instances, however, downdrift sources are invoked as sediment sources for spit elongation as inlets migrate (e.g. Aubrey and Gaines 1982). Once initiated, spits are responsive to variations in the rate of sediment supply and dispersal, both of which may be strongly non-uniform. Their morphological behaviour is further modulated by the morphology of the re-entrant in which they are accumulating. The shape of the spit at any given time is a major determinant of subsequent morphological change.

Spits form from a variety of materials of varying grain sizes from fine sand to boulders. Gravel-sized grades, often reflecting glacial diamict sources, dominate in higher latitudes while coarse carbonate material (coral rubble) builds spits in the tropics under infrequent high energy events. Sand-sized material is, however, the most common component of spits globally.

Once formed, spits evolve in response to wave energy and sediment supply. Sections may develop planform equilibrium in response to wave energy distribution. When starved of sediment, cannibalisation of existing sediment may cause narrowing and even breaching of the spit. Some of the most important work in this regard was presented in conceptual models by Carter and Orford (1993) working on gravel spits and barriers where slow relaxation times enabled spit evolution to be observed more readily.

With their strong reliance on longshore sediment inputs, spits are highly susceptible to disruption by human interference. Sea defences (seawalls, breakwaters and groins) placed on eroding source areas, interruption of longshore drift by groynes, or interference in inlet bypassing system by jetty construction pose serious threats to the integrity of spits.

Changing climate and sea levels are also likely to prompt future changes in spit morphology. Spits may form and elongate or break down entirely in periods of sea level rise depending on local factors. Under stable sea level phase, they may become sediment-starved as sediment sources are depleted. Under such conditions they may evolve toward swash-aligned equilibrium forms that involve cannibalization of existing sediment and major changes in plan and profile.

Spits tend to be best developed in areas dominated by sea waves (short-period waves that approach the coast at a high angle). As such they are common in relatively sheltered sea environments (North Sea, Mediterranean, Caribbean, Caspian) and less common in swell-dominated areas, where waves are more fully refracted when they reach the shoreline.

This volume contains a series of contemporary studies of spits from around the world ranging from high latitudes of southern Argentina (Bujalevsky et al.) to the tropical Caribbean coast of Columbia (Anfuso et al.) and West Africa (Anthony). The studies reported here involve a mix of sand and gravel spits. Stephan et al. and Regnauld et al. focus on gravel spits of Atlantic France, while Burningham and Bujalevsky et al. describe gravel spits in England and Argentina, respectively. Most studies are concerned with a single spit, although Sabatier and Anthony and Furmańczyk and Musielak present regional appraisals of spit morphology and behaviour on the Rhone delta (Mediterranean Sea) and Polish coast (Baltic Sea), respectively.

Many of the chapters are focused on the historical scale of spits, deduced from cartography and aerial photographs (e.g. Anfuso et al.; Crisà et al.; Taaouati et al.), while some authors have utilised historical sources (Burningham) and archaeological data (Crisà et al.; Del Rio et al.). The role of extreme events, fairweather coastal processes and longer-term geomorphological evolution are reported in studies using ground-penetrating radar (Tillmann, Buynevich et al.). Larson et al. provide a state-of-the-art review of contemporary engineering models of spit evolution.

Linkages between spits and adjacent sedimentary environments are stressed in several of the papers. Kelley et al. describe spit-tidal flat interactions in Maine, USA, noting an important supply from algal-rafted gravel. Buynevich et al. describe an important aeolian dune component on Baltic sea spits linked to historical climate variability while Costas et al. assess links between ebb delta and spit development in an integrated approach that links the topographic evolution of subaqueous and terrestrial landforms.

Spit evolution is often affected by anthropogenic interventions (Anfuso et al.; Del Rio et al.), many of which are associated with the management of river basins (Crisà et al.), especially dam construction (Anthony and Sabatier and Anthony). Others are influenced by progressive reduction in sediment supply under natural conditions, leading to spit disintegration (Devoy) or cyclic changes in morphology (Vespremeanu-Stroe and Preoteasa).

While some authors (Sabatier and Anthony; Regnauld et al.; Anfuso et al.) provide explicit suggestions for future spit management, all of the findings reported in this volume have implications for the future management of spits under near-future climate and sea-level change. The high degree of local influence on spit evolution points to a need for detailed local knowledge to be combined with generic knowledge of spit-forming and spit evolutionary processes when making management decisions.

Environmental Sciences Research Institute
University of Ulster
Coleraine, UK
J. Andrew G. Cooper
Derek W.T. Jackson

Dipartimento di Fisica e di Scienze della Terra
University of Messina
Messina, Italy
Giovanni Randazzo

References

Ashton A, Murray AB, Arnoult O (2001) Formation of coastline features by large-scale instabilities induced by high-angle waves. Nature 414:296–300
Aubrey DG, Gaines AG (1982) Rapid formation and degradation of barrier spits in areas with low rates of littoral drift. Mar Geol 49:257–277
Carter RWG, Orford JD (1993) The morphodynamics of coarse clastic beaches and barriers: a short- and long-term perspective. J Coast Res Spec Issue 15:158–197
Evans OF (1942) The origin of spits, bars, and related structures. J Geol 50:846–865.
Johnston DW (1919) Shore processes and shoreline development. John Wiley & Sons, Incorporated, New York, p 584
Schwartz ML (1972) Spits and bars, Benchmark papers in geology. Dowden, Hutchinson & Ross Inc., Stroudsburg
Zenkovich VP (1959) On the genesis of cuspate spits along lagoon shores. J Geol 67:269–277

Chapter 1
Evolution of Sandspits Along the Caribbean Coast of Colombia: Natural and Human Influences

G. Anfuso, Nelson Rangel-Buitrago, and Iván Darío Correa Arango

Abstract This work deals with the evolution of Bocas de Ceniza-Puerto Caimán, Galerazamba, Isla Cascajo and Punta Canoas coastal sectors, located along the 120 km-long coastline between the Magdalena River mouth and Cartagena de Indias, on the Caribbean coast of Colombia. Comparisons of coastline morphology from reliable ancient charts, modern bathymetric surveys and remote sensing data, show major changes (in some cases at kilometre-scale) related to the rapid erosion and formation of offshore sandy shoals, spits and beaches. These sediment bodies are linked to sediment supply from the Magdalena River. In 1935, after the emplacement of two jetties at the river mouth, sediment was channelled offshore and erosion ensued on the western part of the Magdalena delta. Spits and sandy shoals rapidly migrated down drift. As a result, a spit at Puerto Colombia – present on the 1935 and 1947 aerial photograms – progressively diminished and merged with the coastline between 1953 and 1959. South of this location, a new spit formed before 2000 and it presently shelters a marina at Puerto Velero.

At Galerazamba, a 5 km-long spit at a high angle to the coastline, was present until 1864. It was replaced by a new spit, broadly parallel to shoreline, at some point before 1947. This feature was much smaller than the one previously observed and it has migrated down drift until the present. High rates of accretion were also observed around Isla Cascajo, a rocky island that caused the development of a rapidly growing tombolo. At Punta Canoas, a spit formed between 1947 and 1961 and, then migrated southwards.

G. Anfuso (✉)
Departamento de Ciencias de la Tierra, Facultad de Ciencias del Mar y Ambientales, Universidad de Cádiz, Polígono Río San Pedro s/n, 11510 Puerto Real, Cádiz, Spain

Área de Ciencias del Mar, Universidad EAFIT, Carrera 49 N° 7 Sur – 50, Medellin, Colombia
e-mail: giorgio.anfuso@uca.es

N. Rangel-Buitrago
Facultad de Ciencias Básicas, Programa de Física, Universidad del Atlántico, Km 7 Antigua vía Puerto Colombia, Barranquilla, Atlántico, Colombia

I.D. Correa Arango
Área de Ciencias del Mar, Universidad EAFIT, Carrera 49 N° 7 Sur – 50, Medellin, Colombia

G. Randazzo et al. (eds.), *Sand and Gravel Spits*, Coastal Research Library 12,
DOI 10.1007/978-3-319-13716-2_1

1.1 Introduction

Colombia has coasts on the Pacific Ocean and the Caribbean Sea. They are populated by 4.3 million inhabitants, i.e. 11 % of the national population, distributed into 16 and 30 coastal municipalities along Pacific Ocean and Caribbean Sea coasts, respectively (DANE 2010).

The Caribbean littoral is a 1,600 km-long microtidal environment between Panama and Venezuela (Fig. 1.1). The Caribbean coastal margin of Colombia is a geologically complex region where Quaternary interactions among tectonic, tropical climate and oceanographic processes shaped a varied and unstable coastal zone characterized by spits, bars and beaches along the low coastal plains and deltas, alternating with terraced and cliffed coastlines in rocky areas (Correa and Morton 2010; Martinez et al. 2010).

Along the southern Caribbean coast of Colombia, offshore and onshore mud diapiric intrusions are evidenced by weakened rock zones, domes and mud volcanoes that deeply influence coastal geomorphology (Ramirez 1959, 1969; Correa

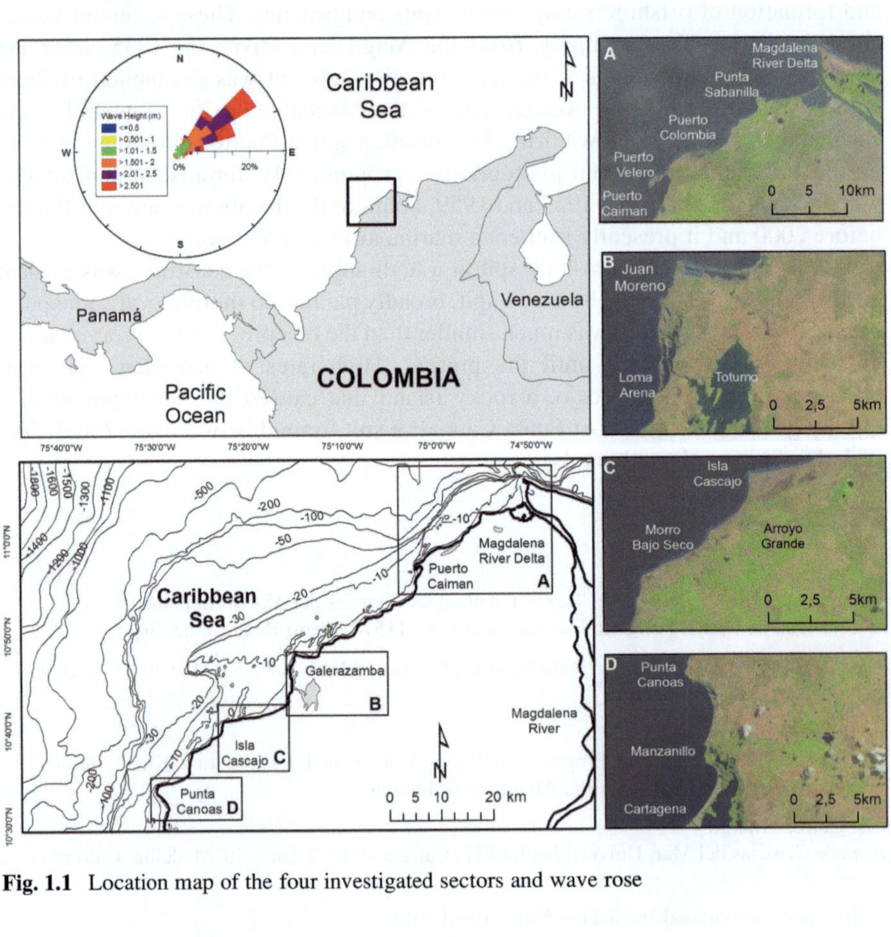

Fig. 1.1 Location map of the four investigated sectors and wave rose

et al. 2007). Presently, some of these mud volcanoes (at Galerazamba, Punta Canoas, Gulf of Morrosquillo and Arboletes) outcrop at or near the coastline and partially affect littoral drift.

Calcareous sediment is common on island beaches (San Andres Islands, San Bernardo and El Rosario archipelagos) and on a few beaches close to Santa Marta and Cartagena de Indias. These sediments are linked to sub-aerial and marine erosion of Plio-Pleistocene to recent coral reefs terraces and living reefs. The supply of coarse terrigenous sediments (sand and gravel) is linked to four large rivers (Atrato, Sinú, Magdalena and Rancheria) and numerous small distributaries draining the Andean region, and to the erosion of granular rocky sectors north of Cartagena de Indias.

Comparisons of coastline morphology from reliable historical charts and modern bathymetric surveys and remote sensing, show substantial, in some cases kilometre-scale changes in morphology of the Caribbean coastline. These changes are most evident in the rapid erosion and accumulation of offshore sandy shoals, spits and beaches (Martinez et al. 1990; Correa 1990). Most cliffed coastlines of this coast are currently in an erosional state depending in part on the presence or absence of frontal intertidal or subtidal depositional features (Correa 1990).

This paper is concerned with the geomorphological change on four of the most rapidly changing sectors of the Caribbean littoral of Colombia, e.g. Bocas de Ceniza-Puerto Caimán, Galerazamba, Isla Cascajo and Punta Canoas (Fig. 1.1).

1.2 Study Area

The four sedimentary systems are located in the Departments of Atlántico (Bocas de Ceniza-Puerto Caiman sector) and Bolivar (Galerazamba, Isla Cascajo and Punta Canoas sectors). They are developed along a 120 km-long coast extending from the Magdalena River mouth to Cartagena de Indias, on the Caribbean coast of Colombia (Fig. 1.1). This coast has two large commercial and touristic cities, e.g. Cartagena de Indias and Barranquilla (DANE 2010) and is the most developed area of the Colombian continental Caribbean.

The first significant human intervention in this zone was the construction in 1893 of a 1,200 m long pier at Puerto Colombia port which was the most important one on the Caribbean littoral of Colombia until the emplacement in 1935 of Bocas de Ceniza jetties to stabilize the main distributary channel of the Magdalena River delta – the entrance to the new port of Barranquilla (Raasveldt and Tomic 1957).

These were followed in the last three decades by the construction of many defense structures (groins and seawalls) and big developments including residential buildings, condominiums and tourist resorts for both large-scale and exclusive use. Most of these developments are presently (or will be in the near future) at risk from the prevailing erosion, so the evaluation of magnitude and causes of coastal change in this area is an essential issue for coastal management.

All the investigated sand bodies, three sandy spits and a tombolo, are composed of terrigenous sand supplied essentially by the Magdalena River which is the largest fluvial system in Colombia. It is 1,600 km long and drains a 257,438 km^2 basin, e.g. a great portion of the Colombian Andes, and has a sediment yield of 560 t km^{-2} year^{-1} (Restrepo and López 2008).

A secondary source of sediments for this stretch of coast is erosion of cliffs north of Cartagena de Indias (Correa 1990).

The Caribbean coast of Colombia is a tropical environment with maximum precipitation of 2,500 mm/year. Seasonal variations in precipitation usually show two rain periods (winter seasons, e.g. April–May and October–November) and two dry periods (summer seasons, e.g. December–March and July–September).

Tides are of the mixed semi-diurnal type, with maximum amplitudes of 60 cm (Andrade 2008; Correa and Morton 2010). Winds mean velocity is lower than 12 m/s, and higher values are associated with winds blowing from NE in the December–March period. Lower values are observed in September–November and are generally associated with winds from E. Surface sea currents are associated with the Caribbean current -flowing during almost all the year from East to West- and the Darien or Colombia current, which flows from Panama northeastward (Thomas 2006).

Significant wave height generally fluctuates between 1 and 2 m, while the average peak period is about 7 s (see wave rose in Fig. 1.1). Most of the year (November to May), the Caribbean wave system is dominated by the presence of swell waves from the NE. For the reminder of the year ower and less frequent waves from NW, WSW and even SW occur. According to INVEMAR (2006) and Restrepo et al. (2012) seasonal variations in wave approach directions are associated with changes in significant wave height values, with the lowest records occurring each year between August and October ($\leq$1.5 m), and the most energetic conditions occurring from November to July, with significant wave heights sometimes exceeding 2 m. Consequently, associated net longshore sand drift has a dominant south-westward component, minor reversal occurring during the rainy periods when southerly winds become dominant in some sectors and set up short, high-frequency waves able to cause significant erosion along cliffed mud coastlines (Correa and Morton 2010).

Erosive events are also related to the impact of hurricanes and cold fronts (Ortiz 2012; Ortiz et al. 2013). Hurricanes, usually originating in the Caribbean area from June to November may affect the Caribbean coast of Colombia with strong winds, heavy rains and storm waves. Cold fronts, occurring during January, February and March, cause strong swell waves which impact may be increased by trade winds blowing from ENE. They usually hit the coast for about 48 h and have an average occurrence of six events per year (Ortiz et al. 2013).

1.3 Methodology

Satellite images and aerial photogrammetric flights were used to reconstruct linear coastline evolution and morphological variations of four sand bodies along the Caribbean littoral of Colombia during different intervals from 1935 to 2013, i.e. the medium to long-term period (Crowell et al. 1993).

Following the methods described by Jiménez et al. (1997), Leatherman (1983), and Pajak and Leatherman (2002), aerial photos were scanned, geo-referenced, and computer rectified to eliminate scale and distortion problems (Chuvieco 2000; Lillesand and Kiefer 1987; Moore 2000). Ground Control Points (GCP_s) for photo registration were obtained from the geo-referenced 2013 satellite image and all information was presented in Projected Coordinate System UTM Zone 18. Taking into account the smooth topography of the study area, a polynomial transformation was applied in the registration process (Chuvieco 2000). The number of GCP_s used varied from one photograph to another (from 9 to 15 units) and their position was located in unequivocal places (Thieler and Danforth 1994).

The error due to document distortion (Moore 2000) was resolved and controlled in the geo-referenced documents by visual observation, achieved by comparing the registered photographs with the base map and deriving the root mean square error (RMSE).

Given that the investigated area is a microtidal environment, shoreline position was defined as the instantaneous water line position at the moment of the photo (Pajak and Leatherman 2002; Boak and Turner 2005).

Wave height effects were not considered because no storm conditions were observed in any of the photographs. In addition, effects of seasonal variation and influence of individual storms on shoreline evolution have limited importance because of considered time span and vast recorded morphological changes (Dolan et al. 1991).

In order to quantify spits evolution, it was used the ArcGis 9.3 extension Digital Shoreline Analysis System (DSAS), v. 3.2 USGS Woods Hole, Massachusetts (Thieler et al. 2005) which uses, as an input, a series of shoreline positions referenced to an arbitrary baseline. In this study, transects perpendicular to the baseline were generated at 100 m intervals and the DSAS allowed calculation of sand shoals surfaces.

Last, wave propagation was carried out using SMC (González et al. 2007) program in order to obtain prevalent current direction vectors along the investigated area.

1.4 Results and Discussion

1.4.1 Coastal Evolution

The evolution of the investigated littoral was strongly influenced and linked to Magdalena River delta evolution, river sedimentary supplies and, to a less extent, the influence of neotectonic processes (Hoover and Bebout 1984; Correa 1990).

During the Pleistocene-Holocene period the Magdalena River delta complex grew toward the WSW (Martínez et al. 1990) and great amounts of sediments moved through the narrow shelf and into a canyon feeding the Magdalena Fan (Shepard et al. 1967; Hoover and Bebout 1984).

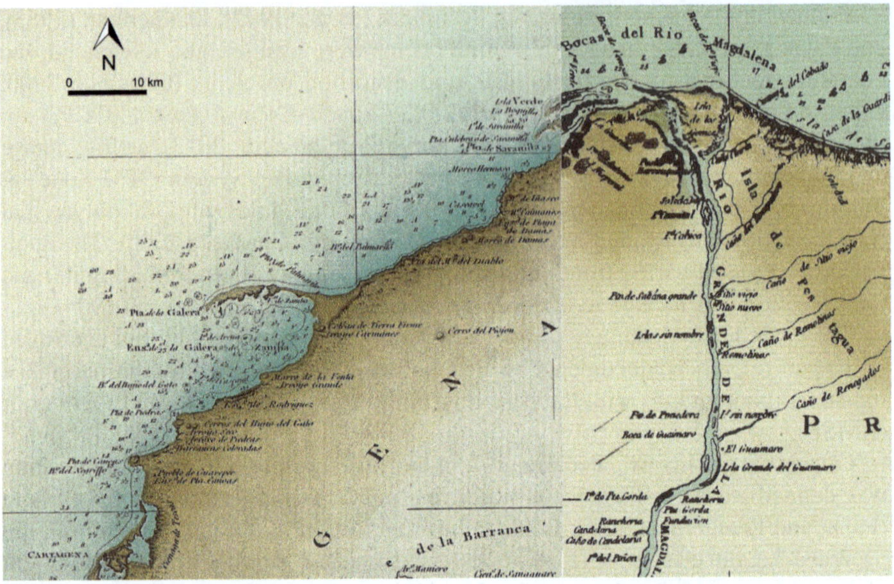

Fig. 1.2 Magdalena River delta reproduced in the historic map of Brigadier Fidalgo dated between 1792 and 1812. Notice the large spit observed at Galerazamba broadly normal to the shoreline

Around 1800, the delta front had a symmetrical form with the main river distributary channel discharging on the west side (Fig. 1.2).

Heavy erosion of the western delta arm started in 1935 when the construction of two 7 km-long jetties was completed. These structures were emplaced on the western side of the Magdalena River delta to stabilise and reduce infilling problems at river mouth entrance – the stabilised channel of the Magdalena River in fact constituted the main entrance to Barranquilla port. After the jetties were constructed it became the most important port on the Caribbean coast of Colombia.

As a result of jetties construction, the western part of the delta in successive decades recorded hundreds of meters of retreat and vast areas of mangrove swamps and coastal lagoons at Boca de Ceniza were eroded (Correa 1990; Matinez et al. 1990).

Probably, this retreat and reconfiguration of the offshore sand bodies is at least in part associated with temporal sediment deficits as a result of several submarine slides. The first was immediately after the construction of the jetties in August 1935, and involved sinking along 484 m of the western jetty and depth changes at the river mouth from 5 to 20–30 m (Koopmans 1971). Ten years later, in November 1945, another submarine slump destroyed 208 m of the western jetty and a new slump was again recorded in July 1963 (Koopmans 1971).

From the Holocene to present, sediments supplied by the river created large sandy shoals, spits and tombolos and constituted the primary sedimentary source for downdrift beaches. Such sedimentary features migrated from NE to SW as far as

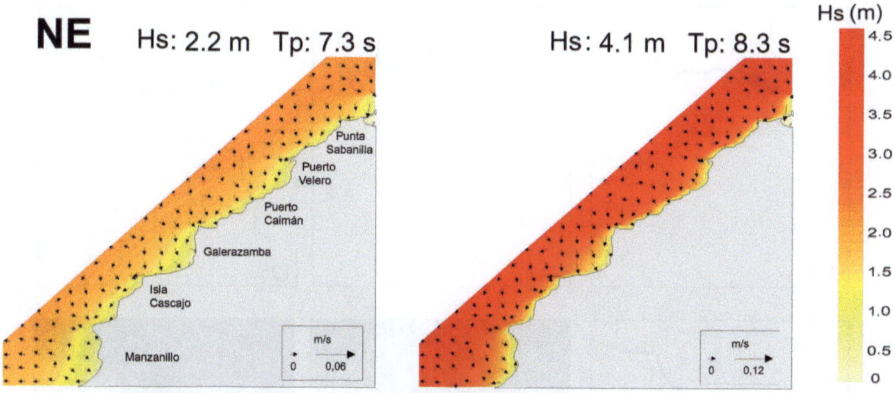

Fig. 1.3 Wave heights and vectors of associated currents for the investigated area

Cartagena de Indias bypassing several rocky headlands and various human and natural structures (Raasveldt and Tomic 1957; Correa 1990). The southward component of sandbody migration was related to diffracted (at jetties) and refracted swell wave fronts originally approaching from NE and ENE which give rise to alongshore drift clearly SW directed (Fig. 1.3). The westerly component was instead linked at places to specific bathymetric conditions (Figs. 1.1 and 1.3) and perhaps to shorter sea waves related to west to east winds with velocities >1 m/s and to eastern sea surface current (e.g. a component of the Caribbean current, Thomas 2006). The specific evolution and behaviour of each of the investigated sectors will be separately discussed from North to South.

1.4.1.1 Bocas de Ceniza-Puerto Caimán

This coastal sector is located immediately on the western side of Magdalena River delta, hence sedimentary features were the largest and most dynamic (Fig. 1.4).

1935 aerial photographs recorded a well developed sand spit that was hundreds of meters wide and >10 km long, with a surface area of c. 4.4 km^2. The spit originated from the mainland south of Punta Sabanilla and enclosed a coastal lagoon (Fig. 1.4).

By **1947** the spit presented smaller dimensions (c. 3.3 km^2) but similar form and orientation (Fig. 1.4). Both the 1935 and 1947 features formed an angle of approximately 40° with the shoreline. This is extremely high since spits are usually almost parallel to shoreline (Carter 1988; Calliari 1994). Such singular orientation was attributed by Martinez et al. (1990) to wave refraction and shoaling processes linked to offshore uplifting of mud diapiric intrusions. Those authors stated that sand spits were temporarily and eventually broke and migrated, but the reason was unknown.

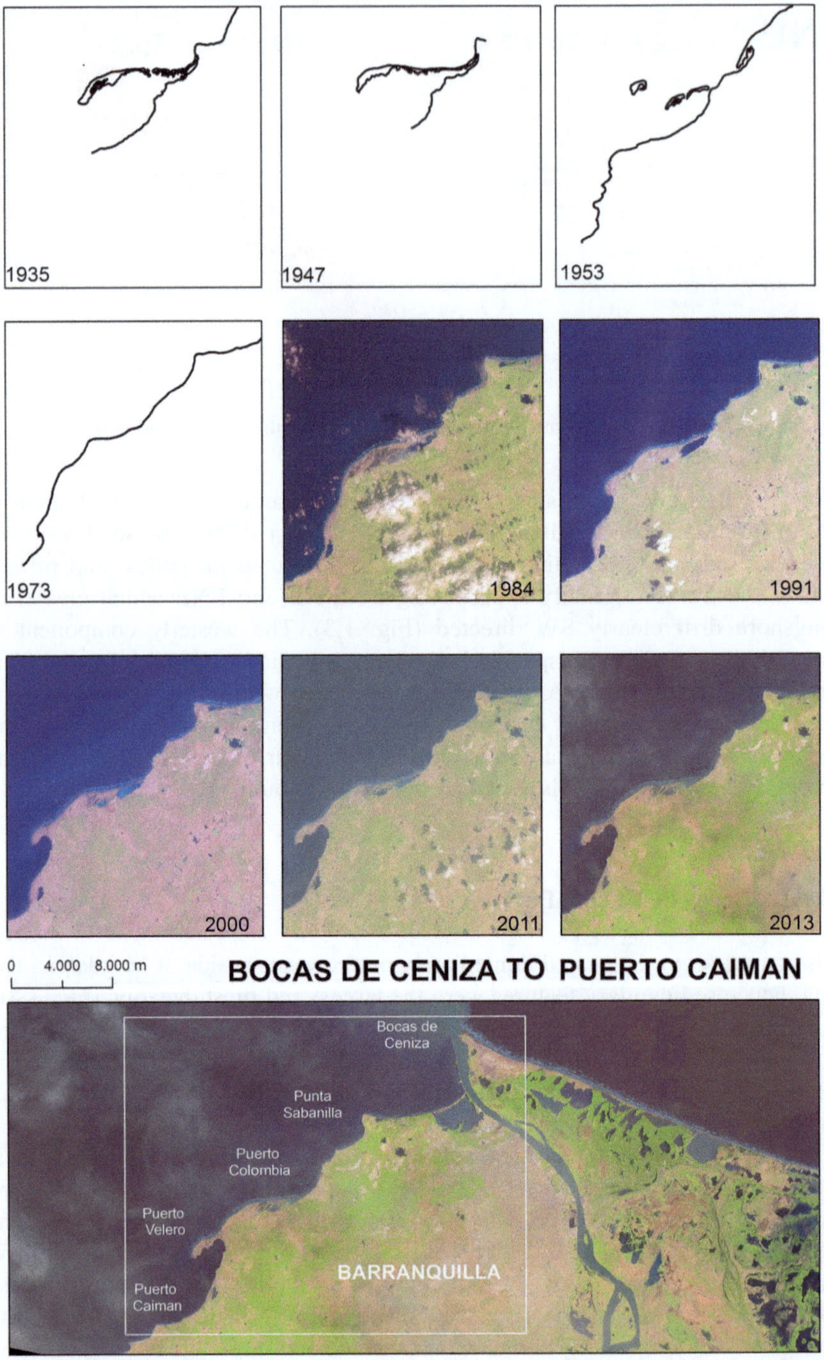

Fig. 1.4 Evolution of Bocas de Ceniza – Puerto Caimán sector between 1935 and 2013

The spit morphology, orientation and migration is related to wave propagation patterns, influenced by jetties at the Magdalena River mouth, and to the bathymetric characteristics of specific areas (Figs. 1.1 and 1.3).

Predominant wave fronts, approaching from NE, experience important refraction and diffraction at the river delta acquiring, south of it, an almost shore parallel orientation as also observed by Ortiz et al. (2014).

As a result of wave action, the spit formed in the 1930s moved to the SW without recording important morphological variations. Immediately south of the delta, the feature probably started to impinge in a shallow area, e.g. a wide zone between −10 and –20 m, and began to adapt its form to the bathymetric contours linked to the existence of a rocky seafloor (Correa 1990, see bathymetric map in Fig. 1.1). Migrating further to the SW, at Punta Sabanilla, the spit migration (and consequently its form) was again strongly influenced by bathymetric lines, specifically the 10 m contour, which runs broadly parallel to coastline until Punta Sabanilla, where it passes close to the headland, and than acquires a very different orientation, i.e. it elongates westward, assuming an oblique orientation with respect to the shoreline, similar to the one shown by the spit in the 1935 and 1947 aerial photographs (see bathymetric map and spits orientation in Figs. 1.1 and 1.4). Wave propagation presented in Fig. 1.3 evidenced as, at Punta Sabanilla, longshore currents present a clear westward component and their velocities strongly increase, this way confirming the importance of a westward directed drift at that point, responsible for spit oblique orientation.

In **1953** the spit presented a very different form compared to previously observations. This was because the sandy feature did not work as a spit sensu stricto, e.g. a drift-aligned feature with a fixed end receiving sediments from an updrift source and a free one migrating down drift (Carter 1988). After the jetties construction, the spit no longer received sediments and thus it turned into a swash-aligned sand shoal rapidly changing and freely migrating SW.

Specifically, in 1953, it appeared segmented into three parts. A sandy island was formed about 4 km offshore of Puerto Colombia. The main body of the spit, breached into two parts, was 4.5 km long and 1.1 km^2 smaller than the 1947 one. The fixed spit edge was located at Punta Pradomar, between P. Sabanilla and Puerto Colombia, and the free one about 2 km from the coast (Fig. 1.4).

The rapid migration observed during the 1947–1953 period (respect to previous time interval) and successive breaching between 1947 and 1953, were related by Martinez et al. (1990) to storm impacts but may be also linked to the impact of Hurricane Fox that in 1952, despite only having a category of Tropical depression, directly hit the coast at P. Colombia.

The successive detachment of the spit occurred between 1953 and **1959** probably because of the landward rolling of the spit associated with normal wave conditions and storm effects (Martinez et al. 1990) and/or the impact of Hurricane Katie that hit littoral – but not directly- in 1955 with the category of Tropical storm.

In **1973** the spit merged with large pre-existing sandy beaches forming a new feature of about 4.6 km^2 elongated south-westward. In the following decade the spit, still merged to coastline, continued its alongshore migration and experienced

Fig. 1.5 Increasing beach erosion at Puerto Colombia, October 2012 (**a**) and 2013 (**b**) Flat beach of fine sediments with umbrellas and chairs (**c**) and human constructions (**d**) at Puerto Velero. Long seawall (**e**) and groins constructed (**f**) south of Punta Sabanilla

0.49 and 0.47 km^2 of areal decrease in **1984** and **1991,** respectively and creating erosion problems at P. Colombia where some beaches totally disappeared (Fig. 1.5a, b). In 1991 the spit reached Puerto Velero headland and in **2000** formed a new small feature rapidly migrating SW as observed in **2011** and **2013** images. Presently the spit shelters a water body where in recent years a recreational marina, Puerto Velero, was constructed (Fig. 1.5c, d). The spit is very flat and low and is therefore strongly threatened by flooding events linked to storm waves. Further, the fixed edge, is very narrow and will likely experience breaching in the near future transforming the spit into a migrating sandy island with serious problems for human settlements developed there.

The small spit which formed the coastal lagoon in 1935 south of Punta Sabanilla persisted for several decades because it was protected by the headland and had a sand supply from updrift eroding coasts such as at Bocas de Ceniza (Correa 1990). In 2012 the spit was totally eroded exposing human settlements to potential erosion. This was counteracted by emplacement of a long seawall and numerous groins

(Fig. 1.5e, f). The groins trapped fine sediments and formed wide, smooth and low beaches rich of quartz, feldspar and mica and heavy minerals that gave a dark colour to the newly formed beaches.

1.4.1.2 Galerazamba Sector

Martinez et al. (1990) reported the existence from 1811 to 1864 of a 5 km-long spit broadly normal to the shoreline at Galerazamba (Fig. 1.2). Such an uncommon orientation was attributed by Martinez et al. (1990) and Correa (1990) to the presence of offshore mud volcanoes (Ramírez 1969) and associated wave shoaling processes that interrupted the drift patterns resulting in spit formation at an angle almost normal to the coastline.

Within this study, it was observed that this ancient spit probably developed in a bathymetric setting very similar to the one observed at Punta Sabanilla. Between P. Caimán and Galerazamba, the 10 m bathymetric contour runs very close to shore but at Galerazamba point it suddenly turns westward (see bathymetric map, Fig. 1.1). Such bathymetric settings have influenced spit formation and orientation blocking alongshore migration of a large sandy shoal that assumed an elongated form parallel to the 10 m bathymetric line. Such observations are confirmed by results of wave propagation: predominant wave fronts (approaching from NE) favour the formation of intensive and clearly westward oriented currents at Galerazamba point (Fig. 1.3).

A new spit, broadly parallel to the shoreline, appeared at some point before **1947** south of Punta Juan Moreno (Fig. 1.6). It is visible in recent aerial photographs and satellite images, although with an area of 1.8 km^2 and it was much smaller than the

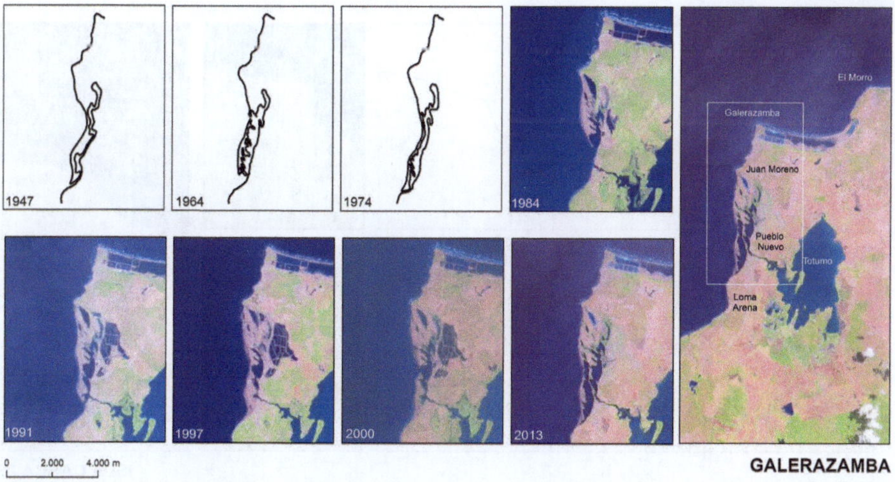

Fig. 1.6 Evolution of Galerazamba sector between 1947 and 2013

ancient one observed until 1864. The modern feature enclosed different small water bodies and migrated and changed its form in last decades (Fig. 1.6).

According to Correa (1990) the spit fulfilled an important protective function for the backing lagoon and human settlements (the small village of Amanzaguapos) until the beginnings of the 1950s. In the 1947–1954 period, erosion processes produced 300 m of coastal retreat (Correa 1990). Spit erosion and narrowing weakened spit protection function and erosion threatened human settlements that were relocated landward. The new village, named Pueblo Nuevo, was founded in 1954 northeast of the previous one.

The spit experienced narrowing in **1964** and **1974** and the south part was totally eroded in **1984**, with the loss of 0.45 km^2; as a consequence Pueblo Nuevo was exposed to erosion processes and the settlement was relocated 100 m eastward in 1985 (Correa 1990).

In following years sandy supplies from alongshore drift favoured spit accretion and southward migration of c. 500 m, with an associated growth of 1.0 km^2 (**1991**, Fig. 1.6).

Southward migration in the following years allowed the free edge to merge with the coastline (**1997**, Fig. 1.6). In successive steps it then disintegrated (**2,000** images) because of the loss of 0.7 km^2, increasing erosion processes at Pueblo Nuevo. In **2013** the spit accreted of c. 0.7 km^2 and migrated down drift about 80 m.

1.4.1.3 Isla Cascajo Sector

Isla Cascajo was a calcareous Plio-pleistocene reef rocky island about 2 km from the mainland as recorded in **1935** photograms (Fig. 1.7). Its distance from the shoreline gradually reduced from 1935 onward because of coastline accretion of

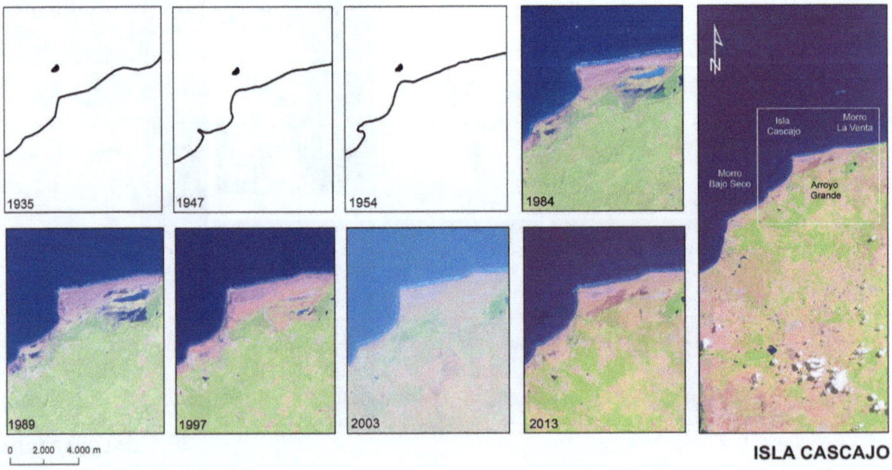

Fig. 1.7 Evolution of Isla Cascajo sector between 1935 and 2013

0.7 and 2.5 km^2 in **1947** and **1954** (Fig. 1.7). Such behaviour was linked to the formation and progressive accretion of a tombolo originated in the shadow area of the island (Fig. 1.7). Accretion processes determined the connection of the tombolo with the mainland in 1981 and an increase of 0.95 km^2 of beach surface in **1984** (Fig. 1.7).

Such accumulative processes continued in following decades and favoured seaward shoreline migration with an increase of 0.53 and 0.54 km^2 respectively in **1989** and **1997** (Fig. 1.7). Accumulation rates diminished in following years and produced an increase of beach surface of 0.24 and 0.03 km^2 in **2003** and **2013.** Further, as evidenced in 1954 photograms, a secondary spit enclosing small lagoons developed landward of a small rocky headland, south of Isla Cascajo. This spit migrated to the SW and merged with the shoreline (1984). Coastal lagoons protected by the tombolo and the spit progressively disappeared because they were artificially drained and naturally infilled by landward migrating dunes.

Martinez et al. (1990) affirmed that sediments accumulated in this coastal sector in recent decades derived from the breaking up of the Galerazamba large spit that was observed in old maps until 1864. It is a plausible hypothesis but hard to confirm. Indeed the lapse time between the dismantlement of ancient Galerazamba spit (1864) and the formation of the tombolo is quite large considering the dynamism of the study area. Alternatively, sediments deposited at Isla Cascajo might have been gradually eroded from updrift areas.

1.4.1.4 Punta Canoas

According to Correa (1990) and Martinez et al. (1990) the spit, not visible in **1947** photograms (Fig. 1.8), was formed at some time before 1961. In **1974** and **1984** photographs it presented a similar form and a length of c. 2 km. In the period between 1984 and **1989** the spit widened, increasing 0.1 km^2, and migrated down drift enclosing new small water bodies (Fig. 1.8). In **1991** a small amount of erosion affected the free (southern) edge of the feature that, in the following years, continued its southward migration until it joined the coastline in **1997** with an accretion of 0.49 km^2. Successively, the northern spit area recorded erosion (-0.59 km^2, in **2003**) and presently (**2013**) a new small spit is reforming (Fig. 1.8).

1.4.2 Morphodynamic Model

A conceptual model of spit and sandbody evolution was presented in Fig. 1.9. It deals with Puerto Colombia spit study case but can be applied to other investigated sectors. The model is broadly based on the works of Forbes et al. (1990, 1995a) y Orford et al. (1991) which described barrier evolution in paraglacial coasts. Such environments are characterised by gravel (and gravel/sand) barriers which size is typically of the order of 10^2–10^4 m in length – depending on the

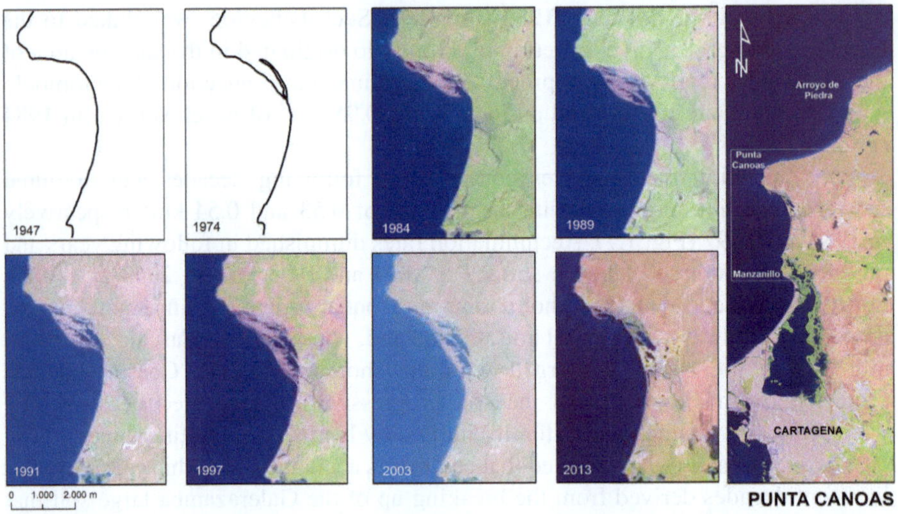

Fig. 1.8 Evolution of Punta Canoas sector between 1947 and 2013

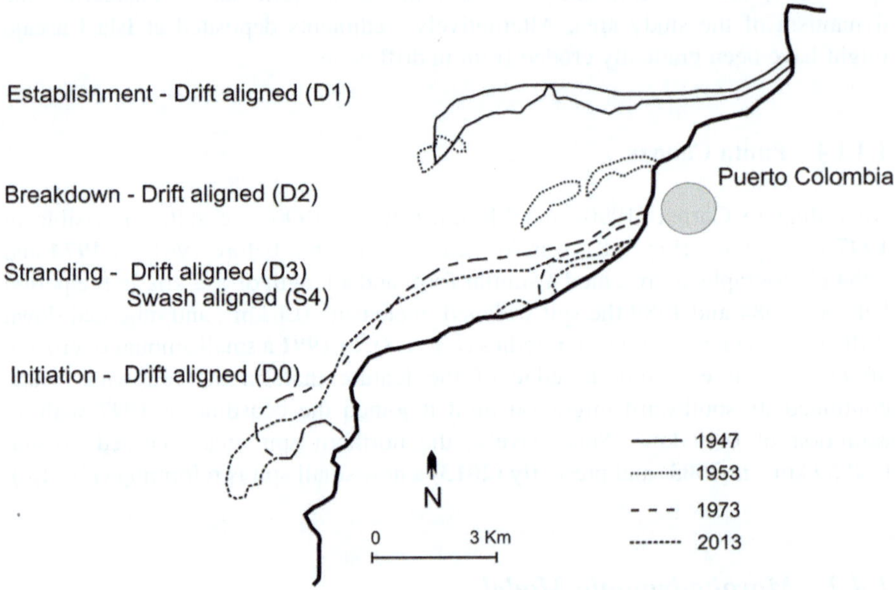

Fig. 1.9 Morphodynamic sketch of investigated sandbody behaviour

rock-irregularities distinctive of many paraglacial coasts, and their origin is linked to the erosion of glacial deposits. Systems formation and evolution take place over time scales of decades to centuries, the lifespan depending in large part on the volume and endurance of the sediment source (Forbes et al. 1995b). Specifically, according to Carter et al. (1989, 1990) and Forbes et al. (1995a), the long-term evolution of coastal systems in paraglacial environments is controlled by several external factors: (i) the geological and physiographic setting; (ii) relative sea-level changes resulting from long-term glacial-eustatic and isostatic adjustments; (iii) time-varying sediment supply, primarily from glacigenic or related deposits; (iv) the climate system, i.e. wave climate and secular changes in related energy inputs and also (v) tides and river discharges.

Sandbody systems analysed in this work presents similar dimensions to the features described by Forbes et al. (1995a) but essentially consisted of sand deposits which give rise to smooth dissipative environments strongly contrasting with the reflective beaches observed in gravel barriers (Carter et al. 1989, 1990). The formation of investigated sandbody and gravel barriers is in both cases linked to the erosion of large sedimentary deposits, respectively coastal sediments of fluvial origin (in this study case from the western arm of the Magdalena River delta) and gravel deposits of glacial origin – as an exempla Carter et al. (1990) in a study case from Nova Scotia, recorded a total sediment release in the order of 10^6–10^7 m^3. Formation of investigated sand bodies takes place at decadal scale so, in contrast to gravel barriers of glacial origin cases, sea level variations are not important in their formation and evolution, achieving major importance sediment supplies and wave climate factors.

Despite mentioned differences in origin and sedimentological characteristics, sandbody systems formation and migration follow the self-organisation, cannibalization and transformation processes close to the ones observed for gravel barriers of glacial origin. The investigated sand bodies (Fig. 1.9) followed the conceptual model presented by Forbes et al. (1990, 1995a) y Orford et al. (1991) which describes as a barrier is initiated, become established, and may eventually range through different stages (transition, stranding, or breakdown), depending on the external conditions (energy and mass inputs) and internal response (morphodynamic feedback) of the system. According to the model, barriers may be initiated from discrete or line sources, and give rise to drift- or swash-aligned embryos (D0 and S0, Forbes et al. 1995a), depending on the local topography (coastline, nearshore bathymetry, and backbarrier accommodation space). In the study case presented in Fig. 1.9 there is no record of the "initiation" stage and it is only possible to presume that the sandbody was originated at some point after jetties construction probably from several minor sand bodies located at river mouth that merged together and migrated downdrift (Fig. 1.2). Later, under conditions of large and increasing sediment supplies from the river delta dismantlement, the sandbody formed a drift-aligned structure ("establishment" stage, D1, Fig. 1.9). Following Ruz et al. (1992), such structures can extend very rapidly in shallow water and this was the case of the investigated spit which rapidly migrated downdrift. According to the model, as supply declines, drift-aligned barriers may experience severe sediment budget deficits, leading to alongshore cell accentuation, barrier breaching, self-adjustment, and breakup. The "breakdown" stage (D2, Fig. 1.9) took place in between 1947 and 1953, when the spit experienced fragmentation and started to migrate landward, i.e. started the transition from a drift-aligned to a swash aligned feature.

In the following stage, "stranding" (D3 and S4, Fig. 1.9), both drift-aligned and swash features are observed. Following to Forbes et al. (1995a), stranded barriers may evolve from swash- or drift-aligned barriers which have lost their seaward anchor or they may develop in place. In such kind of situations, interactions between basement control and longshore cell development can produce a complex pattern of morphodynamic variation alongshore. The volume of sediment remobilised and the degree of self-organization depend on the amount of sediment available and on the erosion and transport potential at a given coastal site. Deficient supplies of material may prevent the development of coherent structures and favour the breakdown of existing forms as well as exaggerated amounts of supplies my alter the self-organisation mechanism in the system because saturate the transport potential. This is often observed because cells develop in long spits in case of sediment starvation (Carter and Orford 1991). As a result, cell formation produces the subdivision of the shoreline into segments wherein the longshore transport potential successively accelerates and decelerates, leading to updrift erosion, downdrift deposition, and the growth of shoreline irregularities. The nearshore wave field is affected by such irregularities and, as a consequence, the wave approaching direction and surf dynamics change creating energy gradients associated with drift- to swash-alignment conditions that reinforce the cell growth. Such self-organization mechanism produces barrier fragmentation, cannibalization of the proximal sections and breakup in drift-aligned features where alongshore sediment exchange represents the main sedimentary process but may give rise to thinning and total feature failure in swash-aligned barriers which are dominated by cross-shore sediment movement (Carter et al. 1990; Forbes et al. 1990, 1995a; Orford and Antony 2011).

Last, at Puerto Colombia (Fig. 1.9), an "initiation" stage (D0) was observed because the drift-aligned (D3) barrier in the "stranding" stage formed an initial, small spit which is growing and migrating to the southwest. All described stages for Puerto Colombia sandbody evolution take also place at other investigated places.

1.5 Conclusions

The investigated coastal sectors show great variability in the form and behaviour of large sand bodies linked to important sediment supplies from Magdalena River, that are transported toward Cartagena de Indias, about 120 km downdrift.

Until the construction in 1935 of two jetties at Bocas de Ceniza (Magdalena River mouth), such sedimentary supplies formed sand spits and shoals and maintained wide beaches in down drift areas. Huge sand spits were observed at Galerazamba in the 1800s and at Puerto Colombia in 1935. Formation of these features – at a high angle with shoreline was determined by bathymetric contours patterns to be linked to structural control.

After the emplacement of jetties, fluvial sediment supply to the littoral system was channelled offshore by structures. Huge volumes of sand were presumably lost to deeper areas by the occurrence of at least, three submarine slumps related to delta front instability. Hence, sedimentary inputs to down drift areas were reduced to those associated with erosion of the western delta arm. Hundreds of meters of retreat were recorded in a few decades. Such supplies were not continuous in time and spits started to migrate down drift, e.g. toward the SW, working as "free" sand bodies. As a result, the spit at Puerto Colombia progressively diminished in size and merged with the coastline between 1953 and 1959. South of this location, a new spit formed before 2000 which now shelters a marina at Puerto Velero.

Sediment supply from coastal retreat at Magdalena River mouth also favoured the formation of a spit at Galerazamba, much smaller than the one that had existed until 1864, great accretion at Isla Cascajo, a rocky island that produced the development of a rapidly growing tombolo, and the formation of a spit at Punta Canoas. Sandbodies evolution recorded self-organisation, cannibalization and transformation processes close to the ones observed for gravel barriers of glacial origin by different authors.

Migration of fixed spit edges and the lack of natural replacement of eroded sediments produced great erosion at several places of important tourist interest, e.g. Punta Sabanilla, Puerto Colombia, etc. Such erosive processes were counteracted by the progressive emplacement of protective structures, essentially groins and seawalls. Structures partially interrupted alongshore transport and favoured the deposition of fine suspended sediments that generated – in few years – large sandy beaches, but created erosion in down drift areas.

Acknowledgments This work is a contribution to the Andalusia Research PAI group n° 328 and was carried out under the framework of the convention between the University of Cadiz (Spain) and the Department of Ciencias de la Tierra, Faculty of Science, EAFIT University of Medellin (Colombia).

References

Andrade CA (2008) Cambios Recientes del Nivel del Mar en Colombia. In: Restrepo JD (ed) Deltas de Colombia: morfodinámica y vulnerabilidad ante el Cambio Global. Fondo Editorial Universidad EAFIT, Medellín, pp 103–122

Boak E, Turner I (2005) Shoreline definition and detection: a review. J Coast Res 21(4):688–703

Calliari L (1994) Cross-shore and longshore sediment size distribution on Southern currituck Spit, North Carolina: implications for beach differentiation. J Coast Res 10(2):360–373

Carter RWG (1988) Coastal environments. Academic Press, London, 617 pp

Carter RWG, Orford JD (1991) The sedimentary organisation and behaviour of drift-aligned gravel barriers. Proc Coastal Sediments '91. (Seattle.) Am Sot Civ Eng, New York, pp 934–948

Carter RWG, Forbes DL, Jennings SC, Orford JD, Shaw J, Taylor RB (1989) Barrier and lagoon coast evolution under differing relative sea-level regimes: examples from Ireland and Nova Scotia. In: Ward LG, Ashley GM (eds) Physical processes and sedimentology of siliciclastic-dominated lagoonal systems. Mar Geol 88:221–242

Carter RWG, Orford JD, Forbes DL, Taylor RB (1990) Morphosedimentary development of drumlin-flank barriers with rapidly rising sea level, Story Head, Nova Scotia. Sediment Geol 69:117–138

Chuvieco E (2000) Fundamentos de Teledeteccio'n Espacial. Rialp Ed, Madrid, 567 pp
Correa ID (1990) Inventory of littoral erosion and accretion (1793–1990) between Los Morros and Galerazamba. Dept. of Bolivar, Colombia, pp 129–142, AGID Report n° 13
Correa ID, Morton RA (2010) Caribbean coast of Colombia. In: Bird EFC (ed) Encyclopaedia of the world's coastal landforms. Springer, Melbourne, pp 259–264
Correa ID, Acosta S, Bedoya G (2007) Análisis de las Causas y Monitoreo de la Erosión Litoral en el Departamento de Córdoba. In: CVS – Universidad EAFIT (ed) Convenio de transferencia horizontal de Tecnología No. 30. Corporación Autónoma de los Valles del Sinú y del San Jorge. Fondo Editorial Universidad EAFIT, Medellín, 128 pp
Crowell M, Leatherman SP, Buckley M (1993) Shore-line change rate analysis: long term versus short term data. Shore Beach 61(2):13–20
DANE (National Department of Statistics) (2010) General census of Colombia – Censo general de la Republica de Colombia (report 1). Santa Fe de Bogotá, Colombia. 501 pp
Dolan R, Fester MS, Holme SJ (1991) Temporal analysis of shoreline recession and accretion. J Coast Res 7(3):723–744
Forbes DL, Taylor RB, Shaw J, Carter RWG, Orford JD (1990) Development and stability of barrier beaches on the Atlantic coast of Nova Scotia. Proc. Canadian Coastal Conf. (1990, Kingston.) Natl Res Count Can, Ottawa, pp 83–98
Forbes DL, Orford JD, Carter RWG, Shaw J, Jennings SC (1995a) Morphodynamic evolution, self-organisation, and instability of coarse-elastic barriers on paraglacial coasts. Mar Geol 126:63–85
Forbes DL, Shaw J, Taylor RB (1995b) Differential preservation of coastal structures on paraglacial shelves: holocene deposits of southeastern Canada. Mar Geol 124:187–201
González M, Medina R, González-Ondina J, Osorio A, Méndez FJ, García E (2007) An integrated coastal modeling system for analyzing beach processes and beach restoration projects, SMC. Comput Geosci 33:916–931
Hoover JD, Bebout DG (1984) Submarine fan diversion by tectonic processes – Magdalena fun and slope, south Caribbean. Gulf coast section- SEPM foundation Research- Conference abstracts 5, p 43
INVEMAR (2006) Climatologie de la vitesse et la direction des vents pour la mer territoriale sous juridiction colombienne 8° a 19° N e 69° a 84° W. Atlas ERS 1 et 2 et Quickscat, Colombie. CNRS e UMRS 8591, Laboratoire de Géographie Physique e Programa Geociencias. Instituto de Investigaciones Marinas y Costeras, Santa Marta
Jiménez J, Sánchez-Arcilla A, Bou J, Ortiz M (1997) Analysing short-term shoreline changes along the Ebro delta (Spain) using aerial photographs. J Coast Res 13(4):1256–1266
Koopmans BN (1971) Interpretación de Fotografías Aéreas en Morfología Costera Relacionada con Proyectos de Ingeniería. Ministerio de Obras Públicas-Centro Interamericano de Fotointerpretación, Bogotá, Septiembre 1971. 23 pp
Leatherman S (1983) Shoreline mapping: a comparison of techniques. Shore Beach 51:28–33
Lillesand T, Kiefer R (1987) Remote sensing and image interpretation, 2nd edn. Wiley, New York
Martínez J, Pilkey O, Neal W (1990) Rapid formation of large coastal sand bodies after emplacement of Magdalena river jetties, Northern Colombia. Environ Geol Water Sci 16(3):187–194
Martínez J, Yokoyama Y, Gómez A, Delgado A, Matsuzaki H, Rendón E (2010) Late Holocene marine terraces of the Cartagena region, southern Caribbean: the product of neotectonism or a former high stand in sea-level? J S Am Earth Sci 29:214–224
Moore L (2000) Shoreline mapping techniques. J Coast Res 16(1):111–124
Orford JD, Anthony EJ (2011) Extreme events and the morphodynamics of gravel-dominated coastal barriers: strengthening uncertain ground. Mar Geol 290:41–45
Orford JD, Carter RWG, Forbes DL (1991) Gravel barrier migration and sea-level rise: some observations from Story Head, Nova Scotia. J Coast Res 7:477–488
Ortiz JC (2012) Exposure of the Colombian Caribbean coast, including San Andrés Island, to tropical storms and hurricanes, 1900–2010. Nat Hazards 61:815–827

Ortiz JA, Salced B, Otero LJ (2014) Investigating the collapse of the Puerto Colombia Pier (Colombian Caruibbean Coast) in March 2009: methodology for the reconstruction of extreme events and the evaluation of their impact on the coastal infrastructure. J Coast Res 30(2):291–300. doi:10.2112/jcoastres-d-12-00062.1

Ortiz JC, Otero LJ, Restrepo JC, Ruiz J, Cadena M (2013) Cold fronts in the Colombian Caribbean Sea and their relationship to extreme wave events. Nat Hazards 13:2797–2804

Pajak MJ, Leatherman S (2002) The high water line as shoreline indicator. J Coast Res 18(2):329–337

Raasveldt HC, Tomic A (1957) Lagunas colombianas. Revista de la Academia Colombiana de Ciencias Exactas, Físicas y Naturales. Bogotá 10(40):175–198

Ramírez JE (1959) El Volcán submarino de Galerazamba. Revista de la Academia Colombiana de Ciencias Exactas, Físicas y Naturales. Bogotá 10(41):301–314

Ramírez JE (1969) Los Diapiros del Mar Caribe Colombiano. Memorias I Congreso Colombiano de Geología. Bogotá, 31–39

Restrepo JD, López SA (2008) Morphodynamics of the Pacific and Caribbean Deltas of Colombia. J S Am Earth Sci 25:1–21

Restrepo C, Otero L, Casas C, Henao A, Gutierrez J (2012) Shoreline changes between 1954 and 2007 in the marine protected area of the Rosario Island Archipielago (Caribbean of Colombia). Ocean Coast Manag 69:133–142

Ruz M, Htquette A, Hill PR (1992) A model of coastal evolution in a transgressed thermokarst topography, Canadian Beaufort Sea. Mar Geol 106:251–278

Shepard FP, Dill RF, Heezen BC (1967) Diapiric intrusions in forest slope sediments off Magdalena Delta, Colombia. Am Assoc Pet Geol Bull 52:2197–2207

Thieler E, Danforth W (1994) Historical shoreline mapping: improving techniques and reducing positioning errors. J Coast Res 10(3):549–563

Thieler ER, Himmelstoss EA, Zichichi JL, Miller TL (2005) Digital Shoreline Analysis System (DSAS) version 3.0: an ArcGIS extension for calculating shoreline change. U.S. Geological Survey Open-File Report 2005, 1304 pp

Thomas YF (2006) Climatología marina, presión atmosférica, viento y olas para las aguas territoriales bajo la jurisdicción colombiana, 8°-19° N y 69°-84° W. Datos TOPEX-POSEIDON, CNRS, Francia, 69 p

Chapter 2
Patterns of Sand Spit Development and Their Management Implications on Deltaic, Drift-Aligned Coasts: The Cases of the Senegal and Volta River Delta Spits, West Africa

Edward J. Anthony

Abstract The sand spits associated with the Volta and Senegal River deltas, the two largest river deltas in West Africa, after that of the Niger River, show complex patterns of morphodynamic development while also strongly reflecting the recent impacts of human activities. The large spit of the Volta delta seems to be a direct outgrowth of a natural change in the location of the mouth of the Volta and of a marked reduction in sand supply on the eastern coast of Ghana that largely predated the construction of the Akosombo dam, but which has been strongly aggravated since this dam was completed in 1961. Spit formation has led, in particular, to segmentation of the unique sand drift cell that prevailed on the Bight of Benin coast between the Volta delta mouth and the western confines of the Niger delta. These changes have been associated with strong gradients in longshore drift and fundamental modifications in the dynamics of sand barriers on the Bight of Benin coast and hitherto fed by sand supplied by the Volta River. The spit has prograded massively by the adjunction, in situ, of new individual beach ridges, rather than by undergoing elongation, a pattern of growth that has entailed sand sequestering within the confines of the delta. The distal tip of the spit has recently welded to the shoreline, creating a new barrier-lagoon system, and assuring the resumption of integral sand drift from the mouth of the Volta towards the rest of the hitherto sand-starved Bight of Benin coast. The process should also offer limited respite from erosion to the beleaguered town of Keta, located just downdrift of the former distal tip of the now welded spit.

The Langue de Barbarie spit at the mouth of the Senegal River delta is an outgrowth of strong longshore drift affecting one of the finest examples of a wave-dominated delta. The spit developed jointly with the delta and appears to have inexorably extended downdrift in conjunction with river-mouth diversion and

E.J. Anthony (✉)
Aix-Marseille Univ., Institut Universitaire de France, CEREGE UM 34,
Europôle de l'Arbois, 13545 Aix en Provence cedex 04, France
e-mail: anthony@cerege.fr

G. Randazzo et al. (eds.), *Sand and Gravel Spits*, Coastal Research Library 12,
DOI 10.1007/978-3-319-13716-2_2

migration southwards during the late Holocene, albeit with occasional breaches, the oldest of which are only preserved north of the city of St. Louis. The spit has had a downdrift migration range of about 30 km, beyond which alongshore sand drift and river discharge conditions led to stabilisation of the position of the river mouth. Numerous natural breaches have been identified on the Barbarie spit between 1800 and the present day, no doubt related to years of exceptionally high river discharge in an overall Sahel-influenced climatic context of rather irregular discharge. The recent evolution and potential future demise of the spit reflect the consequences of hasty and short-sighted artificial breaching to solve an impending flooding problem facing the historic city of St. Louis in 2003. An artificial breach through the spit created by engineers in just a few hours to counter a risk of flooding of St. Louis from exceptional river discharge in 2003 has had dramatic consequences on the integrity of the spit, notably by acting as a new river mouth that underwent rapid and significant widening. Part of the spit sand further downdrift of the new mouth is being recycled into river-mouth bars, while the rest of the eroded sand is transported downdrift by longshore currents, including beyond the sealed pre-2003 mouth. This dismantling of the spit has had dramatic consequences on recent settlements and tourist facilities and infrastructure. The erosion is probably a joint result of sequestering of part of the sand load transported by longshore currents in the new widened river mouth and a reinforced tidal prism through this new mouth. A new phase of spit dismantling and the formation of a second mouth a few kilometres downdrift of the new mouth occurred during high river discharge conditions in October 2012, thus illustrating the potential seasonal effect of high river outflow on spit reworking.

2.1 Introduction

Much of the coast of West Africa is characterised by essentially rectilinear wave-dominated sand barriers, with only a small proportion of muddy coast associated with open estuaries, relatively large tidal ranges and interspersed sandy and shelly cheniers between southern Senegal (south of the Gambia) and central Sierra Leone. The distribution of the long stretches of wave-dominated coast and the much more limited predominantly tidal estuarine sector is highlighted by continental shelf width in Fig. 2.1, the latter sector being subject to significant wave energy dissipation over a broad, low-gradient shelf (Anthony 2006). The two sandy sectors on either side of this muddy estuarine coast are under the influence of dominantly long and regular swell and/or shorter-fetch trade-wind waves. In conjunction with abundant fluvial sand supplies during the Late Pleistocene sea-level lowstand on the presently drowned inner shelf, this swell wave regime has resulted in the build-up of numerous sandy barrier systems. In Senegal, these barrier systems commonly consist of dune-decorated spits. Many of the barrier systems on the rest of the West African coast have been prograded ones and some are still progradational,

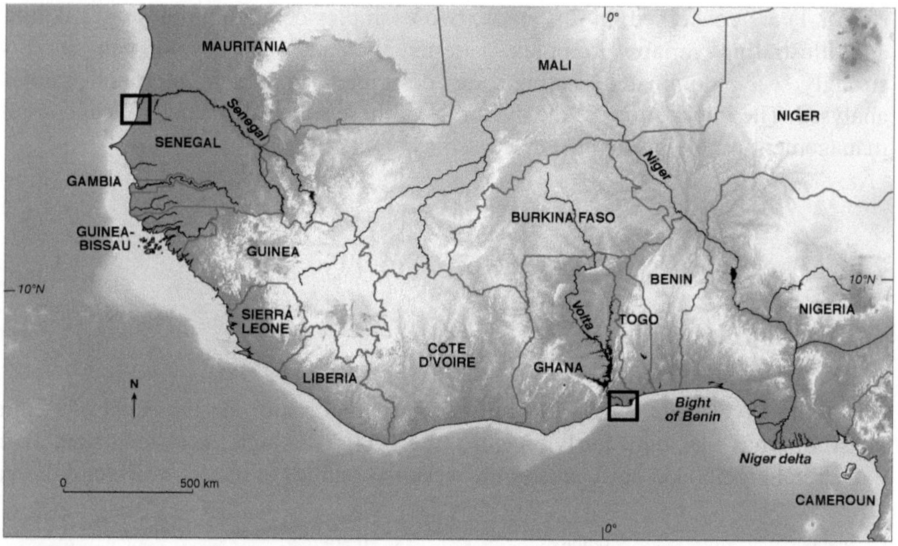

Fig. 2.1 The coast of West Africa. Continental shelf width (clearer hue along the coast) is a fine indicator of the distribution of long stretches of wave-dominated coast (narrow shelf) and the much more limited, predominantly tidal, estuarine sector between Sierra Leone and Guinea-Bissau (broad, low-gradient shelf), subject to significant wave energy dissipation. The two sandy sectors on either side of this muddy estuarine coast are under the influence of dominantly long and regular swell and shorter-fetch trade-wind waves. Abundant sand supplies and strong wave-induced longshore drift have favoured the construction of numerous sand spits, including at the mouths of the Niger, Volta and Senegal River deltas, the last two (*boxes*) of which are briefly described in this chapter

characterised by sequences of wave-formed beach ridges under dominantly 'drift-alignment' patterns (as defined by Davies 1980). Locally, 'swash-aligned' patterns have developed in embayed settings bounded by bedrock headlands, notably in Liberia. A hallmark of this constant wave regime is strong sustained longshore drift responsible for the commonality of spit formation. Some of the prograded barrier systems exhibit, in their distal sector (relative to drift direction), initial successive spit recurves reflecting mid-Holocene closures of a rather irregular coastline with a succession of bays and headlands inherited from the late Pleistocene sea-level lowstand. These spit recurves are succeeded by rectilinear successive beach-ridge extensions in the course of progradation under a high sand supply regime, although long stretches of the West African coast are also characterised by such rectilinear barrier systems with no evidence of spit recurves. This is especially so where a double barrier system exists, a common situation, comprising an inner barrier combining spits and rectilinear ridges, and an outer barrier of rectilinear ridges. In the vicinity of certain river mouths, however, the interaction between fluvial processes and longshore drift has generated more or less complex spit development. These spits are interesting in terms of manifestations of river-mouth morphodynamic patterns and from a coastal management perspective, especially where large deltas, such as those of the Volta and Senegal Rivers

(Fig. 2.1) are concerned. In this study, two examples concerning these two deltas, and illustrating variably complex patterns of sandy spit development in this strongly wave-dominated setting, are described and their morphodynamics analysed. The implications of the dynamics of these spit systems in terms of coastal management problems are also discussed.

2.2 The Volta River Delta Spit

2.2.1 Setting

The Volta River delta (Fig. 2.2) forms the proximal sector, on the Bight of Benin coast, of what was, until fairly recently (the 1960s), a single, very long sand drift cell stretching 500 km eastwards to the western confines of the Niger River delta in

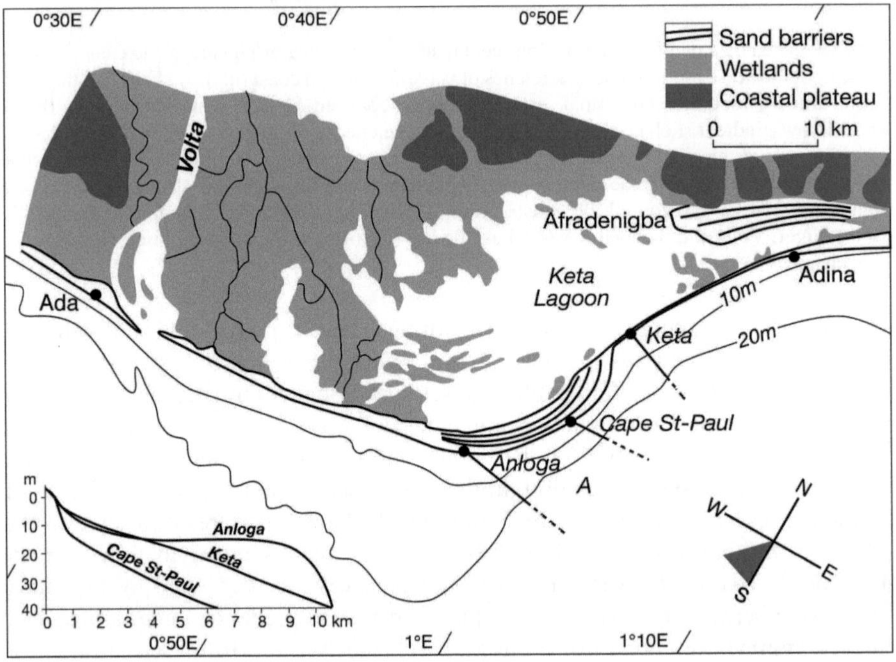

Fig. 2.2 Morphology of the Volta River delta and shoreface, showing the prograded river-mouth spit between Anloga and Keta. Massive spit growth appears to have occurred relatively recently, with regards to the post-mid-Holocene history of sand barriers in the Bight of Benin, in response to a westward shift in the river mouth and increased sequestering of sand by the Volta through enhanced wave refraction in the area south of Keta. This sand trapping has initiated a wave of important erosion of the narrower sand barrier linking the Volta delta to the Adina barrier system. Spit progradation has diminished considerably in response to a decreasing fluvial sand supply but, correlatively, the welding of the distal tip of the spit in the 1990s to the narrow Keta barrier has led to a reduction in erosion of the latter barrier, the southern part of which has been further consolidated by engineering works

Nigeria (Fig. 2.1). The Volta delta has developed in a large salient trap created by left-lateral offsetting on the coast of Ghana probably by the Romanche fracture zone (Kutu 2013), one of the many fracture zones that has offset the West African continental shelf and coast. The delta coast is bounded by a narrow shelf 15–33 km wide, and characterised by a fairly uniform, moderately steep shoreface with a gradient of between 1:120 and 1:150 up to −15 m, considered as the close-out depth for significant wave movements on this coast (Delft Hydraulics 1990; Rossi 1989), and beyond which the inner shelf levels out to a low-gradient (1:350–1:400) plain covered by relict transgressive sands. The wave setting is a cyclone- and storm-free West Coast Swell Environment as defined by Davies (1980), dominated by moderate to high energy ($H = 0.5$–4 m), long period ($T = 10$–15 s) southwesterly swells from the Atlantic. Waves break on the coast after refraction, with angles of 4–9°. Tides impinging on this coast are semi-diurnal, with a mean range of around 1 m. The mean spring tidal range is about 1.95 m. Overall, the constant wave regime of this coast, marked by long, moderate to high waves with low variation in directional approach, and a tendency for the prevalence of relatively steep, reflective beachfaces of medium to coarse sand, result in one of the highest rates of annual unidirectional longshore sand drift in the world, with values of up to 1 million m^3 (Blivi et al. 2002). These peak values are probably exceeded over short stretches of beach along the Volta delta. An important source of this sand has been the Volta River, one of the three largest river basins in West Africa, with the Niger and the Senegal. The Volta River drains a predominantly sandstone catchment that also includes a wide variety of lithologic terranes covering an area of 390,000 km^2, much of it located in the semi-arid Sahel zone of West Africa. The river's liquid discharge varied between a low of 1,000 m^3/s in the dry season and a high of over 6,000 m^3/s in the wet season before completion of the Akosombo Dam in 1961, only 60 km upstream from the sea. Discharge downstream of the dam has been strongly reduced by the decrease in rainfall over the Sahel since 1975 (Oguntunde et al. 2006). The Volta Delta covers an area of around 5,000 km^2. The sand load brought down annually by the river to its delta before dam construction has been estimated at about 1 million m^3 (Delft Hydraulics 1990). Much of this sand was injected into the longshore drift system via a single delta river mouth (Fig. 2.2). Sand supply from the shoreface has also been deemed to be important, especially in the early phases of barrier progradation as shoreface gradients in West Africa adjusted to sea level (Anthony 1995).

2.2.2 *Barrier Dynamics in the Bight of Benin: Prelude to Volta Delta-Mouth Spit Development*

The large spit at the mouth of the Volta delta is tentatively interpreted here, in the absence of absolute dates, to be a relatively recent feature resulting from adjustments between sediment supply from the river, delta dynamics and the strong longshore drift on this coast. The barrier systems that have developed in this sector of the Bight of Benin coast exhibit a relatively complex history, elements of which have been documented by Anthony and Blivi (1999). Much the bight coast exhibits a regressive single or double barrier. This barrier has been shown to be a 'hybrid' system in terms of internal facies composition and plan-view morphology, in that it has evolved from an essentially regressive to a stationary (synonymous with cessation of progradation) system. Once progradation of an inner barrier resulted in the regularisation of what was, hitherto, an indented shoreline, the succeeding phase of coastal development involved the emplacement of the more continuous outer barrier directly linked to the Volta river mouth, suggesting the establishment of a highly efficient drift alignment and transition to an economy of massive sediment sourcing by the Volta that predated the development of the Volta spit. Radiocarbon ages from parts of the barrier front show a phased pattern of progradation from the Volta delta to Benin. Ages of 5,000–6,000 years B.P. from Ada and Anloga near the sea (Fig. 2.2) in the proximal to central sectors of the Volta delta barrier (Streif 1983) suggest little net progradation of this part of the barrier over this period. The Volta delta-mouth barrier zone has thus essentially comprised a 'stationary' barrier sector of active through-drift of sand from the river up to Anloga, beyond which progradation of the distal Volta barrier continued, serving at the same time as a through-drift zone for the rest of the bight coast. This part of the barrier has not been dated. Old maps show that this delta-mouth barrier stretched continuously to Togo, with the zone between the old port of Keta and Adina acting as one of through-drift of sand between the Volta mouth and the barrier in Togo (Fig. 2.2). In spite of the massive sand supply from the Volta to the coast, the rest of the barrier front in neighbouring Togo shows no significant progradation since around 3,800 year B.P., and cessation of progradation occurred over 1,000 years later in Benin, further east (Anthony et al. 1996, 2002). These ages therefore suggest relatively rapid progradation over a fairly short time (5,000–3,000 year B.P.), especially in Togo, once longshore drift conditions became favourable to large-scale eastward advection of Volta sand. The ensuing phase of net long-term longshore stability in Togo and Benin probably stemmed from some sort of equilibrium among shoreline orientation, the nearshore profile and the hydrodynamic regime (Anthony 1995). The final sink for this sand is the eastern end of the Bight of Benin in Nigeria, near the western confines of the Niger delta.

2.2.3 *Inception, Development and Geomorphic Transformation of the Volta Spit*

The inception of a distinct spit along the eastern sector of the Volta delta, in lieu of a former distally shore-attached barrier-lagoon system, is interpreted here as the result of increasing influence of the prograding distal Volta barrier on wave refraction patterns. This has been further compounded by a recent reduction in direct fluvial supply that has induced a wave of erosion of the downdrift barrier and shoreface deposits to fulfill the strong drift requirements. Although the construction of the Akosombo Dam in 1961 and the drastic reduction in sand supply to the Bight of Benin coast it caused (Ly 1980) readily come to mind as the trigger factor of such erosion and barrier readjustments, the evidence pleads for a more complicated scenario. There is clear evidence that coastal erosion antedated dam construction. The barrier sector of Keta, the hinge point between the distal Volta barrier and the Togo barrier, was a once flourishing colonial port, probably located in a sector of stationary barrier that served as a transit zone for Volta delta sand to Togo. Since the mid-1880s, net erosion of this area is probably close to 1 km (Kumapley 1989). The most likely reason for this dramatic erosion is that sand supply from the Volta delta area has progressively become insufficient to compensate for strong drift supply to the rest of the Bight of Benin, probably because the distal delta barrier, between Anloga and Keta had been increasingly sequestering a significant proportion of the river's sand supply to the coast, culminating in the formation and rapid growth of the present Volta delta spit. This may have occurred in conjunction with a shift in the river mouth from the distal barrier sector at about the present area of Anloga, where the shoreface bathymetry appears to suggest the presence of a deltaic lobe, to the present proximal position near Ada (Fig. 2.2). The estimated amount of sand captured annually in this prograding distal part of the barrier has been estimated, for the period 1968–1996, at about 750,000 m^3 a year (Anthony and Blivi 1999). In particular, the necessity to satisfy the strong longshore drift budget towards Togo has resulted in considerable reworking of the barrier downdrift of this prograding distal segment, including the nearshore zone, threatening coastal settlements such as Keta. As erosion around Keta has proceeded, this severely eroded drift acceleration zone (Fig. 2.3) became characterised, by the 1990s, by a narrow (<100 m wide) eroding transgressive barrier subject to overwash and breaching during the summer months of strong swell (Fig. 2.3). A shoreline stabilisation project completed in 2004 (Keta Sea Defence Project) and comprising several groynes and a seawall (Fig. 2.3) have reduced erosion in this sector, which is still estimated at about 5.5 m/year in a recent study (Boateng 2012). Downdrift of this sector erosion is even stronger (Addo et al. 2011), whereas the Volta spit has continued growing, increasingly with a concave seaward plan-view shape due to accretion of successive beach ridges, but with restricted longshore growth of the distal tip. Beach-ridge accretion occurs over subaqueous prograding of lobes of coarse to medium Volta sand over a silty shoreface at a depth of around 10 m (Rossi 1989), and this explains the relatively steep shoreface-to-shelf transition of the prograding sector off Cape St. Paul (Fig. 2.2). Once the initial through-drift configuration in this zone was altered by the sediment budget imbalance evoked above, accretion of the Volta spit forced further wave refraction, amplifying cell segmentation, and resulting in the instauration of a swash-alignment associated with sequestering of sand by this new spit complex. These changes have also induced new patterns of barrier behaviour that include transgressive and regressive dynamics at various locations alongshore. Drift requirements have been satisfied by substantial barrier retreat and shoreface erosion east of the Volta barrier.

Fig. 2.3 Google earth images (*top right* and *bottom left*) and aerial views of the Volta spit hinge point with the shoreline, a particularly fragile drift-pulse sector that has been subject to important erosion in the past. *Top left* photograph taken in 2001 and *top right* 2013 Google extract show the narrow barrier and inlet system that connects the Volta delta spit to the Bight of Benin barrier system at Adina. The barrier is still subject to erosion, especially north of the sector shown here. *Bottom right* photograph (2001) and *bottom left* 2013 Google image show part of the town of Keta at the tip of the Volta spit and engineering works completed in 2004 (Keta Sea Defence Project) aimed at consolidating the spit hinge point and protecting the town. *Broad arrowheads* with dots show locations of photographs on the Google images (Photo credit: Great Lakes Dredge and Dock Company)

The situation has merely been aggravated since the 1960s by the construction of the Akosombo Dam. A recent feature of this spit growth is that the formerly highly eroded proximal zone of the drift acceleration zone, or drift 'pulse', is now being protected by this spit. Anthony and Blivi (1999) considered that this probably heralded a shift towards a less swash-aligned distal spit zone, and predicted eastward 'leakage' of some of the Volta sand, rather than quasi-total sequestering, thus alleviating erosion further downdrift. Recent images indeed show that the distal tip of the spit has now welded to the shoreline, thus resulting in the reconversion of spit to attached barrier, a process probably consolidated by the engineering works in this sector (Fig. 2.3).

2.3 The Senegal River Delta Spit

2.3.1 Context

The Senegal River is about 1,800 km long, the second longest river in West Africa. Reports regarding the area of the Senegal catchment range widely from 288,000 to 450,000 km^2, which is quite surprising, given the modern possibilities of determination of river basin coverage from satellite imagery! The Senegal delta (Fig. 2.4) is a classical text-book example of a wave-dominated delta, characterised by strong longshore drift generated by Atlantic waves from the northwest. Two clear manifestations of such wave domination are the absence of a notable classic deltaic

Fig. 2.4 The Senegal River delta, a fine example of a wave-dominated delta characterised by the Langue de Barbarie spit and river-mouth system subject to strong north-south longshore drift. This Google image shows a much widened river mouth in 2013 that developed very rapidly from an artificial breach created to prevent river flooding of St. Louis in October 2003. The former natural mouth much further south (sector not shown here) has been sealed by longshore drift

'bulge', and the presence of a persistent sand spit, the Langue de Barbarie, an extremely mobile feature, subject to repeated past breaches, associated with phases of delta-mouth migration over a total distance of about 30 km at least since the mid-seventeenth century. Of particular significance in terms of coastal management is the historic and picturesque city of St. Louis (population: 200,000), a UNESCO world heritage city at the proximal delta hinge point of the spit (Fig. 2.4), and protected from the ocean by the latter.

The lower Senegal delta is characterised by high biological productivity, rich agricultural and fishing sectors, a relatively high level of urbanisation and a recent strong bent towards tourism. St. Louis is frequently affected by flooding of the lower Senegal valley in the rainy season (May–October), as the discharge of the river rises dramatically, much of the city lying at an elevation of less than 2.5 m above sea level (Sall 2006). The mean annual liquid discharge of the Senegal River is 676 m^3/s. It varies from a low dry season value of 9 m^3/s in May, to an extreme flood value of 3,320 m^3/s in September. The wet season discharge is thus relatively moderate, given the size of the Senegal basin, and shows a slow rise, typical of large tropical river systems. Interannual variability is extremely high, however, with the mean annual discharge ranging from 250 to 1,400 m^3/s. The river drains a large area of the western Sahel, and its discharge has been particularly affected by droughts in the Sahel since the 1970s.

The Senegal delta coast is characterised by a relatively narrow continental shelf. The coast is dominated by waves that are predominantly from the northwest, and this direction is especially active during the dry season from November to June. Significant wave heights range from 1.5 to 2 m. NW waves impinge on the Langue de Barbarie at angles of 10–20°, thus generating active longshore drift to the south. Waves from the southwest and from the west account for the rest, and are only active during the rainy season between July and October. Wave periods range from 6 to 12 s, reflecting a mix of distant swell and wind waves (Anthony 2006). The tidal regime is semi-diurnal and the range microtidal, comprised between 0.5 m at neap tides and 1.6 m at spring tides.

2.3.2 *The Migration Dynamics of the Langue de Barbarie Spit*

The relatively moderate river discharge, including during the flood season, the permanence of moderate waves propagating across a relatively narrow shelf, and the microtidal regime are three conditions that explain the wave-dominated character of the Senegal River delta and quasi-permanent river-mouth diversion by the Langue de Barbarie. The Barbarie spit is a 100–400 m-wide feature that has fluctuated in length between 10 and 30 km over the last century. The spit is capped by a 5–10 m high dune. The river-mouth depths range from 2.5 m in the rainy season when sand deposition results from high river discharge, to 3.5 m in the dry season, when sediment flushing is assured essentially by tidal currents. A drift volume that decreases from north to south along the spit from 0.70 to 0.60 $\times$ 10^6 m^3 has been calculated by SOGREAH (1994), the gradient attributed to progressively more significant aeolian dune abstraction of sand between the relatively urbanised sector of St. Louis, where dunes have been fixed by vegetation, notably plantations of Filao (*Casuarina equisetifolia*), and the relatively poorly vegetated distal zone. Mild opposite drift towards the north occurs during the short summer period when waves from the southwest dominate.

Rates of spit growth proposed in the literature from the analysis of satellite images and aerial photographs and from field measurements vary widely, from 94 to 700 m a year. It is likely that rates actually vary interannually depending on variations in wave characteristics, river discharge and mouth dynamics coupled with breaching. Spit growth occurs through the classic adjunction of ridges at the distal end, and the process is undoubtedly favoured by the shallow overall depths of the mouth. Gac et al. (1982) conducted a historical analysis of the mobility of the spit and of the corresponding locations of mouth openings. They showed that the spit has varied neither in elevation nor in width since 1800, and that the farthest downdrift position of the mouth of the river, which corresponds to the maximal distal spit extension, did not exceed 30 km, no doubt constrained beyond this point by the conjunction of sand supply taken up by dunes along the spit and river-mouth discharge supplemented by drainage of the southern part of the delta which comprises a shallow infilling basin. Joiré (1947) and Tricart (1961) situated the mouth in the vicinity of St. Louis at about the mid-seventeenth century, but Gac et al. (1982) have identified even earlier mouth scars north of St. Louis. Michel (1980) dated the formation of the spit at between 4000 and 1900 BP. The spit lengthened by 11 km between 1850 and 1900, and its distal tip was located at the turn of the twentieth century 15 km south of St. Louis. It was affected over this period by seven breaches, of which the most important, in 1884, resulted in a new mouth opening that rapidly attained a width of 4 km (Gac et al. 1982). Since 1900, a major coastal management preoccupation has been that of preventing breaches in the vicinity of St. Louis, achieved through dune reinforcement via Filao planting. Between 1900 and the date of their study, Gac et al. (1982) enumerated 13 breaches, 8 of which occurred between 1954 and 1973. The 1973 breach was also the last natural breach of the twentieth century. The absence of breaches over the last quarter of the twentieth century may be explained by the absence of major floods over this period, characterised by the Sahel drought, and by the flow regulatory effects of dams constructed in the lower Senegal valley. The mean reduction in river discharge in Bakel, a hydrographic station located 550 km upstream of St. Louis, attained 25 % between 1970 and 1990, and even diminished by half between 1980 and 1990 (Mahé and Olivry 1995). An anti-salt dam, the Diama dam, was constructed 54 km upstream of St. Louis in 1987, and a hydropower dam, the Manantali dam, constructed in 1987 in the upper basin in Mali, thus resulting in modification of the hydrological regime and in sediment trapping upstream.

2.3.3 Impending Demise of the Spit?

One ancillary function of the two dams constructed on the Senegal River was to alleviate floods in the lower valley, notably in the deltaic sector of St. Louis. The flood control function of these dams has yielded rather mixed results (Mietton et al. 2006), with flooding still being persistent in the critical area of St. Louis, as a result of rainfall anomalies and the potential blocking of freshwater outflow at the Langue de Barbarie inlet by strong wave activity. One such rainfall anomaly occurred in October 2003, resulting in a rapid rise in water level around St. Louis that led to the drastic decision, by the local authorities, to create an artificial breach near the city to alleviate the flooding. In urgency, on the night of October 3, 2003, a 4 m-long, and 1.5 m-deep trench was opened up by engineers in a relatively narrow (100 m-wide) portion of the spit about 7 km south of St. Louis (Durand et al. 2010), enabling a rapid overnight drop in water level of up to 1 m, thus preventing flooding. Following this opening, the tips of the two opposed spits were rapidly eroded, the breach acting as the new river mouth, and attaining a width of 250 m 3–4 days after the opening (Fig. 2.5a), and 800 m 6 months later. The depth of the

Fig. 2.5 Ground photographs showing the initial trench (*top left*) that preceded the artificial breach of the Langue de Barbarie created in October 2003 in order to alleviate flooding of parts of St. Louis. *Top right* photograph shows the breach considerably widened by river and tidal flow a few days later (Photo credit: Service régional de l'Hydraulique, St. Louis du Sénégal). *Bottom* photographs taken in January 2013 show a natural breach that occurred in October 2012 500 m downstream of the artificial breach (Photo credit: Bertil Wilmotte)

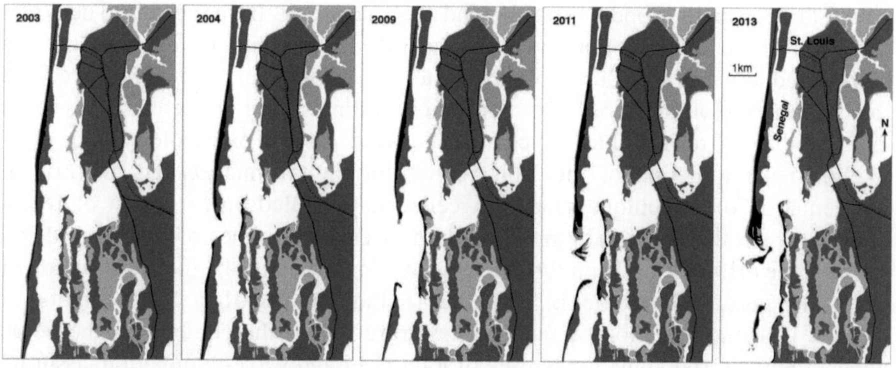

Fig. 2.6 Assemblage of sketches showing changes in the Langue de Barbarie spit and inlet system between March 2003, prior to the October 2003 artificial breach, and 2013. The artificial breach has been followed by significant destabilisation of the spit and by widening of the new river mouth that has exploited this breach. Parts of the eroded spit are being reworked by waves and currents into river-mouth bars and islets. Mouth widening has also led to an increase in tidal influence within the lower Senegal delta and in direct wave attack of parts of the delta plain hitherto protected by the spit

breach had also increased to 6 m, and its width by 1.5 km by 2007 (Bâ et al. 2007). Similar artificial breaches of an extremely mobile spit in Benin, with migration rates of up to 700 m, are commonly operated in order to alleviate flooding, without major consequences on the morphodynamic functioning of the river mouth (Laïbi et al. 2011). In the case of the Senegal River delta spit, the implications of this breach, operated without a consideration of the consequences of subsequent behaviour of the spit and of river-mouth dynamics, has turned out to be dramatic for the numerous villages in the lower delta plain south of St. Louis, and the tourism-based structures that have developed on the Langue de Barbarie in the decade preceding the opening. The spit may also have been lastingly destabilised by this artificial breach, this new mouth characterised by in situ widening rather than migration (Fig. 2.6). Channeling of the Senegal river flow in this new mouth led to closure of the former natural mouth located downdrift. A natural breach 500 m south of the artificial 2003 breach occurred in October 2012, thus creating a new additional mouth (Fig. 2.5b). Following this breach, much of the remaining spit is now being eroded, resulting in several washovers that tend to coalesce, widening sea intrusion pathways, as the sand is transported downdrift by the strong longshore drift. This has led to the dismantling of campsites and other tourist structures. The delta in this eroding sector is now directly exposed to erosion, threatening numerous villages.

2.4 Discussion and Conclusion

The patterns of spit development of the two largest river deltas in West Africa, after the Niger delta, strongly reflect the recent impacts of human activities and, in at least the case of the Senegal delta spit, lack of foresight in engineering practice aimed at delta-plain flood control behind the Langue de Barbarie spit. The Volta delta spit seems to be an outgrowth of a change in river-mouth location and of a marked reduction in sand supply on the eastern coast of Ghana that largely predated the construction of the Akosombo dam. The resulting strong gradients in longshore drift have been responsible for changes in sand barrier morphology and, most likely, in the initiation of the Volta delta spit. Spit formation led, in turn, to segmentation of the unique sand drift cell that prevailed on the Bight of Benin coast between the Volta delta mouth and the western confines of the Niger delta. This unique drift cell has been further segmented in the 1960s by the construction of deepwater ports in Togo, Benin, and Nigeria (Laïbi et al. 2014). The formation of the spit is a manifestation of sand sequestering within the confines of the delta, under a delta morphodynamic regime of joint river and wave-domination, assuring its future integrity in a context of sand deficit. The Volta delta spit has behaved essentially as one of restricted longshore growth, prograding massively instead by the adjunction, in situ, of new individual beach ridges. This pattern of growth has ultimately led to the welding of the distal tip of the spit to the Bight of Benin shoreline, thus creating a new barrier-lagoon system in replacement of the spit. This change should be synonymous with the reestablishment of a single drift cell on the western Bight of Benin coast between Ghana and Togo, thus assuring resumption of integral sand drift from the mouth of the Volta. The process should also offer limited respite from erosion to the hitherto beleaguered town of Keta.

The Langue de Barbarie spit at the mouth of the Senegal delta is altogether different in its inception, dynamics and future evolution. The spit is an outgrowth of strong longshore drift affecting one of the finest examples of a wave-dominated delta. Its recent evolution and potential future demise reflect, however, the consequences of hasty and short-sighted artificial breaching to solve an impending flooding problem facing the historic city of St. Louis. The Barbarie spit developed jointly with the delta and appears to have inexorably extended downdrift in conjunction with river-mouth diversion and migration southwards during the late Holocene, albeit with occasional breaches, the oldest of which are only preserved north of the city of St. Louis. The spit has had a downdrift migration range of about 30 km, beyond which alongshore sand drift and river discharge conditions led to stabilisation of the position of the river mouth. Over time, breaching appears to have been essentially influenced by river discharge spates. Numerous natural breaches have been indentified on the Barbarie spit between 1800 and the present day, no doubt related to years of exceptionally high river discharge in an overall Sahel-influenced climatic context of rather irregular discharge. In contrast, the long phase of river-mouth stability between 1973 and 2003 was associated with a decrease in river discharge related to the Sahel drought, and this phase of stability was marked by the development of tourism-based activities on the spit. The artificial breach through the spit created by engineers in just a few hours to counter a risk of flooding of St. Louis from one such exceptional river discharge season in 2003 has had dramatic consequences on the integrity of the spit, notably by acting as a new river mouth that underwent rapid and significant widening (Fig. 2.6).

It is not clear why the spit is being progressively dismantled. One important question relative to this is that of the net sediment budget. There are no reasons to believe that sand supply from updrift, along the coast of Mauritania, and hitherto feeding the spit, may have diminished, and this does not seem to be a likely explanation for spit demise. A more likely reason may reside in sequestering of sand drifting alongshore from the north within the confines of the new, wide river mouth, a process that may be reinforced by a now larger tidal prism associated with this new mouth. Much of the lower delta plain and the main river channel are now situated over 20 km upstream of the former mouth, between the new enlarged mouth and the anti-salt intrusion Diama dam that confines the tidal prism to the lower delta plain. It is interesting to note that the latest phase of spit dismantling occurred during high river discharge conditions in October 2012, thus illustrating the potential seasonal effect of high river outflow on spit reworking. As sand is being sequestered within the confines of the new mouth, the spit downdrift is being eroded to balance the strong requirements in longshore drift.

References

Addo KA, Jayson-Quashigah PN, Kufogbe KS (2011) Quantitative analysis of shoreline change using medium resolution satellite imagery in Keta, Ghana. Mar Sci 1:1–9

Anthony EJ (1995) Beach-ridge progradation in response to sediment supply: examples from West Africa. Mar Geol 129:175–186

Anthony EJ (2006) The muddy tropical coast of West Africa from Sierra Leone to Guinea-Bissau: geological heritage, geomorphology and sediment dynamics. Afr Geosci Rev 13:227–237

Anthony EJ, Blivi AB (1999) Morphosedimentary evolution of a delta-sourced, drift-aligned sand barrier-lagoon complex, western Bight of Benin. Mar Geol 158:161–176

Anthony EJ, Lang J, Oyédé LM (1996) Sedimentation in a tropical, microtidal, wave-dominated coastal-plain estuary. Sedimentology 43:665–675

Anthony EJ, Oyédé LM, Lang J (2002) Sedimentation in a fluvially infilling, barrier-bound, estuary on a wave-dominated, microtidal coast. the Ouémé River estuary, Benin, West Africa. Sedimentology 49:1095–1112

Bâ K, Wade S, Niang I, Trébossen H, Rudant JP (2007) Cartographie radar en zone côtière à l'aide d'images multidates RSO d'Ers-2: application au suivi environnemental de la Langue de Barbarie et de l'estuaire du fleuve Sénégal. Teledetection 7:129–141

Blivi A, Anthony EJ, Oyédé LM (2002) Sand barrier development in the Bight of Benin, West Africa. Ocean Coast Manag 45:185–200

Boateng I (2012) An application of GIS and coastal geomorphology for large scale assessment of coastal erosion and management: a case study of Ghana. J Coast Conserv 16:383–397

Davies JL (1980) Geographical variation in coastal development, 2nd edn. Longman, London, p 212 pp

Delft Hydraulics (1990) National and regional aspects of coastal erosion in the Bight of Benin. Project 6607.43.94.155, European Development Fund, Brussels

Durand P, Anselme B, Thomas YF (2010) The impact of the opening of the breach in the Langue de Barbarie (Saint-Louis du Senegal) in 2003: a change of flood hazards nature? Cybergeo: Eur J Geogr. http://cybergeo.revues.org/23017

Gac JY, Kane A, Monteillet J (1982) Migrations de l'embouchure du fleuve Sénégal depuis 1850. Cahiers ORSTOM Série Géologie 12:73–76

Joiré J (1947) Amas de coquillages du littoral sénégalais dans la banlieu de Saint-Louis. Bulletin de l'Institut Français de l'Afrique Noire 9:170–340
Kumapley NK (1989) The geology and geotechnology of the Keta basin with particular reference to coastal protection. In: Proceedings, KNGMG Symposium on Coastal Lowlands, Geology and Geotechnology, pp 311–320
Kutu JM (2013) Seismic and tectonic correspondence of major earthquake regions in southern Ghana with Mid-Atlantic transform-fracture zones. Int J Geosci 2013(4):1326–1332
Laïbi RA, Gardel A, Anthony EJ, Oyédé LM (2011) Apport des séries d'images LANDSAT dans l'étude de la dynamique spatio-temporelle de l'embouchure de l'estuaire des fleuves Mono et Couffo au Bénin, après la construction du barrage de Nangbéto sur le Mono: dynamiqye d'une embouchure fluviale. Revue Télédétection 10:179–188
Laïbi RA, Anthony EJ, Almar R, Castelle B, Senechal N, Kestenare E (2014) Longshore drift cell development on the human-impacted Bight of Benin sand barrier coast, West Africa. J Coast Res Spec Issue 70:78–83
Ly CK (1980) The role of the Akosombo Dam on the Volta river in causing erosion in central and eastern Ghana (West Africa). Mar Geol 35:323–332
Mahé G, Olivry JC (1995) Variations des précipitations et des écoulements en Afrique de l'Ouest et central de 1951 à 1989. Sécheresse 6:109–117
Michel P (1980) The southwestern Sahara margin: sediments and climate change during the recent Quaternary. Palaeoecol Afr 12:297–306
Mietton M, Dumas D, Hamerlynck O, Kane A, Coly A, Duvail S, Baba MLO, Daddah M (2006) Le delta du fleuve Sénégal. Une gestion de l'eau dans l'incertitude chronique. Actes du Colloque international "Incertitudes et Environnement - mesures, modèles, gestion", Arles, France, 23–25 Nov 2005, 12 p
Oguntunde PG, Friesen J, van de Giesen N, Savenije HHG (2006) Hydroclimatology of the Volta River Basin in West Africa: trends and variability from 1901 to 2002. Phys Chem Earth 31:1180–1188
Rossi G (1989) L'érosion du littoral dans le Golfe du Bénin: un exemple de perturbation d'un équilibre morphodynamique. Zeitschrift für Geomorphologie NF Suppl Band 73:139–165
Sall M (2006) Crue et élévation du niveau marin à Saint-Louis du Sénégal: impacts potentiels et mesures d'adaptation. Thèse de Doctorat. Université du Maine. 332 p
SOGREAH (1994) Etudes de faisabilité et d'avant projet sommaire de l'émissaire delta. Rapport final, 70 p
Streif H (1983) Die Holozäne Entwicklung und géomorphologie der Küstenzone von Ghana. Essener Geografische Arbeit 6:1–27
Tricart J (1961) Notice explicative de la carte géomorphologique du delta du Sénégal. Mémoires B.R.G.M. 8, 137 p

Chapter 3
El Paramo Transgressive Gravel Spit, Tierra del Fuego, Argentina

Gustavo Gabriel Bujalesky and Gustavo Gonzalez Bonorino

Abstract The northeastern Atlantic coast of Tierra del Fuego is located in the extra-Andean lowlands of the island. This coast experiences a macrotidal regime, moderate energy waves and intense westerly winds. Extensive beaches and other littoral deposits are composed of gravel and coarse sand. This area has been free of ice since 1.8 Ma B.P. Glacigenic deposits were re-worked by littoral processes that formed beaches during sea level highstands of the Pleistocene. Península El Páramo (El Páramo spit) is a 20 km-long gravel spit barrier that closes partially the San Sebastián Bay. It has formed during the last 6,000 years. The growth of the spit has taken place under limited sediment supply. Its elongation is the result of erosion and sediment recycling, resulting in a landward retreat.

The Atlantic beach is composed of gravel and coarse sand and exhibits a reflective morphodynamics. Regularly spaced washover channels develop on the crests of the Atlantic beach of El Páramo spit. These channels are active under storm waves or swell during spring tides. The washover channels of the Atlantic beach of El Páramo spit were formed as the result of subharmonic edge waves generated by incident waves with periods of 1 l to 17 s. The modal spacing (90 m) can be related to subharmonic edge waves generated by incident waves with periods of 15 s. The waves in the bay rework the sediments of the washover fans at the inner side of the spit, building a bayside beach ridge and plugging the backbarrier mouth of the washover channels. This process helps preserve the spit crest.

A fossil gravel beach ridge at El Páramo spit shows regular and periodic spaced washover channels. These are attributed to a single episodic event of high-energy incident waves with a period of 6 s. Such waves are generated within San Sebastián Bay by extreme south-westerly winds, during maximum high tide. Although resonance or standing waves could also be generated by a strong earthquake, or by a tsunami wave entering the bay.

G.G. Bujalesky (✉)
Centro Austral de Investigaciones Científicas (CONICET), Av. Houssay 200, V9410CAB Ushuaia, Tierra del Fuego, Argentina
e-mail: bujalesky@gmail.com

G. Gonzalez Bonorino
CONICET-CADIC, UTN-FRRG, Av. Houssay 200, V9410CAB Ushuaia, Tierra del Fuego, Argentina

G. Randazzo et al. (eds.), *Sand and Gravel Spits*, Coastal Research Library 12,
DOI 10.1007/978-3-319-13716-2_3

3.1 Introduction

El Paramo is a Holocene spit developed on the Atlantic coast of Tierra del Fuego, Argentina (Fig. 3.1a). Like many spits on paraglacial coastlines (e.g., Forbes and Taylor 1987), El Paramo is attached to a receding headland underlain by Pleistocene till that supplied gravel and sand for its construction. The hydraulic setting has, however, given El Paramo spit a distinctive morphological and sedimentological character.

The El Paramo spit evolved under high-energy Atlantic wave climate combined with a macrotidal regime, and, more significantly, facing the wide, glacially-carved embayment of San Sebastian Bay, that provided a long fetch to the strong and persistent westerly winds typical at these latitudes. Atlantic longshore currents transported sediment southward to the spit terminus, and also into the bay domain by overwashing across the spit and by wave refraction around the spit end. In turn, wave-induced longshore currents on the bay flank of the spit were competent to remobilize gravel-sized clasts and thus preclude the deposition of fine-grained sediment between among the recurved ridges. The near-equal transport capacity for gravel of the Atlantic and the bay longshore currents have been responsible for the continued elongation of El Paramo. This paper describes the processes involved in shaping the El Paramo spit and speculates on the reasons for its persistence without breaching.

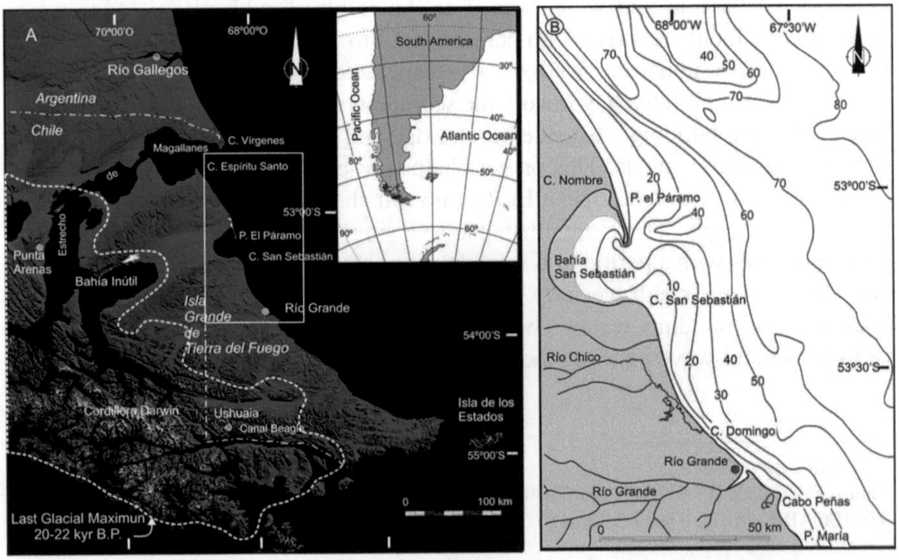

Fig. 3.1 (a) Shaded relief map of the Isla Grande de Tierra del Fuego. *Dotted line* marks the extent of ice cover during the Last Glacial Maximum (Modified from Coronato et al. 2004); *rectangle* shows location of panel b. (b) Coastal area around El Paramo spit; depths in meters referred to spring low tide level; location shown in panel a. In both panels *c* stands for Cape

3.2 Geologic and Tectonic Setting

The island of Tierra del Fuego spans, from southwest to northeast, the southernmost stretch of the Andean orogen and foreland (Fig. 3.1; Lodolo et al. 2003). The orogen consists of severely deformed and partly metamorphosed, Cretaceous and older marine sedimentary and volcanic rocks, unconformably overlain by moderately deformed Tertiary shallow marine and continental strata. These rocks underlie highlands attaining 3,000 m a.s.l. In the foreland, subhorizontal Cretaceous to Tertiary shallow marine and continental deposits underlie fluvially-dissected lowlands. This varied substrate is partly covered by dominantly Pleistocene glacial deposits that, in the Andean orogen are lodged in glacially scoured troughs, and over the foreland area form extensive sheets that locally underlie eroding coastal cliffs. Holocene deposits occur as fluvial valley fills and peat bogs, and as extensive tidal flats, strandplains and spits, along the Atlantic coast (Bujalesky et al. 2013).

Related to continuing Andean orogenesis in the Holocene, Tierra del Fuego has experienced differential tectonic uplift at rates decreasing from about 3 mm/year in the southwest to virtually zero along the Atlantic shoreline (Bujalesky et al. 2013). By the Last Glacial Maximum (ca 20 ka BP) glacier ice fronts had largely retreated from the Atlantic shelf to the Andean highlands (McCulloch et al. 2000; Araya-Vergara 2001). Only a few outlet glaciers reached the sea, flowing along glacially-scoured depressions presently occupied by the Strait of Magellan, the Inutil Bay-San Sebastian Bay corridor, and the Beagle Channel. About 12 ka BP these depressions were flooded as the rising sea level outpaced tectonic uplift. The sea entered deeply into a much more extensive San Sebastian Bay, carving cliffs on the sides of the glacial scour. Shortly after, however, around 7 ka BP, glacio-eustatic sea-level rise slowed down to match tectonic uplift along much of the Atlantic shore. The late Holocene was thus a time of relatively stable sea level that favoured the development of numerous accretionary coastal landforms, among which El Paramo spit. El Paramo spit partly encloses San Sebastian Bay and interacts with tidal sedimentation within the bay (Fig. 3.1b).

3.3 Dynamic Setting

The British Meteorological Office reported ship observations on wave hydraulics offshore from San Sebastian Bay between 1949 and 1968 (IMCOS Marine Limited 1978). These data indicate a very low frequency of wave heights higher than 3.5 m. Strong northeasterly gales with a return period of 50 years generate wave heights of 12 m with periods around 11.5 s. A 1 year record of nearshore observations 60 km north from San Sebastian Bay indicates a maximum significant wave height of 3.4 m, with maxima up to 7 m for north to northeasterly winds with 50 year return period, and a maximum significant period of 12.9 s, with maxima of 17.5 s (Compagnie de Recherches et d'Etudes Oceanographiques and Geomatter 1985).

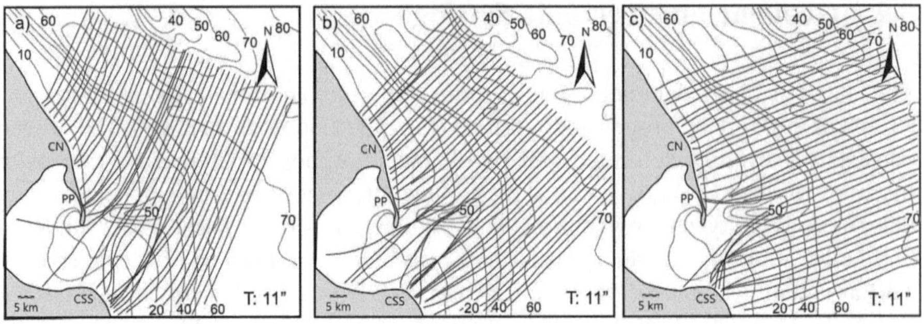

Fig. 3.2 Wave refraction patterns at San Sebastian Bay referred to high spring tide level and incident waves with periods of 11 s from the (**a**) north-northeast, (**b**) northeast, and (**c**) east-northeast and east (Computations as per Elliot 1990). *CN* Cape Nombre, *PP* El Páramo Peninsula, *CSS* Cape San Sebastián. Depths in meters

In turn, the waters of San Sebastian Bay are swept by SW-NW winds attaining velocities in excess of 120 km/h with 50 year return periods (IMCOS Marine Limited 1978; Compagnie de Recherches et d'Etudes Oceanographiques and Geomatter 1985). For these wind conditions and a 29-km-long fetch, significant spectral wave heights attain 2.74 m, with maximum periods of 5.9 s (calculations done with the Automated Coastal Engineering System, of Coastal Engineering Research Center 1992). On the upper intertidal zone of the bay beach plunging breakers may reach heights of up to 1.5 m and periods of up to 5 s. The mean tidal range at Bahía San Sebastián is 6.6 m (maximum recorded range is 10.4 m), with flood and ebb currents reaching velocities of 2 knots (Servicio de Hidrografía Naval 1981, 2011).

The influence of submerged topography on El Paramo spit was explored through wave refraction modeling (Elliot 1990). The runs were carried out for ordinary swell conditions, high spring-tide levels and incident waves with periods of 11 s. The shore platform off Cape Nombre induces wave convergence on the central sector of El Páramo spit. A 36-m-deep channel bounds the spit to the south. The channel is likely a Pleistocene glacifluvial feature but may have been re-excavated by tidal currents (Fig. 3.2).

3.4 Spit Morphology and Composition

El Paramo spit is a 20-km-long, narrow landform attached to the Cape Nombre headland (Fig. 3.3a). The headland is part of a 50-km-long coastal stretch extending to the Strait of Magellan, lined by cliffs of Pleistocene till and Tertiary sandstones up to 90 high. The shoreline is currently receding episodically; at Cape Nombre the cliff face was stable from February 20th 1987 until February 20th 1988. At that time, strong swells at spring high tide level were accompanied by breaker heights >3 m and periods of 9–12 s. This event caused cliff retreat of 3.7 m in a few days

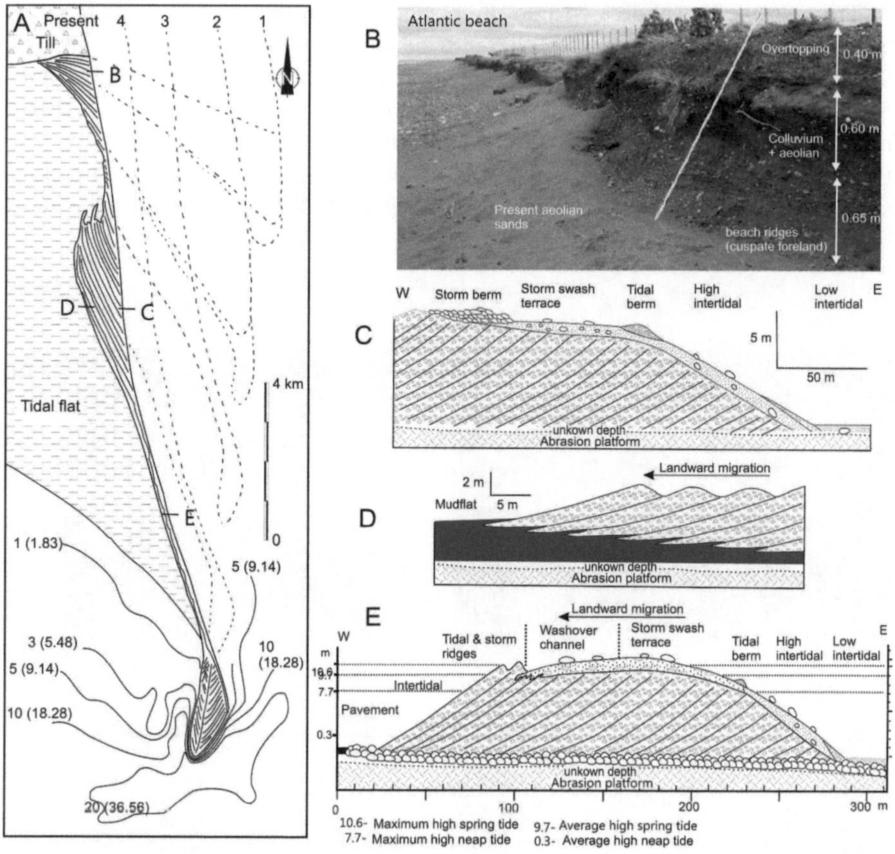

Fig. 3.3 (**a**) Evolution of El Páramo spit. Numbers *1*, *2*, *3*, and *4* approximately indicate successive plan-view reconstructions of El Páramo spit, starting as a cuspate foreland attached to a former coastal cliff. Depths in meters referred to spring low tide level. The location of panels (**b**–**e**) are indicated. (**b**) Wave-cut erosive scarp on the Atlantic beach exposes beach ridges of the cuspate foreland. Ridge crests are approximately 1 m above the present storm berm, and are covered by colluvium derived from the paleocliff at Cape Nombre interlayered with eolian deposits; modern overtopping gravel caps the sequence. (**c**) Atlantic beach facies where recycling of fossil gravel bay-domain ridges occurs. (**d**) Bay beach facies composed of medium to coarse openwork gravel. (**e**) Middle section of the spit facies, washover channel domain where by-pass of sediment supply to the bay beach takes place and constitutes a significant factor in landward migration

(Bujalesky 1990). In addition to these episodic events, continuous percolation and freeze-thaw processes on these unconsolidated deposits contribute to their erosion. Cliff recession has left behind a wide shore platform cut in the Pleistocene deposits, that is partly exposed at lowest tides.

The northernmost gravelly beach ridges, at the base of a palaeocliff, are concave toward the bay area (Fig. 3.3a). They are interpreted as a remnant of a cuspate foreland that predated spit formation; erosive faces in these ridges on the Atlantic beach show crests 0.6–1.0 m above the present Atlantic storm berm (Fig. 3.3b).

The spit can be subdivided into three segments. In the north, over 200 gravel beach ridges form a strandplain 1,200 m wide and 8 km long that accreted into San Sebastian Bay (Fig. 3.3c, d); in plan view the ridges are convex toward the bay area. Spit terminations indicate northward growth. Interference with tidal flat expansion is reflected in the diminishing northward extension of the ridges from older to younger. One of the fossil beach ridges, located 50 m east of the bay shoreline, shows numerous washover channels showing evidence of eastward palaeoflow (Fig. 3.4a).

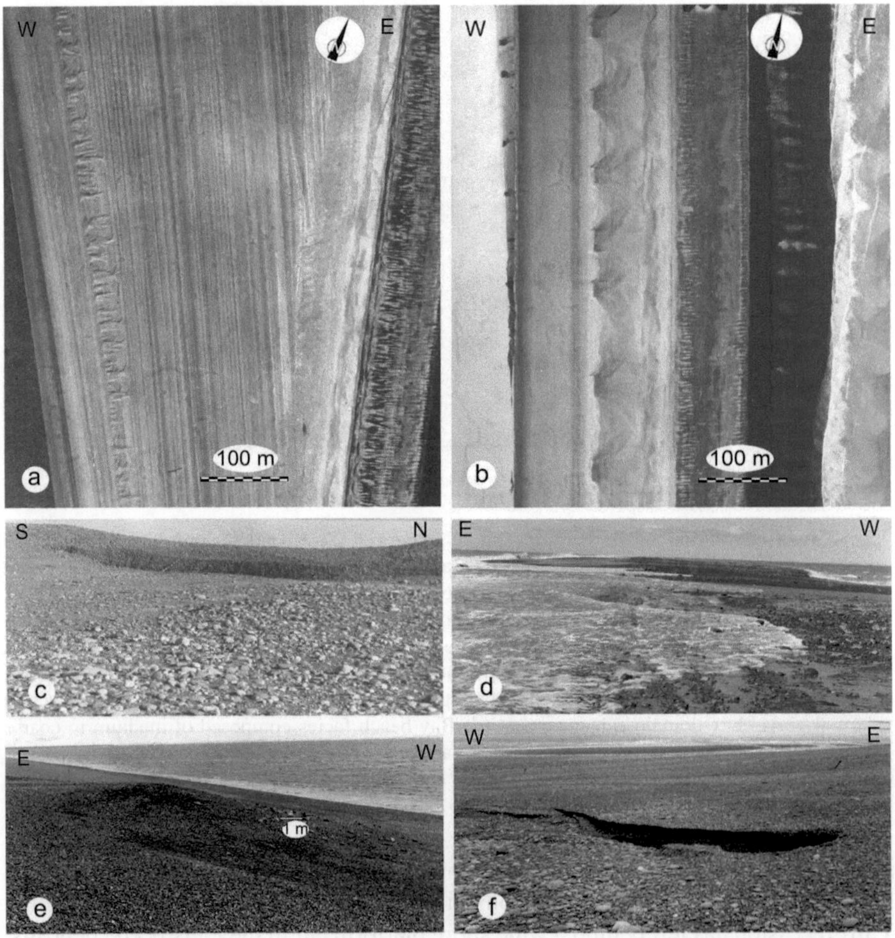

Fig. 3.4 Washover channels of El Paramo spit. (**a**) Fossil washover lineament and gravel beach ridges of the northern segment of the spit. (**b**) Central section of the spit. (**c**) Plugged bay discharge mouth of the central section of the spit. (**d**) Overwashing in the central section of the spit (02/20/1988). (**e**) A 400 m^3 washover fan on the bay beach (02/20/1988). (**f**) A 0.7 m deep gully carved by overwashing discharge on the bay beach of the spit (02/20/1988)

The second segment is 7 km long and narrow; its width varies from 50 m at high tide to 200 m at low tide (Fig. 3.3e). About 55 washover channels occur along this segment, sloping towards the bay (Fig. 3.4b). The channels are funnel-shaped, narrowing toward the bay from 25 to 95 m (mean width 70 m) to 5 to 60 m (mean width 25 m). The depth of the channels reaches 0.7 m. Associated with the overwash channels are gravelly washover fans that reach the bay beach (Fig. 3.4e). Shortly after overwashing, however, bay waves relocate the fan gravel onto high beach levels, plugging the channels (Fig. 3.4c).

The third spit segment, at its southern end, shows a lozenge-shaped strandplain with a maximum width of 900 m and a length of 2 km (Fig. 3.5a). The strandplain comprises three distinctive sets of gravel ridges. The larger set consists of ridges accreting to the south and southeast that are erosively truncated by Atlantic wave action. An intermediate-sized set of ridges, interspersed with set A along an

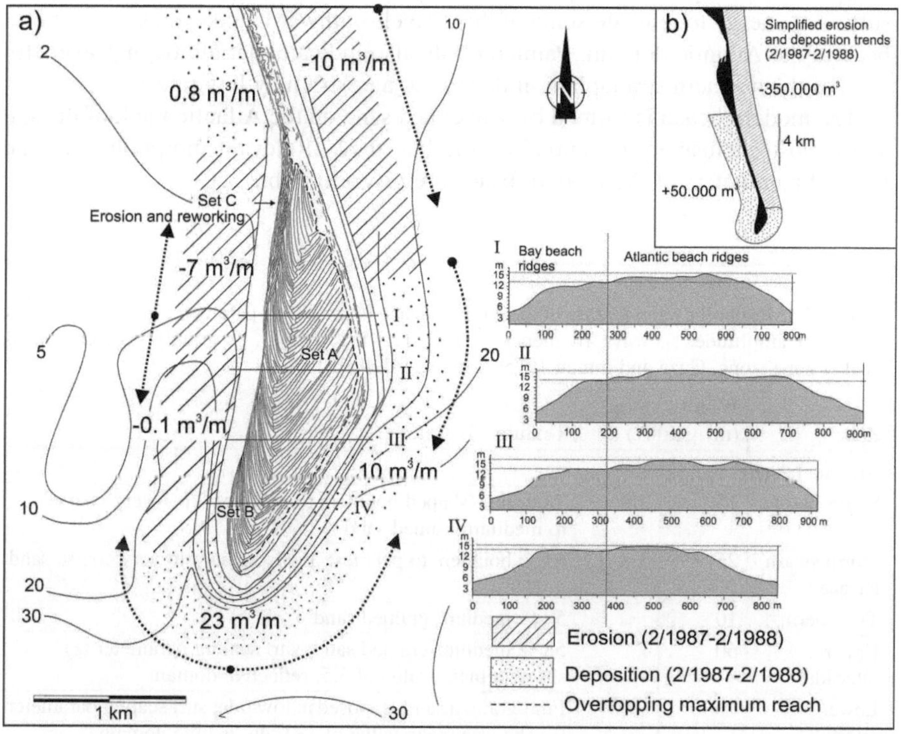

Fig. 3.5 (**a**) Geomorphology, sediment budget and gravel transport (measured between February 1987 and February 1988) in the southernmost segment of El Páramo spit. Deposition and overtopping dominate in the southeastern end of the spit, meanwhile tangential strong sediment transport operates in the bay domain. These processes are evidenced in the crest height difference of the beach ridges of the sets A and B (sections I–IV). (**b**) Simplified erosion and deposition trends along the present beach of the spit

irregular central suture, reflects accretion to the south-southwest and west (set B in Fig. 3.5a). Ridge geometries suggest contemporary erosion of set B in the north-west. The third, smallest, set of ridges (set C in Fig. 3.5a) erodes into set B. The ridges of set C are built from amalgamated swash bars and beach ridges accreting to the north-northwest and west.

The present-day sediment budget and circulation of gravel on the southernmost strandplain, was studied by topographic beach profiling and using painted pebbles and cobbles as tracers. Beach profiles repeated 12 months apart in the Austral summer, revealed erosion at a mean rate of 10 m^3/m of beach length where beach ridges of set A are truncated, shifting southward to accretion of 23 m^3/m of beach length near the spit end. On the bay side, slightly negative values indicate non-deposition to erosion, and revert to accretion as the zone of overwashing is approached (Fig. 3.5). Painted clasts clearly showed that southbound Atlantic gravel is delivered to the bay domain by wave refraction around the spit point, as is typical of other recurved spits (Bujalesky 1990), but also that bay waves build sufficient energy to relocate some of this gravel southward, and even send part of it back to the Atlantic domain. Painted clasts also indicate that although gravel still reaches the northern strandplain, it does so at an extremely low rate.

The modern beach is formed by gravel and sand on the Atlantic flank of the spit and almost exclusively by gravel on the bay flank. Relevant morphological and textural parameters of the modern beach are given in Table 3.1.

Table 3.1 Morphology and texture of the modern beach (surf scaling parameter: $\varepsilon = a\,\omega^2/g\,\tan^2\beta$; a: breaker amplitude, ω: wave frequency $-2\pi/T$, T: wave period-, g: gravity acceleration, β: breaker zone slope, Guza and Inman 1975; Guza and Bowen 1977)

Zone	Width (m)	Slope (°)	Texture
Atlantic beach			
Storm berm	15	3.5	72 % disc-shaped boulders to pebbles; 28 % very coarse to medium-grained sand
Storm swash terrace	25	2	64 % boulders to pebbles; 36 % medium to very coarse sand
Tidal berm	10		58 % medium-grained sand
Upper intertidal	60	7	58 % medium-grained sand; surf scaling parameter (ε) shows a mean value of 3.5; reflective domain
Lower intertidal	70	*ca* 0	Fine-grained sand; exposed at low tide; surf scaling parameter (ε) shows a mean value of 183; dissipative domain
Bay beach			
Storm ridge	10		Disc-shaped pebbles (mean size 28 mm)
Tidal ridge	10		Cylindrical and spherical pebbles (mean 23 mm)
Intertidal	65	4–6	Spherical and cylindrical pebbles (mean 22 mm)
Pavement	10		Boulders and cobbles

3.5 Overtopping and Overwash Processes

On February 20th, 1988, El Paramo was affected by north-northeasterly swell with periods of 9–12 s, that at a high tide level of 10.5 m gave rise to plunging breakers up to 3 m high (Fig. 3.4d). The tidal berm of the Atlantic beach behaved as a bar, where the waves broke, and immediately behind it a shore-parallel channel developed. Furthermore, small return channels with a 50 m spacing cut the crest of the tidal berm. Overtopping was observed in the northern and southern reaches of the central, narrow, spit segment, the run-up extending 20 m landward.

Overwashing of Atlantic waters and sediment was more common than overtopping, however. Saturation of the core of the spit increased the washover discharge. Four washover channels became active simultaneously; two of them lay side by side. Overwash flows breached the bay berm ridge and transported significant quantities of gravel toward the bay, forming washover fans 400 m^3 in volume (Fig. 3.4e). The bay beach face was gullied by the joint action of the washover flows and the seepage discharges (Fig. 3.4f).

Bujalesky (1997) analyzed the spacing of washover channels in the central section of El Paramo spit and in a fossil beach ridge following the statistical methodology of Orford and Carter (1984). Distances between the channels axes were obtained from aerial photographs (scales 1:5,000 and 1:10,000) taken by the Argentine Servicio de Hidrografía Naval in 1970–1971. The statistical tests carried out for the washover channels spacing of both sets support the hypothesis of uniform spacing. Tests of randomness also indicated that the washover channel distribution does not adjust to the Poisson theoretical model, consisting of uniform distributions of regular pattern (Table 3.2; Bujalesky 1997).

To test for spatial periodicity of the washover channels (Orford and Carter 1984), a Markov Chain analysis was performed in order to determine if certain spacing is probabilistically dependent of a precedent state (Bujalesky 1997). In developing the transition matrices each interval was considered as a determinate state (e.g.: 20–29 m spacing state). Then the number of times that each state was followed by each of the others was established. Overall, the independence hypothesis proved to be statistically valid. One exception occurred with the fossil washover channels, for which the 10 m interval size allowed the statistical independence hypothesis to be rejected. If k is the number of steps through which the memory persists until forgotten (Davis 1986). The k number for the mentioned matrix was equal to 10. The largest probability (0.49; $k = 10$) was evidence for 15–24 m spacing (Bujalesky 1997).

3.6 Discussion

3.6.1 Overwash Processes

Edge wave processes can explain the occurrence of rhythmic littoral forms (Bowen and Inman 1969). Guza and Davis (1974) demonstrated that on reflective beaches the edge waves are subharmonic (the period is twice the period of the incident

Table 3.2 Stem and leaf diagrams and statistics of the washover channels spacing

Atlantic beach			Fossil beach ridge (bay domain)		
Unit = 1 m; 1|2: 12 m			Unit = 1 m; 1|2: 12 m		
Depth			Depth		
2	4	55	6	0o	578888
3	5	0	16	1*	0000000002
10	6	0000555	(18)	1o	5555555555555578
18	7	00000255	30	2*	000000000022
23	8	00577	17	2o	555
(12)	9	000000005555	14	3*	000
13	10	0000005	10	3o	5
4	11	00000555	9	4*	000
3	12	0			
3	13		Outliers		45, 50, 50, 55, 62, 65
2	14	5			
	15	0			
Outliers		165			
N:55			N:64		
Mean:	90.20 m		Mean:	21.33 m	
Standard deviation: 24.61 m			Standard deviation: 13.55 m		
Coefficient of variation: 0.27			Coefficient of variation: 0.64		
Median: 90 m			Median: 17.5 m		
Lower fourth: 70 m			Lower fourth: 12.75 m		
Upper fourth: 105 m			Upper fourth: 25 m		
Skewness: 0.57			Skewness: 1.51		
Kurtosis: 0.61			Kurtosis: 1.72		

Modified after Bujalesky (1997)

wave, $Te = 2\ T_i$) and operate with zero mode, that is, their troughs and crests lie, respectively, below and above the mean water level. For mode $n = 0$ the following formula is valid (Huntley and Bowen 1979):

$$L_e = \frac{g}{2\pi} T_e^2 \tan \beta \tag{3.1}$$

where, L_e is the wavelength of the edge wave, g is the acceleration of gravity, T_e is the edge wave period ($Te = 2T_i$), and β is the beach slope. Orford and Carter (1984) considered that the regular and periodic spacing pattern of washover channels is controlled by subharmonic edge waves that reinforce the uprush and mould beach cusps on the upper levels of the beach, favouring washover channel formation. During the overwash event of February 1988 at El Paramo spit, wave periods of 9, 11 and 12 s were measured. By Eq. 3.1, these incident wave periods correspond with subharmonic edge wave periods of 18, 22 and 24 s and wavelengths of 110, 93 and 62 m. The two adjoining washover channels that operated simultaneously would have been subjected to edge waves forced by 11 and 12 s period incident waves.

Table 3.3 Relationship between washover channels spacing, cero mode subharmonic edge waves length (Le) and period (Te) and incident wave period (Ti, modified after Bujalesky 1997)

	Spacing (m)	Le (m)	Te (s)	Ti (s)
Atlantic beach washover channels: beach slope (β) = 7°; tan β = 0.123				
Minimum	45	90	21.7	10.8
Lower fourth	70	140	27.0	13.5
Median	90	180	30.7	15.3
Upper fourth	105	210	33.1	16.6
Maximum not outlier	150	300	39.6	19.8
Outlier	175	330	41.7	20.8
Fossil beach ridge (bay domain): beach slope (β): 10°; tan β: 0.176				
Minimum	5.0	10.0	6.0	3.0
Lower fourth	12.8	25.5	9.6	4.8
Median	17.5	35.0	11.3	5.6
Upper fourth	25.0	50.0	13.5	6.7
Maximum not outlier	40.0	80.0	17.1	8.5
Outlier	65.0	130.0	21.7	10.9

Table 3.3 shows the relationship between channel spacings, subharmonic edge wavelength and period, and incident wave period, for both the active (Atlantic beach) and the fossil (bay domain) washover channels. The maximum wave period measured for the northern Atlantic coast of Tierra del Fuego was 17.5 s. This fact indicates that spacings larger than the upper fourth percentile (ca 105 m) are unlikely to be related to subharmonic edge wave dynamics. Incident waves with periods of 11 and 12 s that are able to cause overwash on the spit have a return period of 6–24 months. The recurrence interval of incident waves with periods of 15–17 s is more than 2 years.

In the case of the fossil washover channels, the median spacing (17.5 m) corresponds to incident waves with periods of 5.6 s and would be compatible with the maximum estimated wave periods (5.89 s) from 124 km/h winds with a return period of 50 years. No washover channels were observed on the other fossil beach ridges indicating that overwash from the bay domain is a rare event with a probability of occurrence much less than that of the extreme winds. The significant result of the Markov Chain analysis and the larger probability (0.49) of the row vector of the equilibrium matrix for the spacing 15–24 m indicate that these channels originated from a unique storm event. Considering the seismotectonic setting of Tierra del Fuego, however, an origin of the overwash processes in the bay domain by resonance or standing waves generated by a strong earthquake, or by a tsunami wave entering the bay cannot be ruled out.

3.6.2 Spit Development

The continued growth of gravel barriers is conditioned by the sediment volume that reaches their distal points (Carter et al. 1987). Elongation is maintained when the extreme part of the barrier is within the wave transport capacity and sediment is available. With greater distances from the distal point of the barrier, the possibility that a clast reaches it decreases. Then, the sediment along the barrier is recycled causing a thinning of its proximal part (cannibalisation, Carter et al. 1987). Gravel barriers that develop in areas with limited sediment supply tend to migrate landward, even under stable sea level conditions, and they turn into transgressive features. This results in a landward gravel transport, due to the beach reflective behaviour, high infiltration rates and washover processes (Orford and Carter 1982, 1984; Carter and Orford 1984, 1991; Carter et al. 1987, 1989; Forbes and Taylor 1987).

Tidal flat development in San Sebastian Bay likely started after relative sea level stabilized, by the mid-Holocene. Initially, a cuspate foreland developed in the north, attached to a paleocliff carved during sea-level highstand. The greater part of this early foreland has been cannibalized and the detritus relocated in the spit that grew from the tip of the cuspate foreland (Bujalesky and González Bonorino 1991). Speculatively, it is postulated that while the spit was short, littoral drift and wave refraction brought Atlantic sediment into the bay domain, from where bay waves picked it up and transported it northward to accrete the wide strandplain convex to the west. As the spit lengthened, overwashing of gravel into the bay domain became relatively much more significant. Erosion along the Atlantic beach of El Paramo spit has been compensated by deposition along the bay beach due to overwash and sediment transport around the spit end (Bujalesky and González Bonorino 1991). The bay waves in turn redistribute the sediments supplied by overwash and reconstruct the upper beach ridge, giving stability to the beachface. This process preserves the central section of the spit under successive overwash episodes.

Formation and elongation of El Páramo spit has resulted from an unusual combination of hydraulic processes. In general terms, southward-driven gravel on the Atlantic shoreline is partly washed over into the central bay beach, from where westerly-wind-generated waves transport it northward to the accreting northern set of recurved ridges. Remaining gravel accretes to the southernmost spit platform and is partly transported around the spit end by wave refraction. When wind provenance rotates to the northwest, bay-side gravel at the spit end is partly reverted southward and southeastward, resulting in the non-deposition to slightly erosive behaviour. In turn, alternating directions of longshore transport around the spit terminus give rise to the zig-zag junction between ridge sets A and B. The younger set C may have developed in response to an increase in the supply of gravel to the bay through overwashing across the central, slender spit segment. This is the first recognizable sign of change in the main body of the spit system after several thousand years of growth but the likely consequences are unknown. Spit elongation to the south may be progressively checked for two reasons: the deepening sea bottom as the spit platform dives into the 30-m-deep channel, and the strengthening of the erosional power of the ebb tidal flow due to constriction of the bay outlet.

3.7 Conclusions

El Páramo spit has exhibited progressive thinning during its evolution which gives it a very mature stage character. During the Holocene, the growth of the spit took place under limited sediment supply. The progressive elongation was sustained by erosion and sediment recycling (cannibalization) at the seaward side, resulting in a significant landward retreat.

The active and fossil washover channels are uniformly spaced, and have been generated by edge waves; return periods for overwash events range from 6 to 24 months. The modal spacing of 90 m relates to incident waves with periods of 15 s.

The washover channels on the Atlantic beach do not show periodicity, which suggests that they are the consequence of a superposition of incident waves with different periods. It is unlikely that the Atlantic washover channels spacings longer than 105 m resulted from subharmonic edge waves because that would imply incident waves with periods longer than 17 s, the maximum recorded. These large spacings might be associated with height variations over the central section of the spit crest.

The washover channels on a fossil bay domain beach ridge of the northern section of the spit would have been originated due to incident waves with periods of 5.6 s as consequence of extreme south-westerly winds, during a maximum high tide level. However, taking into account the seismotectonics of Tierra del Fuego, an origin of the overwash processes in the bay area by resonance or standing waves generated by a strong earthquake, or by a tsunami wave inflowing the bay cannot be ruled out.

Acknowledgements The authors wish to thank the great number of colleagues and friends that shared field and laboratory work with me. CONICET (PIP 4283, PICTR 2002-67, PIP 05/06-6200, PIP 09/11 0533, PIP 12-14 0395) provided the financial support for the Coastal Geology Research of Tierra del Fuego.

References

Araya-Vergara JF (2001) Formas deposicionales submarinas en el perfil longitudinal del estrecho de Magallanes, Chile. Cienc Tecnológicas Mar 24:7–21

Bowen AJ, Inman DL (1969) Rip currents, 2. Laboratory and field observations. J Geophys Res 74:5479–5490

Bujalesky G (1990) Morfología y Dinámica de la Sedimentación Costera en la Península El Páramo, Bahía San Sebastián, Isla Grande de la Tierra del Fuego. PhD thesis, Universidad Nacional de La Pata, 188 pp

Bujalesky G (1997) Patrón espacial y dinámica de canales de sobrelavado de la costa atlántica septentrional de Tierra del Fuego. Rev Asoc Geol Argentina 52(3):257–274

Bujalesky G, González Bonorino G (1991) Gravel spit stabilized by unusual (?) high-energy wave climate in bay side, Tierra del Fuego, southernmost Argentina. In: Kraus N, Gingerich K, Kriebel D (eds) Coastal sediments '91. Proceedings of a special conference on quantitative approaches to coastal processes, Seattle, Washington, vol 1. American Society of Civil Engineers, New York, pp 960–974

Bujalesky G, González Bonorino G, Abascal L (2013) Holocene coastal environments and processes in subantarctic/temperate cold Tierra del Fuego, Argentina-Chile

Carter RW, Orford JD (1984) Coarse clastic barrier beaches: a discussion of the distinctive dynamic and morphosedimentary characteristics. In: Greenwood B, Davis RA (eds) Hydrodynamics and sedimentation in wave dominated coastal environments. Mar Geol 60:377–389

Carter R, Orford JD (1991) The sedimentary organisation and behaviour of drift-aligned gravel barriers. In: Kraus NC, Gingerich KJ, Kriebel DL (eds) Coastal Sediments '91: proceedings of a special conference on quantitative approaches to coastal processes, Seattle, Washington, June 25–27, vol 1. American Society of Civil Engineers, New York, pp 934–948

Carter RW, Orford JD, Forbes DL, Taylor RB (1987) Gravel barriers, headlands, and lagoons: an evolutionary model, Proceedings Coastal Sediments '87, 1776–1792. American Society of Civil Engineers, New York

Carter RW, Forbes DL, Jennings SC, Orford JD, Taylor R, Shaw J (1989) Barrier lagoon coast evolution under differing relative sea-Level regimes: examples from Ireland and Nova Scotia. Mar Geol 88:221–242

Coastal Engineering Research Center (1992) Automated coastal engineering system. Waterways Experiment Station, Vicksburg

Compagnie de Recherches et d'Etudes Oceanographiques and Geomatter (1985) Campagne meteo-oceanographique, site Río Cullen/Hydra, periode de Fevrier 1984 a Fevrier 1985. Technical Report CREO/1249, Total Austral, Buenos Aires, Argentina, 399 pp

Coronato A, Meglioli A, Rabassa J, Ehlers J (2004) Glaciations in the Magellan Straits and Tierra del Fuego, Southernmost South America. In: Gibbard P (ed) Quaternary glaciations–extent and chronology, part III, Developments in Quaternary Science. Elsevier, Amsterdam, pp 45–48

Davis J (1986) Statistics and data analysis in geology, 2nd edn. Wiley, New York, 646 pp

Elliot AJ (1990) Wave refraction in shallow water. University College of North Wales, Unit for Coastal and Estuarine Studies, Report U90-7, Gwynedd, 27 pp. (Unpublished)

Forbes DL, Taylor RB (1987) Coarse-grained beach sedimentation under paraglacial conditions, Canadian Atlantic coast. In: Fitzgerald D, Rosen P (eds) Glaciated coasts. Academic, New York, pp 51–86

Guza RT, Davis RE (1974) Excitation of edge waves by waves incident on a beach. J Geophys Res 79:1285–1291

Guza RT, Inman DL (1975) Beach cusps and edge waves. J Geophys Res 80:2997–3012

Guza RT, Bowen AJ (1977) Resonant interactions for waves breaking on a beach. In: Proceedings of the 15th coastal engineering conference, pp 560–579

Huntley DA, Bowen AJ (1979) Beach cusps and edge waves. In: Proceedings of the 16th coastal engineering conference, pp 1378–1393

Imcos Marine Limited (1978) Meteorological and oceanographic study: offshore Tierra del Fuego. Technical Report 78/111, Total Austral, Buenos Aires, 17 pp, 4 app

Lodolo E, Menichetti M, Bartole R, Ben-Avraham Z, Tassone A, Lippai H (2003) Magallanes-Fagnano continental transform fault (Tierra del Fuego, southernmost South America). Tectonics 22(6):1076. doi:10.1029/2003TC001500

McCulloch RD, Bentley MJ, Purves RS, Hulton NRJ, Sugden DE, Clapperton CM (2000) Climatic inferences from glacial and palaeoecological evidence at the last glacial termination, southern South America. J Quat Sci 15:409–417

Orford JD, Carter RW (1982) Crestal overtop and washover sedimentation on a fringing sandy gravel barrier coast, Carnsore Point, southeast Ireland. J Sediment Petrol 52:265–278

Orford JD, Carter RW (1984) Mechanisms to account for the longshore spacing of overwash throats on a coarse clastic barrier in southeast Ireland. Mar Geol 56:207–226

Servicio de Hidrografía Naval (1981) Derrotero Argentino. Parte III: Archipiélago Fueguino e Islas Malvinas. Armada Argentina, Publicación H. 203, 304 pp

Servicio de Hidrografía Naval (2011) Tablas de marea para el año 2011. Puertos de la República Argentina y puertos principales de Brasil, Uruguay y Chile. Armada Argentina, Publicación H 610, 494 pp

Chapter 4
Gravel Spit-Inlet Dynamics: Orford Spit, UK

H. Burningham

Abstract The Orford Spit and Ness complex on the gravel-dominated shoreline of Suffolk, eastern England comprises over 350 ha of beach ridge shoreline that diverts the channel of the Alde estuary c. 12 km southward. Centuries of progradation have formed an extensive foreland at Orford Ness, but south of this, Orford Spit has exhibited significant phases of growth and retreat which are in part linked to the dynamics of Orford Haven, the inlet and ebb-delta system at the mouth of the Alde/Ore estuary. Examination of an extended history (500 years) of this system shows a definitive reduction in scales of change over the centuries, and implies a decrease in sediment availability and transport, and a shift from sediment sink to sediment source. Spit dynamics are instrumental in the development and decay of transient lagoons and estuary-oriented spits which rely on breaching, ebb delta bypassing and changes in protection facilitated by the shifting spit.

4.1 Introduction

Spits are wave-formed sedimentary structures that develop primarily within natural embayments where littoral sediment transport supplies the longshore extension of a sub-, inter- and supra-tidal bank system. The longshore planform of spits is generally smooth, reflecting their drift-aligned formation. Sediment delivered to the distal portions of the spit is usually reworked locally to form a succession of ridges and recurves that may (or may not) be linked to pulses in supply and/or shifts in transporting energies. Spits can comprise a range of sediment grades, though are typically composed of sand and/or gravel. This sedimentology is largely a product of geography (gravel is generally a feature of paraglacial, mid-high latitude coastlines), but can impart significant controls on the morphology and dynamics of spits and their associated sedimentary environments.

Here, the morphology and historical development of the gravel-dominated Orford Spit, which extends for several kilometres on the east coast of England

H. Burningham (✉)
Department of Geography, University College London, London, UK
e-mail: h.burningham@ucl.ac.uk

G. Randazzo et al. (eds.), *Sand and Gravel Spits*, Coastal Research Library 12,
DOI 10.1007/978-3-319-13716-2_4

and forms the updrift margin to the Alde/Ore inlet, is described. Orford Spit has arguably one of the most well documented histories of any coastal landform, primarily due to its connection to Orford Ness, a gravel-dominated cuspate foreland. The work of Carr (1965–1986) dominates this literature, and his final work on Orford Spit (Carr 1986) reported on 30 years of monitored change at the mouth of the Alde/Ore. In this paper, he noted that "*In the light of nineteenth century experience it is tempting to suggest that over the next decade or so Orford spit will extend rapidly southwards while sowing the seeds of its own destruction*". This current work offers a timely re-evaluation – 25 years on – and re-assessment of the historical morphodynamics of Orford Spit.

4.2 Regional Setting

The Suffolk coast, southeast England comprises a series of estuarine valleys that connect with the open coast through sediment-filled inlets. Except for the breaks that these inlets form, the open coast beach provides a near-continuous sediment pathway along the length of Suffolk. Geology is dominated by the shelly sands of Pliocene – Pleistocene Coralline and Red Crags that are underlain by London Clay (Carr and Baker 1968; Funnell 1996; Kendall and Clegg 2000), which is exposed in places on the adjacent southern North Sea shoreface. Coastal sediments are largely sorted by depositional environment: fines (silts and clays) are found almost exclusively within estuaries, gravel (colloquially referred to as shingle) dominates the littoral shoreline and sand is found extensively across the shoreface, and variably across the foreshore and inlet shoals.

Large gravel deposits exist on the inlet-adjacent shorelines associated with the Alde/Ore, Deben and Stour/Orwell estuaries, although the sedimentary system of the Alde/Ore is by far the most extensive (Fig. 4.1). The main tidal channel of the Alde estuary is diverted 12 km south from the town of Aldeburgh by a gravel ridge foreland and spit (Orford Ness and Orford Spit respectively). This extended estuarine channel is referred to as the River Ore, which connects with the Butley river just north of the final connection to the North Sea at Shingle Street. The consequence of this diversion is that the Alde/Ore system has a total tidal length of over 30 km. Combined with a mean spring tidal range of 2.8 m at Shingle Street, this equates to a spring tidal prism of around 26×10^6 m^3. Tidal currents maintain a permanent inlet near Shingle Street (known as Orford Haven) which comprises a dynamic ebb delta that is surveyed annually for navigation purposes.

The wave climate of Suffolk is bimodal (Fig. 4.1). Northeasterly waves have a median significant wave height of 0.9 m and account for 47 % of waves recorded at the West Gabbard Buoy: southwesterly waves have a median height of 1.1 m and account for 36 % of the record (2010–2012 inclusive) (Carr and Baker 1968). The 99th percentile wave height is around 3 m irrespective of direction. Inshore (Sizewell) the modes shift eastward but bimodality is retained with northeasterly waves accounting for 46 % of the record ($H_{s50} = 0.6$ m, $H_{s99} = 2.6$ m) and southerly waves 41 % of the record ($H_{s50} = 0.7$ m, $H_{s99} = 2.2$ m).

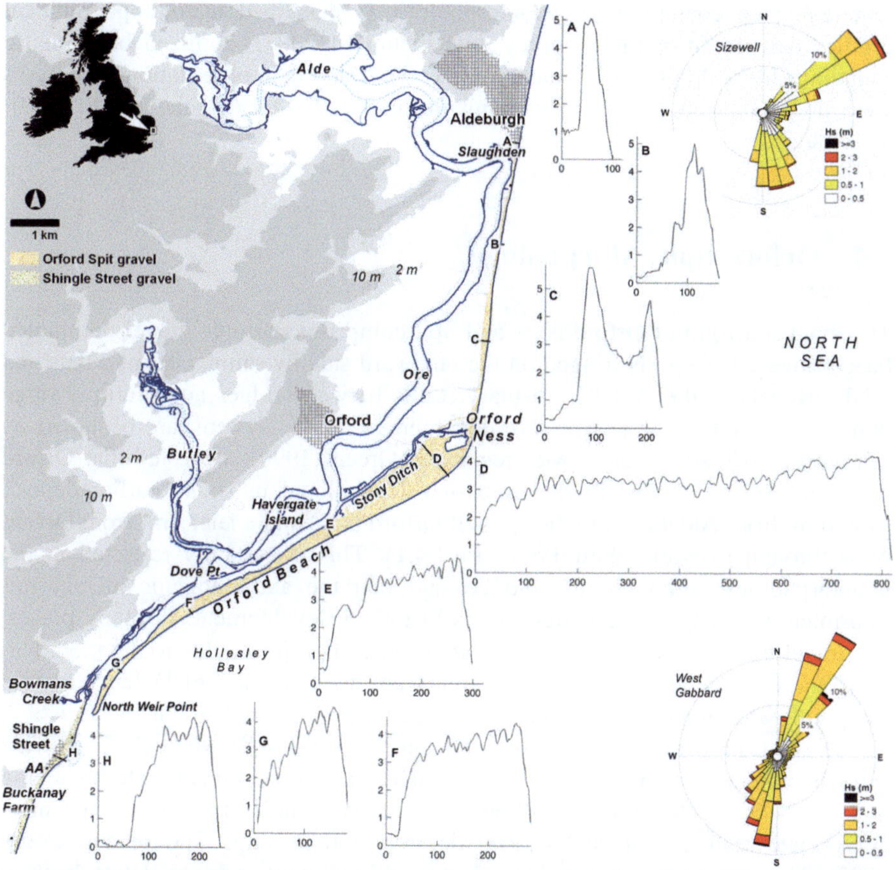

Fig. 4.1 Site map and geomorphological summary of Orford Spit and Orford Ness, illustrating locations referred to in the text (black markers south of Shingle Street show locations of Martello Towers and The Beacons bungalow). Topographic transects (**a–h**) are derived from Lidar survey data (2008). Wave roses summarise the wave climate recorded at Sizewell (c. 14 km north of Orford Ness) and West Gabbard (c. 36 km offshore of Orford Ness) over the period 2010–2012 inclusive (CEFAS)

4.3 Method

Published accounts of the historical evolution of Orford Ness and Spit (e.g. Steers 1926; Carr 1965, 1986) are revisited here in the context of more recent data sources and analytical tools. The relative position of the terminus of Orford Spit was assessed from early historical maps and charts, Ordnance Survey maps, aerial photography, satellite imagery and Lidar data. Most of these analyses were undertaken within a GIS, but where geographic control was lacking (i.e. in maps and charts produced prior to the nineteenth century), positions were derived relative to consistently represented locations (such as local villages and stationary geomorphic features). The contemporary geomorphology is described using topographical profiles derived from Lidar survey data (acquired 2008), digitised bathymetric soundings from Trinity House navigation surveys and field observations. Analyses were undertaken in ArcGIS (including the use of DSAS (Thieler et al. 2009)) and Matlab.

4.4 Orford Spit: Morphology

The main spit form of Orford Ness and Spit comprises multiple gravel-dominated beach ridges. The spit is fringed on the landward side by saltmarsh, mud flats and tidal channels of the Alde/Ore estuary. Lulls between ridges and multiple ridge units, which reflect the progressive development of the system, are occupied by saltmarshes of varying ages. McGregor and Green (1990) suggested that Orford Ness and Spit could be divided into two depositional units: an earlier deposit extending from Aldeburgh to the apex of Orford Ness and a later unit from Orford Ness through to North Weir Point (Fig. 4.1). This is certainly reflected in the geomorphology, although it could be said that the anthropogenic history has imprinted as much on the system as its historical development. In many places, the gravel beach ridge structures are somewhat compromised due to the modifications and damage associated with 70 years use as a Ministry of Defence atomic weapons research facility and bombing range (McGregor and Green 1990).

The spit is narrowest at Slaughden, just south of Aldeburgh, where the 25 m wide ridge is protected by groynes and seawall. This section is vulnerable to breaching and has been the focus of modelling and strategy studies to evaluate the most appropriate management and coastal defence options (e.g. Guthrie and Cottle 2002). This part of the system has a history of sediment recharge dating back to at least 1965 and undertaken on a near-annual basis in the 1990s (Green and McGregor 1990; Guthrie and Cottle 2002). Most of this northern section comprises a single, tall beach ridge (which was until relatively recently artificially steepened), supplemented in places with storm berms high on the foreshore (Fig. 4.1a–c). The spit is widest just south of the apex of Orford Ness, where there are 30+ ridges (Fig. 4.1d), and from here to North Weir Point, beach ridges are smaller than their northern counter parts, reaching 3.5–4.5 m above sea level. This southern section broadly tapers toward North Weir Point, but is variable in width, specifically where the Alde/Ore channel meanders along the landward spit boundary in contrast to the smooth seaward shoreline. Ridge structure is notably consistent throughout this section (Fig. 4.1d–g).

Orford Spit reaches a terminus at North Weir Point, where the Alde/Ore connects with the North Sea through the ebb delta of Orford Haven (Fig. 4.2a). The inlet comprises a number of gravel and sand shoals through which the main ebb channel passes: some smaller marginal flood channels dissect this ebb delta (Burningham and French 2007). On the south side of Orford Haven, the hamlet of Shingle Street is located on a series of gravel beach ridges (Fig. 4.2b) comparable

Fig. 4.2 Photographs of (**a**) view of North Weir Point from Shingle Street across the ebb delta shoals of Orford Haven (with Orford Castle visible in the distance); (**b**) gravel beach ridges at Shingle Street; (**c**) modern saltmarsh peat exposed on the seaward foreshore of the Shingle Street gravel; and (**d**) small gravel spit extending into the Alde/Ore inlet

in structure to those present on the southern section of Orford Spit (Fig. 4.1h). Reclaimed land and infilled wetlands lie landward of this gravel foreland.

Sedimentologically, the Orford Ness and Spit gravel is largely flint (>90 %) with some sandstone (<3.3 %) and quartz (<3 %), and falls primarily in the fine to very coarse gravel size range (Carr 1970; McGregor and Green 1990). This material can be directly linked to Pleistocene deposits exposed in cliffs at Dunwich (the Westleton Beds), approximately 15 km north of Orford Ness (McGregor and Green 1990). As Dunwich has experienced a long history of cliff erosion (Robinson 1980; Pye and Blott 2006), it is generally supposed that this recession has supplied the gravel accumulation at Orford Ness (Fuller and Randall 1988; McGregor and Green 1990). Sand is not commonly found in the sub-aerial spit, but does occur in patches and stratigraphic lens within the foreshore.

4.5 Spit Development: A 500 Year History

The earliest map evidence of Orford Spit is from c. 1539 depicting the coast from Yarmouth in the north to Stour/Orwell estuary entrance in the south (British Library 2013a). Orford Ness is a prominent feature on this map, as is Orford Castle, which is located just inland of the Alde estuary mouth and distal point of Orford Spit. This map is widely used as evidence of a considerably shorter spit, and a gauge against which the growth of the spit has historically been measured (Redman 1864; Steers 1926). Although the date is likely imprecise, there is general agreement with written records that Orford town was relatively close to the open sea during the Middle Ages when it operated as a port (Steers 1926). Historical development of the spit can really only be assessed in terms of the relative position of North Weir Point – the terminus of Orford Spit. Due to navigation needs, this feature is generally better described or represented than other features of the Spit or Ness, certainly until the modern mapping and surveying era.

A map c. 1575 (unattributed), equivalent in detail to that from 1539, locates Orford Haven half way between Dove Point and Shingle Street (British Library 2013b). Appleton's 1588 map is very similar, which suggests that the spit had extended by almost 5 km in half a century. This may be a fair reflection of a particularly dynamic period in the history of Orford Spit, but placing confidence in the historical record that these older maps provide is challenging as many maps published before the twentieth century are copies of earlier maps, and the chronology isn't always clear. The two states (short or long spit) represented in the earliest maps continues to be illustrated in maps across the sixteenth to eighteenth centuries, with evidence also of an intermediate state (Fig. 4.3a). As previously noted, it is very likely that the later maps which show a shorter spit are simply copies of earlier maps; for example, a Blaeu map from 1645 is redrawn from a 1579

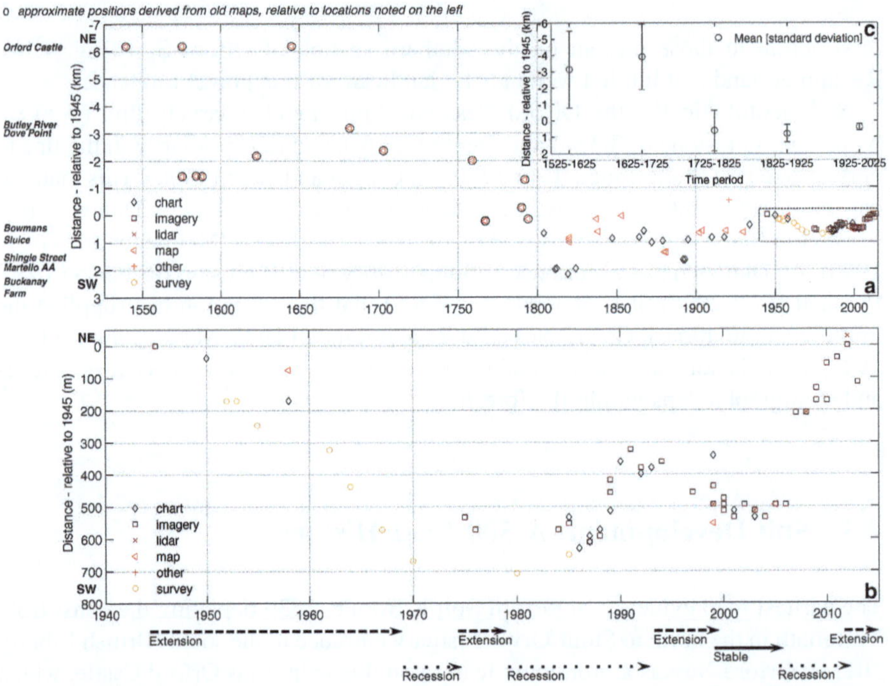

Fig. 4.3 Historical changes in the position of North Weir Point showing the (a) long-term history, (b) post-1940 and (c) century-scale magnitudes of change. Positions are referenced to the location of North Weir Point in 1945, as previously used by Carr (1986) from which several of the nineteenth and twentieth century positions are obtained

map credited to Saxton. The exact temporal sequence is uncertain, but comparing these approximate locations with the more recent history does confirm that Orford Spit extended southward quite significantly (by over 6 km) over the period 1500–1800.

Although maps from the nineteenth century benefited from increased accuracy and geographic control, lags between surveys and publication continue to imprint noise on the time series (Fig. 4.3a). It is fair to note though that Orford Spit was in its most extended form during the early 1800s, and that the envelope of variability over the most recent 200 years has been substantially less than during the preceding 300 years (Fig. 4.3). From its maximum extent around 1810–1820, the spit spent much of the mid-nineteenth century in a process of retreat or was stationary (Redman 1864), which contrasts with the predominance of growth prior to this time. The spit reached a minimum length c. 1850–1860, after which it extended southward again until 1893 when November storms breached the southern part of the spit, causing breakdown and some recession over the following 10 years (Cobb 1957; Steers 1927). Progradation reportedly recommenced shortly after, but mapping evidence for this is inconclusive (Fig. 4.3a).

The first aerial photography survey available dates to 1945, and this places North Weir Point around 700 m north of its breached position in the early 1900s, suggesting that the accounts of growth by Steers (19277) and Cobb (1957) were small-scale and short-lived. Since 1945, the evidence for change has been significantly enhanced by (i) the topographic monitoring undertaken by the Nature Conservancy (1956–1985) and (ii) the onset of airborne and satellite image acquisition (since 1975). The time-series produced from these and other data is detailed and displays a strong signal, the most striking aspect of which is the decreasing envelope of variability, particularly when comparing pre- and post-1945 periods (Fig. 4.3b). Considered at the century timescale, the main shift from shorter spit and large scale change, to longer spit and smaller-scale change occurs toward the end of the eighteenth century (Fig. 4.3c). The spit length achieved by 1800 seems to represent a limit to growth as subsequent centuries cluster around the same average length, whilst century-scale magnitudes of change continue to decline.

The long-term picture is one of episodic, but significant spit growth until the early 1800s when the spit experienced smaller-scale and shorter periods of recession followed by growth. Between c. 1810 and 1945, the North Weir Point cycled through recession, growth and recession, each phase taking around 40 years. The cycles have continued throughout the last 70 years, but in general, each part of the cycle reduces in time scale and magnitude of change. Certainly, the rapid growth that Carr (1986) expected in the late 1980s and 1990s did not occur; in fact this period was dominated by recession with only relatively small-scale (<200 m) growth toward the end of the 1990s. The post-1980 spit has been largely recessional and North Weir Point is currently located 600 m to the north of its 1980 position.

Analysis of shoreline position along the entire spit, from published OS maps and aerial photography reveals a spatially variable pattern of recession and advance (Fig. 4.4). Pye and Blott (2009) showed that the Walberswick-Dunwich gravel beach-barrier (c. 14 km north of Aldeburgh) has progressively receded over the last

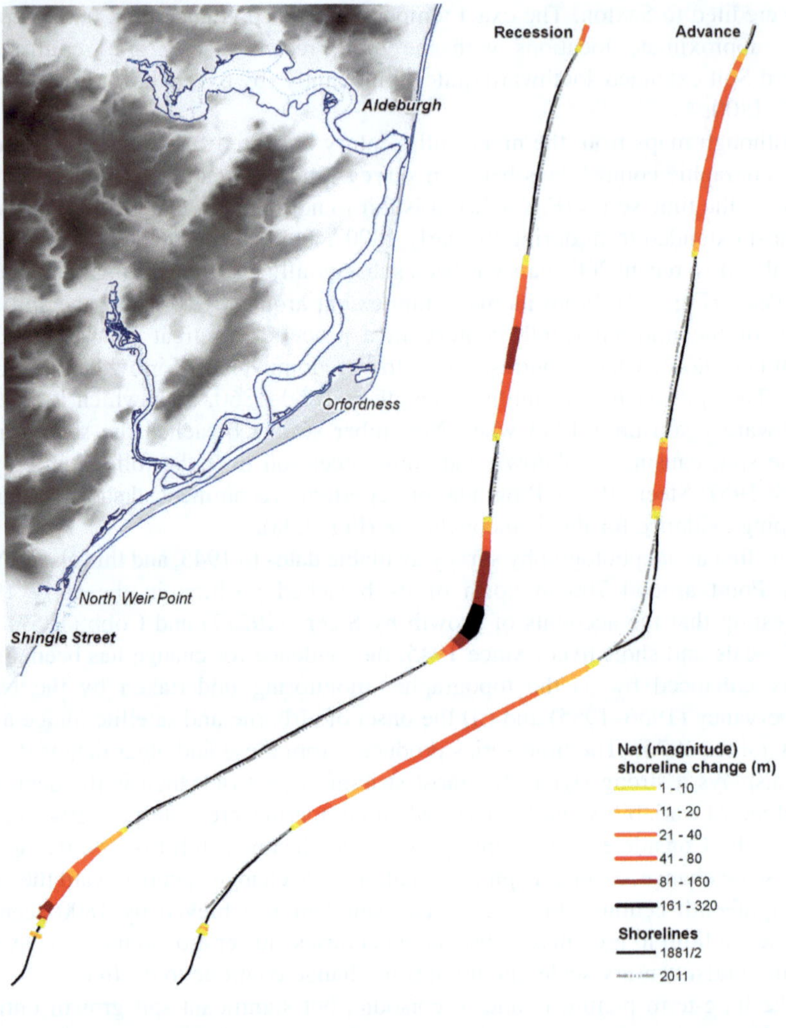

Fig. 4.4 Shoreline change analysis of Orford Spit, illustrating the spatial variation in net change (recession and advance) between 1881 and 2011

500 years, with an average erosion rate of 0.3–1.3 myr^{-1} over the last 160 years, which is thought to be slower than the rate during preceding centuries. Orford Spit does not present a similar 'barrier rollover' behaviour. Although it is not possible to map accurate shorelines from the earlier maps, it does seem that most of the 1500–1800 history was dominated by spit extension. More recently, recession has taken place along discrete sections of shoreline, which are divided by sections of progradation (Fig. 4.4): 56 % of the Aldeburgh to North Weir Point shoreline has prograded over the 1881–2011 period, with an average seaward shift in the mean high water of 37 m (0.28 myr^{-1}). The remaining 44 % of shoreline has eroded,

with an average recession of 63 m (0.48 myr^{-1}). Over this same time frame, the total gravel deposit of Orford Spit has decreased in area from c. 3.9×10^6 m^2 in the late 1800s to the present area of 3.5×10^6 m^2. A large portion of this decrease is associated with the damage caused by years of defence-related activity during the mid-twentieth century, where gravel ridges have been removed from the landward margin of the spit. Overall, these figures suggest a general decline in sediment volume, but this is not expressed in a simple landward shift in the barrier. Washover fans akin to those present along the Walberswick-Dunwich barrier are not evident on Orford Spit, and the spit has experienced reshaping (primarily in the form of reduced curvature in the shoreline) rather than rollover. This is particularly evident at Orford Ness where the maximum recession has occurred along the apex (1.7–1.8 myr^{-1}).

4.6 Spit-Inlet-Delta Dynamics

As noted, North Weir Point and the broader inlet system of Orford Haven have received much attention throughout the 1950–1980s. The terminus of Orford Spit was surveyed on a near annual basis by the Nature Conservancy, and reported extensively by Carr (1965, 1969, 1970, 1972, 1986) and others (e.g. Cobb (1957); Kidson et al. (1958); Kidson and Carr (1959); Kidson (1961); Kidson (1963); Randall (1973)). During this period, Carr described how Orford Spit extended through the accumulation of gravel at North Weir Point in the form of progradational ridges, which forced the infilling and down-drift shift of the Alde/Ore tidal channel and erosion of the Shingle Street foreland. In his most recent reporting of this system Carr (1986) presented a cycle of development that described a form of sediment bypassing similar to the inlet migration and spit breaching model defined by Fitzgerald (1988). This model is a fair description of the longer-term behaviour of this gravel spit, ebb delta and inlet, but spit-inlet dynamics over the last 40 years do not align quite so well. Certainly, spit recession is now a far more common process than spit breaching.

Morphological changes in the ebb delta imply sediment is delivered to intertidal banks on the northern/updrift portion of the delta (attached to North Weir Point) causing southerly growth and extension of the ebb shoals (Fig. 4.5a). This continues until a point when the ebb channel route is so significantly constrained to the south and against the Shingle Street shoreline that it re-routes to the north, cutting through the North Weir Point banks. This switch from an extended southerly path to a shortened easterly path (breaching the ebb delta), occurred in 1998–1999 and again in 2011–2012, a period of 13 years between ebb delta breaches. The late 1990s ebb delta breach immediately followed a short period of extension in Orford Spit after more than a decade of progressive recession (Fig. 4.5b). It is possible that this recession was reflecting a more general decline in sediment supply to the inlet system, which would have also led to some breakdown in the ebb delta. The short period of spit extension at this time could have been facilitated by an increase in

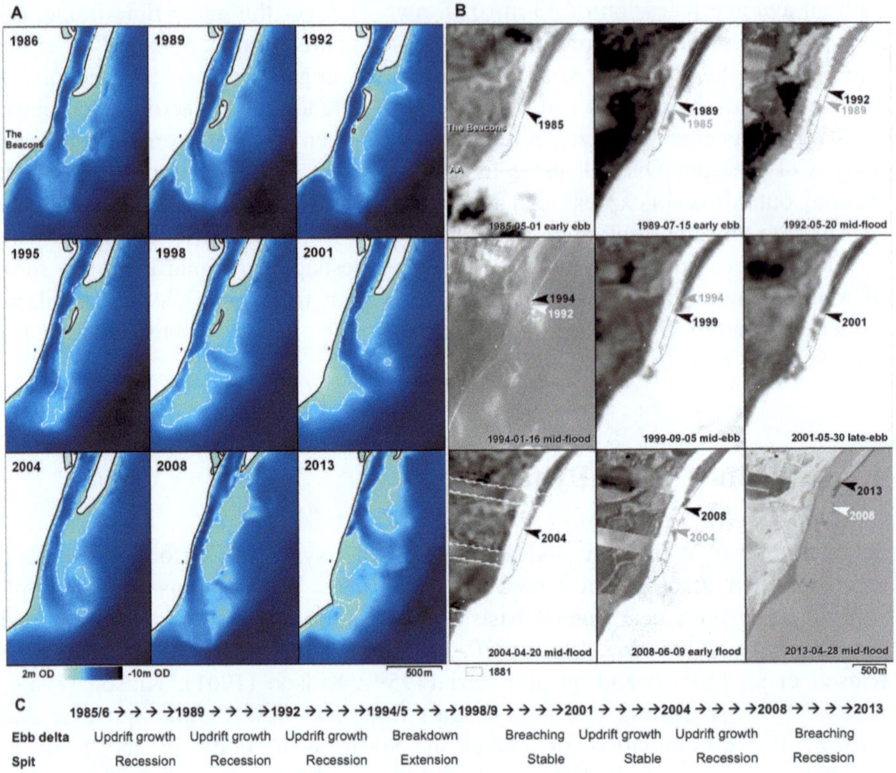

Fig. 4.5 Inlet (**a**) and spit (**b**) dynamics over the 1986–2013 period, and summary of behaviour (**c**). For geographical reference, markers show the position of The Beacons bungalow and Martello Tower AA

the rate of sediment supply, but this was perhaps insufficient to strengthen the ebb delta. The ebb delta breach in 2011–2012 again followed a period of recession that was dominated by 300 m of erosion during the winter of 2006/2007. This erosion was most likely associated with the 31 October–1 November 2006 storm surge that was also responsible for significant breaching and overwash of the Walberswick-Dunwich barrier (Pye and Blott 2009). There is no explicit evidence of a spit breach at this time, and if any form of breaching occurred, speedy reworking of the breached spit across the ebb delta followed, as no spit remnants are visible in 2007 aerial photography. It is worth noting that storm surge events are not particularly well associated with change in this spit system; for example the 1953 storm surge that caused extreme flooding throughout southeast England (Steers 1953) had little impact on North Weir Point (Fig. 4.3). Progressive shortening of Orford Spit continued from 2007 at around 40 myr^{-1} until 2012. It is therefore apparent that the inlet migration and spit-breaching model of bypassing has not applied over the last 40 years despite clear evidence that the system has undergone at least one bypassing cycle.

Although the dominant morphological changes within the inlet region are those associated with spit growth/decay and ebb delta shoal migration and breaching, smaller scale changes are evident along the Shingle Street shoreline. The Shingle Street beach ridge system is a product of the inlet bypassing process whereby breaching of the ebb delta and shift in ebb jet position further north supplies large volumes of sediment to the down drift (Shingle Street) foreshore. An additional consequence of this shoal migration is the development of lagoons within the gravel foreland. Past expressions of these lagoons have been described extensively by Cobb (1957), but these transient features are also now recognised as important habitats described as *percolation lagoons* in the EU 1150 Coastal lagoon habitat. Lagoons were well developed at Shingle Street in the 1940–1950s, and are reliant on the supply of gravel in the form of substantial banks that migrate onshore and attach to the downdrift shoreline in the later stages of the bypassing cycle. The onshore and alongshore movement of shoals and sediment ensure that the lagoons themselves are temporary features but, as with the varying scales of change recognised in Orford Spit itself, the lagoons at Shingle Street appear to be increasingly transient, as illustrated by the most recent lagoon development (Fig. 4.6). The ebb delta at Orford Haven was breached during the winter of 1998–1999, with a shift in ebb jet position just to the north of The Beacons bungalow and closure of the previous route along the Shingle Street shoreline. Closure of this previous channel was achieved through the onshore migration of the downdrift ebb delta shoal, and its attachment to the Shingle Street shoreline. This process facilitated the development of a lagoon within the gravel ridge foreland (Fig. 4.6). The lagoon was intact from spring 2001 and persisted for around 30 months, but was infilled by the end of 2003. Between 2000 and 2004, there was no appreciable change in the position of North Weir Point.

Features within the Shingle Street gravel retain evidence of decadal-scale inlet dynamics, and the sediment transport patterns within the inlet region are ostensibly more complex than suggested when considering the full extent of the spit system. Decadal-scale barrier rollover is evident along the Shingle Street shoreline, where the gravel ridges have migrated over the backbarrier sediments, exposing modern saltmarsh stratigraphy within the foreshore (Fig. 4.2c). The tidal flat and saltmarsh within the Bowmans Creek area have had substantial protection by either Orford Spit or the mid-twentieth century lagoonal foreland of Shingle Street, allowing saltmarsh sedimentation through to the near-present, as shown in the presence of plastic litter within these exposed sediments. In addition to landward transport of gravel, small gravel spits have extended northward from Shingle Street into the narrow estuary (Figs. 4.2d and 4.7), clearly showing that the inlet sediment transport processes are operating in multiple directions. These small spits are very responsive to changes in the position of North Weir Point and extent of Orford Spit. As the ebb delta shoals would obstruct low tide wave processes, and that the small spits occupy the upper foreshore and supratidal, these Shingle Street features are evidence of northward sediment transport by high tide wave processes, presumably from a southerly direction.

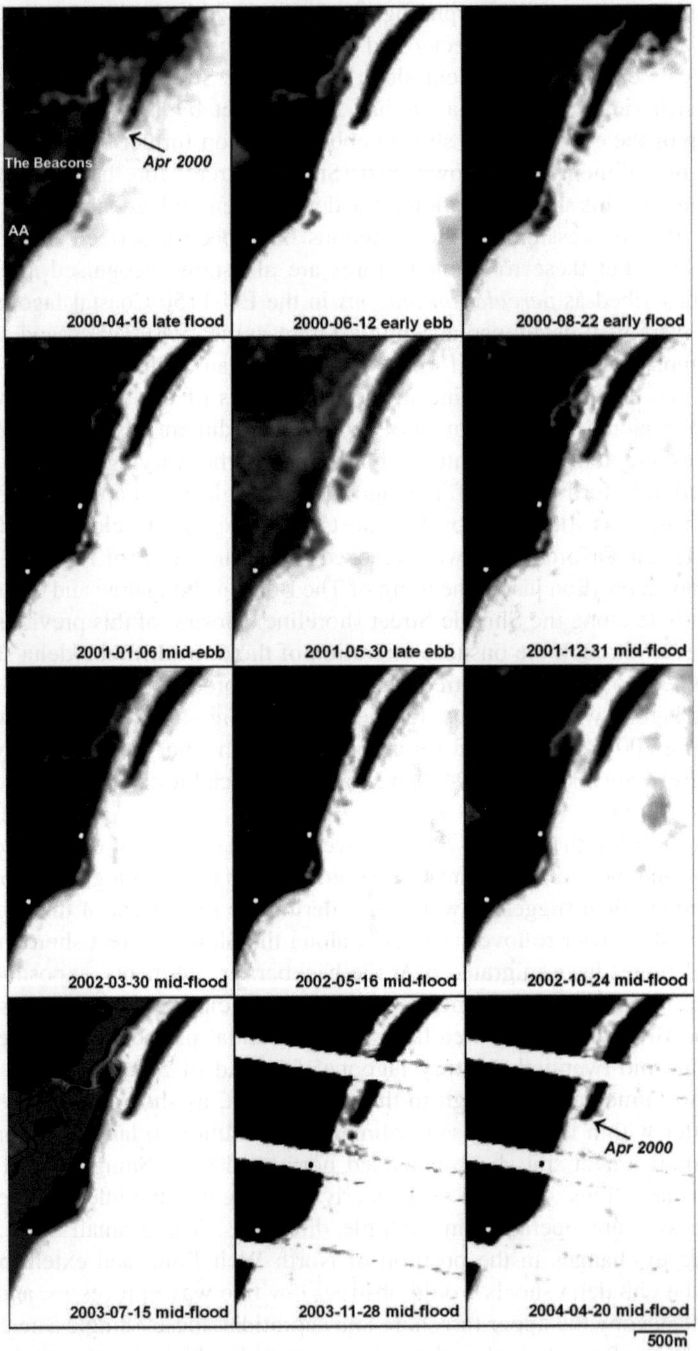

Fig. 4.6 Onshore bank migration, lagoon development and lagoon decay at Shingle Street (2000–2004) illustrated through LandSat imagery. For geographical reference, markers show the position of The Beacons bungalow and Martello Tower AA

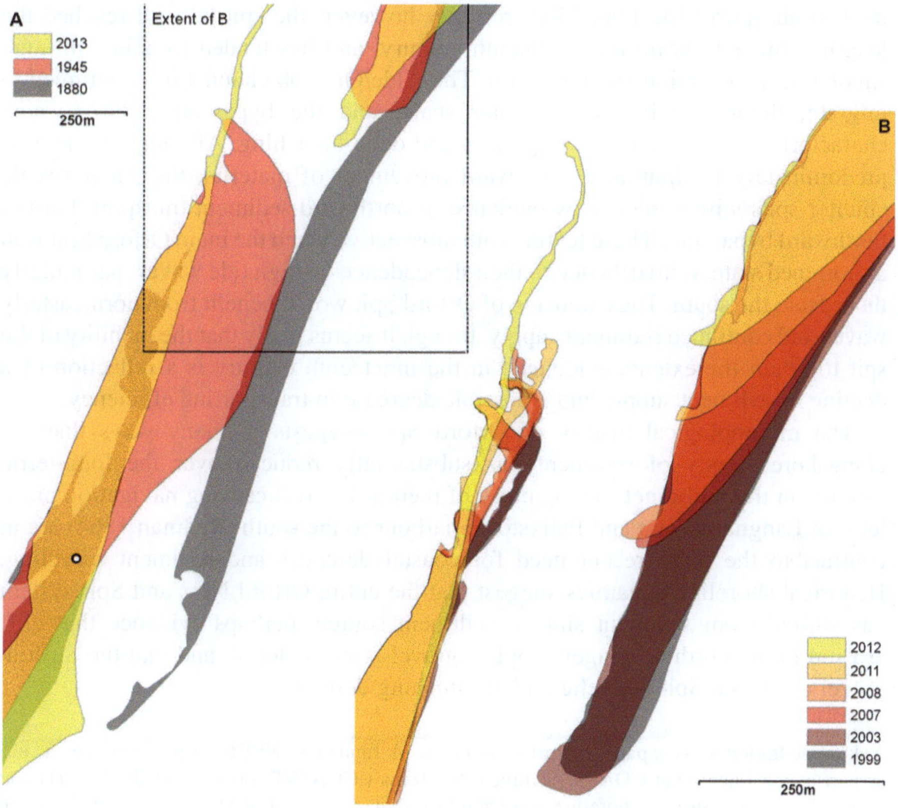

Fig. 4.7 Shoreline change and estuary-directed spit development at Orford Haven

4.7 Conclusions

The gravel spit and inlet system associated with Orford Spit has exhibited significant shifts in scales of development over the last 500 years. Throughout this history, sediment supply and spit-inlet dynamics are non-linear, but there is a significant downscaling and dampening of system dynamics over this 500 year period. The 1500–1800 period seems to be characterised by century- and km-scale growth, presumably facilitated by a plentiful, though likely episodic, sediment supply. The spit length reached in the early 1800s appears to present a morphological limit to spit extension on this coast which was only achieved again in the late 1800s. The post-1800 period represents a change from progressive growth to cyclical change where periods of growth and decay are evident. During this time, the extended spit is known to have breached significantly.

During the 1800s to mid-1900s, the inlet system migrated large distances as a result of accumulation and progradation of North Weir Point, and bypassing of sediment was primarily achieved through spit breaching and re-positioning of the inlet to an updrift location. More recently however, the spit has not reached the length achieved during the nineteenth century, and has tended to adjust through smaller scale accretion and recession. The Alde/Ore ebb channel has continued to migrate, though again on a smaller scale, and the bypassing cycle is now characterised by ebb channel migration and delta breaching. Although the inlet is predominately facilitating a southward movement of material, the smaller-scale counter-spits within inlet show evidence of northward sediment transport, but not northward bypassing. These features are most active when the main Orford Spit is in a shortened state, primarily due to their dependence on high tide waves, particularly those from the south. The extension of Orford Spit would benefit from northeasterly waves and continued sediment supply, though it seems likely that the inability of the spit to attain the extents evidenced in the nineteenth century is a reflection of a decline in sediment supply and a possible decrease in transporting efficiency.

The morphological history of Orford Spit suggests in many ways that the alongshore supply of sediment has substantially reduced over the long-term. Reports in the mid-nineteenth century of plentiful gravel causing navigation problems at Languard Point and Felixstowe harbour to the south (Redman 1864) are in contrast to the more recent need for coastal defences and sediment recycling. Historical shoreline dynamics suggest that the entire Orford Ness and Spit system has shifted from sediment sink to sediment source, perhaps evidence that cliff erosion to the north no longer supplies gravel-sized material, and that the Suffolk gravel of Orford Spit is a relic and diminishing deposit.

Acknowledgments This paper has arisen out of work funded by NERC (http://www.nerc.ac.uk) as part of the Integrating COAstal Sediment SysTems (iCOASST) project (NE/J005541/1) and Crown Estate (on Suffolk shoreline and shoreface dynamics). The iCOASST project draws on additional support, including provision of datasets, from the Environment Agency (http://environment-agency.gov.uk) as an embedded stakeholder. I am also grateful to Trinity House for the provision of navigation surveys.

References

British Library (2013a) A coloured chart of the coast of Suffolk, from Orwell Haven to Gorlston, near Yarmouth – dated 1539. http://www.bl.uk/onlinegallery/onlineex/unvbrit/a/001cotaugi00001u00058000.html

British Library (2013b) A coloured chart of the coast of Suffolk, from Bawdseye to Thorpe, including Orford and Aldboroug – dated 1575–1600. http://www.bl.uk/onlinegallery/onlineex/unvbrit/a/001cotaugi00001u00064000.html

Burningham H, French JR (2007) Morphodynamics and sedimentology of mixed-sediment inlets. J Coast Res SI50:710–715

Carr AP (1965) Shingle spit and river mouth: short term dynamics. Trans Inst Br Geogr 36:117–129

Carr AP (1969) The growth of Orford Spit: cartographic and historical evidence from the sixteenth century. Geogr J 135:28–39

Carr AP (1970) The evolution of Orfordness, Suffolk, before 1600 AD: geomorphological evidence. Zeitschrift für Geomorphologie Neue Folge 14:289–300

Carr AP (1972) Aspects of spit development and decay: the estuary of the River Ore, Suffolk. Field Stud 4:633–653
Carr AP (1986) The estuary of the River Ore, Suffolk: three decades of change in a longer- term context. Field Stud 6:43–58
Carr AP, Baker RE (1968) Orford, Suffolk: evidence for the evolution of the area during the quaternary. Trans Inst Br Geogr 45:107–123
Cobb RT (1957) Shingle Street, Suffolk. Report of the Field Studies Council, 1956–57, 31–42
FitzGerald DM (1988) Shoreline erosional-depositional processes associated with tidal inlets. In: Aubrey DG, Weishar L (eds) Hydrodynamics and sediment dynamics of tidal inlets. Springer, New York, pp 186–225
Fuller, Randall (1988) The Orford Shingles, Suffolk, UK–classic conflicts in coastline management. Biol Conserv 46:95–114
Funnell BM (1996) Plio-pleistocene palaeogeography of the southern North Sea Basin (3.75-0.60 Ma). Quat Sci Rev 15:391–405
Green CP, McGregor DFM (1990) Orfordness: geomorphological conservation perspectives. Trans Inst Br Geogr 15(1):48–59
Guthrie G, Cottle R (2002) Suffolk coast and estuaries coastal habitat management plan. Final Report, English Nature/Environment Agency, 209 pp
Kendall AC, Clegg NM (2000) Pleistocene decalcification of Late Pliocene Red Crag shelly sands from Walton-on-the-Naze, England. Sedimentology 47:1199–1209
Kidson C (1961) Movement of beach materials on the east coast of England. East Midland Geogr 16:3–16
Kidson C (1963) The growth of sand and shingle spits across estuaries. Z Geomorphol 7:1–22
Kidson C, Carr AP (1959) The movement of shingle over the sea bed close inshore. Geogr J 125:380–389
Kidson C, Carr AP, Smith DB (1958) Further experiments using radio-active methods to detect the movement of shingle over the sea bed and alongshore. Geogr J 124:210–218
McGregor DFM, Green CP (1990) Composition of Orfordness shingle. Proc Geol Assoc 101(3):259–263
Pye K, Blott SJ (2006) Coastal processes and morphological change in the Dunwich-Sizewell area, Suffolk, UK. J Coast Res 22(3):453–473
Pye K, Blott SJ (2009) Progressive breakdown of a gravel-dominated coastal barrier, Dunwich–Walberswick, Suffolk, U.K.: processes and implications. J Coast Res 25(3):589–602
Redman JB (1864) The east coast between the Thames and the Wash estuaries. Min Proc Inst Civil Eng 23:186–256
Robinson AHW (1980) Erosion and accretion along part of the Suffolk coast of East Anglia, England. Mar Geol 37:133–146
Steers JA (1926) Orford Ness; a study in coastal physiography. Proc Geol Assoc 37:306–325
Steers JA (1927) The east Anglian coast. Geogr J 69:24–48
Steers JA (1953) The east coast floods, January 31-February 1 1953. Geogr J 119:280–295
Thieler ER, Himmelstoss EA, Zichichi JL, Ergul A (2009) Digital Shoreline Analysis System (DSAS) version 4.0 – an ArcGIS extension for calculating shoreline change. U.S. Geological Survey, Open-file report 2008–1278

Chapter 5
Aeolian Sand Invasion: Georadar Signatures from the Curonian Spit Dunes, Lithuania

Ilya V. Buynevich, Albertas Bitinas, and Donatas Pupienis

Abstract Aeolian landforms may occupy large portions of coastal spits and often consist of multiple generations defined by periods of stability and reactivation. Where earlier phases of aeolian activity are masked by subsequent deposition, continuous high-resolution (200 MHz) ground-penetrating radar (GPR) images are used to reconstruct their dynamic history. Subsurface profiles from younger segments of the Great Dune Ridge massif along the northern Curonian Spit, Lithuania reveal several phases of dune migration, which generally correspond to climatic deterioration during the Little Ice Age. Exacerbated by regional deforestation, 15–30-m-high dunes have invaded coastal settlements, causing their repeated relocation. Along the study transect, an optical age of the younger slipface strata is consistent with the record of village entombment by drifting sand by AD 1797. This study demonstrates that in the absence of mappable paleosols, a combination of geophysical, luminescence-dating, and documentary databases offers a means of reconstructing the history of human-landscape interaction, with potential implications for the future stewardship of natural and cultural resources along spit coastlines.

5.1 Introduction

Positioned near the land-water interface, coastal dunefields have been active throughout various phases of the Holocene and are sensitive to major reorganization in climatic forcings and geological processes along marine and lacustrine margins (Hesp and Thom 1990; Carter 1991; Pye 1993; Jimenez et al. 1999; Clarke et al. 2002; Hesp 2002; Murray-Wallace et al. 2002; Loope et al. 2004; Maia

I.V. Buynevich (✉)
Temple University, Philadelphia, PA, USA
e-mail: coast@temple.edu

A. Bitinas
Klaipėda University, Klaipėda, Lithuania
e-mail: albertas.bitinas@jmtc.ku.lt

D. Pupienis
Vilnius University, Vilnius, Lithuania
e-mail: donatas.pupienis@gf.vu.lt

G. Randazzo et al. (eds.), *Sand and Gravel Spits*, Coastal Research Library 12,
DOI 10.1007/978-3-319-13716-2_5

et al. 2005; Pedersen and Clemmensen 2005; Clarke and Rendell 2006; Hart and Peterson 2007; Bristow et al. 2009; Buynevich et al. 2010a; Dobrotin et al. 2013). Understanding the effects of environmental and anthropogenic changes on coastal landscapes is particularly important in a current regime of rapid shifts in climate, sea level, and sediment supply, combined with ever increasing population pressures along sandy coasts.

One of the fundamental issues regarding the evolution of aeolian landscapes is determining whether a particular dunefield represents a single generation or developed through multiple phases of activity, stability, and reactivation. The latter sometimes leads to sand invasion into coastal settlements (Sherman and Nordstrom 1994; Rick 2002; Buynevich et al 2010a; Girardi and Davis 2010). Therefore, understanding long- and short-term dune dynamics is an important aspect of regional coastal management practices (Žilinskas et al. 2001; Povilanskas et al. 2009). Many earlier studies of long-term dune dynamics have been limited to exposures and sediment cores, inability to trench into the older part of the sequence below the water table, and challenges in establishing regional correlation of aeolian and intervening organic horizons (Bakker et al. 1990; Klijn 1990; Sevink 1991; Junger et al. 2000; Halfen and Johnson 2013).

In recent decades, advances in high-resolution subsurface imaging technology, such as ground-penetrating radar (GPR), allow rapid collection of continuous records, with decimeter-scale resolution and penetration depths of 5–10 m common for freshwater-saturated coastal successions (van Heteren et al. 1998; Neal and Roberts 2000; Jol and Bristow 2003; Pedersen and Clemmensen 2005; Buynevich et al. 2007a, 2009, 2010a). Greater penetration (10–30 m) has been obtained in unsaturated dune sands (Schenk et al. 1993; Clemmensen et al. 2001; Neal and Roberts 2001; van Dam et al. 2003; Havholm et al. 2004; Bristow 2009). GPR allows for rapid imaging of dune stratigraphy, both above and below the water table, and provides information about apparent dip angles of lateral accretion and truncation surfaces, thickness of aeolian units, depth and extent of buried soil (paleosol) horizons, distribution of buried structures, and relative chronology of dune formation, migration, and deflation (Botha et al. 2003; Havholm et al. 2004; Bristow et al. 2005; Clemmensen et al. 2007; Bristow 2009; Buynevich 2009; Girardi and Davis 2010; Dobrotin et al. 2013).

The Curonian Spit along the southeast coast of the Baltic Sea presents an ideal opportunity to investigate the signatures of reactivation of aeolian activity (Fig. 5.1). It is located north and east of the sites where dune dynamics and its links to climate change (e.g., storminess and North Atlantic Oscillation regime) have been investigated in recent years (van der Meulen 1990; Seppälä 1995; Wilson et al. 2001; Clarke et al. 2002; Clemmensen et al. 2007; Forman et al. 2008). Furthermore, this sand spit is the home to famous “wondering dunes” that invaded tracts of maritime forests and a number of coastal settlements in historic times (Mager 1938; Sherman and Nordstrom 1994; Gudelis 1998; Buynevich et al. 2007b). The aim of this paper is to present the first subsurface evidence for historical dune reactivation along a section of the northern Curonian Spit, Lithuania (Fig. 5.2),

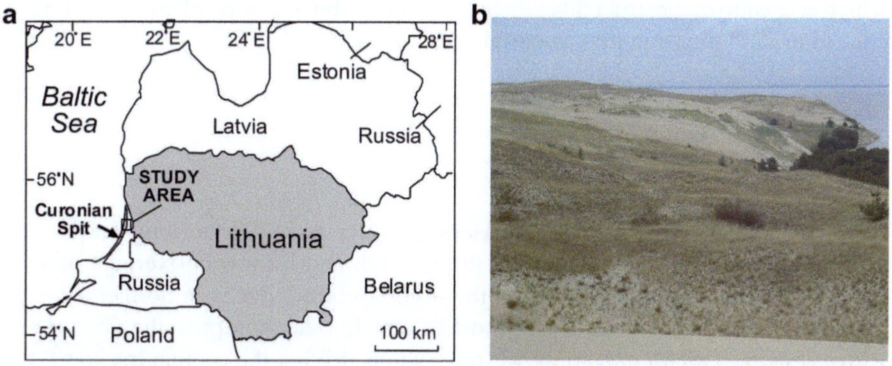

Fig. 5.1 Study area. (**a**) Location of the Curonian Spit along the southeast Baltic Sea coast. (**b**) Aerial photograph of the Great Dune Ridge along the Lithuanian part of the spit

Fig. 5.2 Examples of Curonian settlements buried by migrating dunes. (**a**) Map of historical villages along the northern half of the spit (database: archival materials summarized by Mager 1938; Gudelis 1998; Bučas 2001). (**b**) Photograph of a partially buried house along the lagoon shoreline (From Gudelis 1998). (**c**) Inscription on the monument at the site of the Old Karvaičiū II settlement ("1797 – burial by thick sand", in Lithuanian)

which is used to reconstruct dune thickness and establish independent chronological control of sand invasion into a coastal settlement.

5.2 Physical Setting

Curonian Spit, a UNESCO World Heritage Site, is a 100-km-long barrier spit divided between Russian Federation in the south and Lithuania (Neringa municipality) in the north (Fig. 5.1). It has the highest coastal dunes in northern Europe (more than 65 m above sea level), most of which today comprise the Great Dune Ridge (Fig. 5.2). With prevailing westerly winds driving the aeolian transport, this landward (eastern) part of the spit is dominated by both active and stabilized Holocene dunes (Gudelis 1998; Žilinskas et al. 2001; Povilanskas et al. 2009). Due to the width (1–3 km) and height of the dunes preventing breaching and overwash by storms, the entire spit has been migrating landward during the Holocene due to aeolian processes (Gudelis 1998; Žilinskas et al. 2001; Buynevich et al. 2007a; Povilanskas 2009; Dobrotin et al. 2013).

Deflation of relict dunes on the northern part of the spit exposes a number of mappable paleosol horizons, with the youngest dating back to the beginning of the Little Ice Age (Fig. 5.2b; Gudelis 1998; Buynevich et al. 2007a; Gaigalas and Pazdur 2008; Dobrotin et al. 2013). The geometry and extent of these paleosols allow reconstruction of dune morphology and migration direction similar to some examples of relict dunes along the Polish coast (Borówka 1995). The earlier, cursory studies in this region suggest that at least four paleosol horizons are regional in extent and that charcoal within paleosols may be related to burning of coastal heathlands as early as 2,000 years ago (Moe et al. 2005). The first set of optical dates from several sites place the latest aeolian activity phase at <500 yBP (Bitinas 2004; Dobrotin et al 2013).

The history of human-landscape interaction along the southeast Baltic Sea coast dates back to at least the mid-Holocene and the impact on the landscape became evident during medieval times (twelfth to fifteenth centuries AD; Mager 1938; Gudelis 1998). The scale and speed of geological impact on local population have reached their peak during the fifteenth to nineteenth centuries (Gudelis and Michaliukaitė 1976; Bučas 2001). During this time, a number of communities were established along the Curonian lagoon seeking protection from the Baltic winds behind the high dunes. Only a few small areas remained forested by the end of eighteenth century. Historical documents suggest that this time was also marked by several mobilization episodes of large dunefields, triggered largely by land clearance. During this period, a number of Curonian villages were buried by migrating dunes (Fig. 5.2a; Gudelis 1998; Povilanskas and Chubarenko 2000; Bučas 2001; Povilanskas 2009; Buynevich et al. 2011). In addition, dune migration triggered extrusion of lagoonal muds (Bitinas et al. 2009; Buynevich et al 2010b) and likely generated micro-tsunamis through occasional massive grainflows into the Curonian Lagoon (Gudelis 1998).

5.3 Methods

The internal stratification of Holocene and recent dunes was investigated using high-resolution, continuous ground-penetrating radar imaging. We used a digital Geophysical Survey Systems Inc. SIR-2000 GPR system with a 200 MHz monostatic antenna (Fig. 5.3; for technical aspects, see van Heteren et al. 1998; Neal and Roberts 2000; Jol and Bristow 2003; Buynevich et al. 2009). Penetration of up to 15 m and vertical resolution of 0.16–0.20 m was typical for unsaturated sand. The two 50–60-m-long profile segments chosen as examples were extracted from a 2-km-long traverse of the Curonian Spit (Fig. 5.3a). GPR images were analyzed using Radan v6.0 software package, which involved standard post-processing routines, including surface normalization that accounts for changes in topography (Figs. 5.3 and 5.4).

To groundtruth key bounding surfaces and assess textural variations within the upper part of the imaged sections, geophysical data were groundtruthed by rotary and Edelman hand augers, with penetration depth of 4–5 m (Fig. 5.3c). For

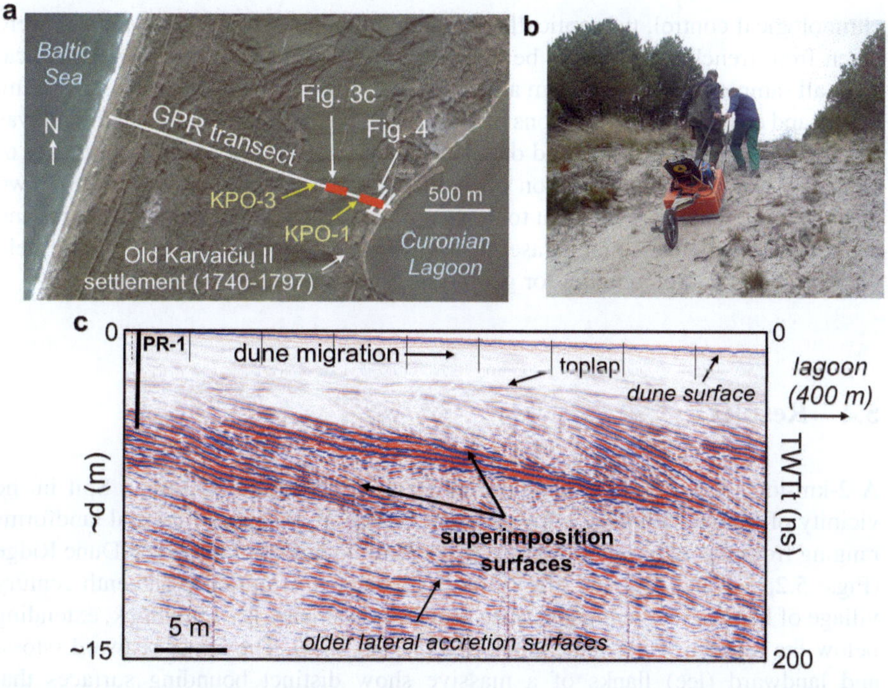

Fig. 5.3 Example of a historically active dune north of the modern Preila settlement: (**a**) Location of the GPR transect and optical samples (KPO) at the site of a former settlement. (**b**) Geophysical data collection with a 200 MHz antenna (*upper slipface*). (**c**) Georadar profile across a recently stabilized dune (see Figs. 5.2a and 5.3a for location; PR-1 – hand auger). Note the gently dipping (<15°) older accretion surfaces separated by a superimposition bedding plane from the steeper (30–31°) paleo-slipfaces above

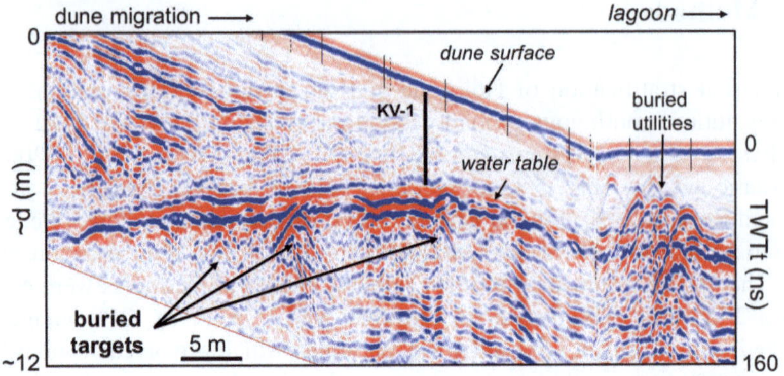

Fig. 5.4 Georadar profile along the slipface of the dune. Hyperbolic diffractions at right show buried utilities along the road. Similar radar signatures beneath 6–7 m of dune sand likely represent part of a historically buried settlement or forest The water table is slightly higher than the present lagoon water level (see Fig. 5.3a for transect location; KV-1 – hand auger). The end of the GPR profile is within 100 m of the monument (Fig. 5.2c)

chronological control, two optically stimulated luminescence (OSL) samples were taken from trench walls 50 cm below the ground surface. Guided by geophysical data, all samples were taken from above the key superimposition surfaces, avoiding toplap and ensuring upper sections of slipface strata are sampled. Aeolian sand was collected in lightproof cores and dated at the OSL Laboratory at the University of Nebraska-Lincoln (for discussion of methodology, see Huntley et al 1985). Two dates presented here are relevant to the historical record of landscape dynamics and are part of a larger regional dataset that will be published elsewhere. A hand-held GPS provided georeferencing for geophysical and chronological datasets.

5.4 Results

A 2-km-long sea-to-lagoon geophysical transect across the Curonian Spit in the vicinity of Preila, Lithuania allowed high-resolution imaging of coastal landforms ranging from recent beach sediments to aeolian sequences of the Great Dune Ridge (Figs. 5.2, 5.3 and 5.4). At the location adjacent to a buried eighteenth-century village of Karvaičiū, the imaged dune sequence is at least 12–15 m thick, extending below the modern water table (Figs. 5.2, 5.3 and 5.4). The upper seaward (stoss) and landward (lee) flanks of a massive show distinct bounding surfaces that separate gently landward-dipping reflections in the lower part of the record (10–13° apparent migrated angle) from steep (31–34°) sigmoidal-oblique reflections above (Figs. 5.3c and 5.4). The latter either have a toplap contact or continue upward to the dune surface. In the landward segment, a number of high-amplitude diffractions are located at or above the water table (Fig. 5.4).

Due to approximately W-E profile orientation, the apparent dip of the GPR reflections is close to a true depositional dip. A hand auger taken at the seaward segment penetrated 4.3 m of unsaturated fine-medium-grained dune sands with no visible heavy-mineral component, which suggests that slight textural variations (grain fabric and water retention; Bristow 2009) within the cross-stratified aeolian sediments are responsible for the upper third of the imaged sequence. At the landward section, a 4-m-deep rotary auger terminated at the water table (Fig. 5.4). Two optical samples of the upper slipface strata (KPO-3 and KPO-1) yielded downwind-younging ages of 415 ± 60 and 215 ± 30 years before present (present = AD 2013), respectively. These dates complement the chronology of historically documented burial of the coastal village – Old Karvaičiū II (AD 1740–1797; Fig. 5.2c).

5.5 Discussion

Where exposures are lacking or organic-rich horizons have not developed, geophysical surveys provide the only means of assessing the internal architecture of aeolian deposits. The key discontinuity (bounding) surfaces imaged in GPR profiles represent superimposition (higher-level reactivation) surfaces (Figs. 5.3 and 5.4) indicative of changes in depositional regime. The dip angle of the upper (younger) lateral accretion surfaces (paleo-slipfaces) approach the angle of repose and similar to modern active and vegetated slipfaces.

Along the landward flank of the dune, high-amplitude hyperbolic diffractions represent point-source signal return due to buried objects (Buynevich 2009). Whereas modern utilities are responsible for a cluster of targets near the road (lagoon-ward part of the record; Fig. 5.4), those covered by 4–6 m of dune sand with mature trees likely represent remains of buried trees or man-made structures left during the relocation of the Old Karvaičiū II settlement. The fact that it is a second location of the "Old" village indicates that it existed in the older location, prior to moving to the study site and subsequently relocating father south due to complete sand burial by 1787 (Fig. 5.2c). Optical dates are consistent with historical records of settlement movement, with a rapid dune migration within a 200-year period (ca. AD 1600–1800), culminating in village abandonment at this site and a number of other parts of the Lithuanian half of the spit (Fig. 5.5).

Abrupt climate shifts have been recognized as an important driving force of aeolian system dynamics in both coastal and inland dunefields (Klijn 1990; van der Meulen 1990; Lancaster 1997; Clemmensen et al. 2001; Clarke et al. 2002). Climate deterioration during the Little Ice Age, likely exacerbated by deforestation, was the likely cause of the latest massive dune migration (Wilson et al. 2001; Hurrell et al. 2003; Wiles et al. 2003; Clemmensen et al. 2007; Forman et al. 2008; Povilanskas 2009; Buynevich et al. 2010c). Prior to this period, the regional destabilization of forested landscapes resulting from natural (lightning) and man-made fires, have been the likely causes of devegetation (Fillon 1984; Seppälä 1995; Moe et al. 2005; Gaigalas and Pazdur 2008).

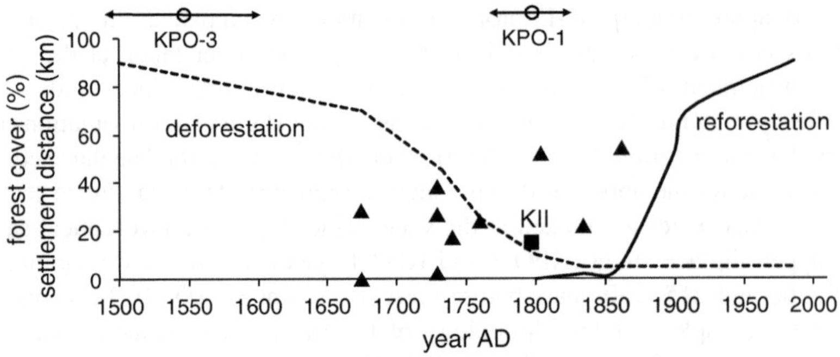

Fig. 5.5 Summary of forest cover and settlement burial along the northern Curonian Spit (Data from Gudelis 1998; Bučas 2001; Moe et al 2005). KII – Old Karvaičiū II village

Burial of lagoon-side Curonian settlements by drifting sand began shortly after regional deforestation (Fig. 5.5; Gudelis 1998; Moe et al. 2005). In contrast to other parts of the Great Dune Ridge that reveal 5,000 years of dune migration and deflation during the Little Ice Age (Buynevich et al. 2007a; Gaigalas and Pazdur 2008; Povilanskas 2009; Dobrotin et al. 2013), the Preila dune has migrated landward more than 1 km during the LIA. New geological evidence for increased wind activity just prior to LIA (Buynevich et al 2010a) is consistent with findings from inland dunefields during the Medieval Warm Period (Sridhar et al. 2006; Halfen and Johnson 2013).

Reforestation efforts that started more than a century ago have stabilized portions of the Great Dune Ridge (Fig. 5.5), however such efforts are in conflict with the long-term translation of the spit via dune migration (Povilanskas et al 2009; Dobrotin et al. 2013). Future research efforts will address the role of natural and anthropogenic factors in driving century-scale landscape dynamics along the southeast Baltic Sea coast.

5.6 Conclusions

Geophysical profiles from the younger portions of the Great Dune Ridge, Curonian Spit, Lithuania reveal distinct phases of aeolian deposition that spanned only several centuries. In contrast to other parts of the dune massif that took more than 5,000 years of dune migration and recent deflation (Buynevich et al. 2007a; Gaigalas and Pazdur 2008; Dobrotin et al. 2013), the Preila dune has moved more than 1 km in a few centuries during the Little Ice Age. This migration culminated in the burial and abandonment of the lagoon-side village by AD 1797. Within 15–30-m-thick dune sequences lacking such exposures, lateral accretion surfaces indicate the direction of sediment transport before and after the stability period, or a change in the direction of dune migration. In some cases, the upper sections of large dunes have been deflated exposing slipface-angle deposition following periods of stabilization (paleosols). However rapid dune migration (or organic matter consumption in low-latitude dune sites) often precludes soil formation. In such instances, integrated geophysical, OSL, and historical databases are crucial for reconstructing the landform dynamics, vegetation history, and cultural context of coastal dunefields.

Acknowledgments This study was funded by the Ocean and Climate Change Institute (Woods Hole Oceanographic Institution) and the National Geographic Society. We thank Anton Symonovich, Aldona Damušytė, Nijole Strakauskaite, Nikita Dobrotin, Dainius Michelevičius, and the Geological Survey of Lithuania for field support, the Curonian Spit National Park for access to study sites, and Ronald Goble for optical dating analyses. Daniel Wiese helped with depth determination of selected GPR targets and Heather Rychlak assisted with manuscript preparation.

References

Bakker TWM, Jungerius PD, Klijn JA (1990) Dunes of the European coasts. Catena Suppl 18:227

Bitinas A (2004) The age of the aeolian sediments in Lithuania. Geologija 45:1–5

Bitinas A, Buynevich IV, Damušytė A, Gadeikis M, Kabailienė M, Pupienis D (2009) A phenomenon of "dune tectonics": Curonian Spit, southeastern Baltic Sea coast. GSA Northeastern Section Abstracts with Programs, Portland, ME, 41:86

Borówka RK (1995) Dunes of the Łeba Barrier – their history and dynamics of present-day aeolian processes. In: Rotnicki K (ed) Polish coast: past, present, and future. J Coast Res Spec Issue 22:247–251

Botha GA, Bristow CS, Porat N, Duller G, Armitage SJ, Roberts HM, Clarke BM, Kota MW, Schoeman P (2003) Evidence for dune reactivation from GPR profiles on the Maputuland coastal plain, South Africa. In: Bristow CS, Jol HM (eds) Ground penetrating radar in sediments, vol 211. Geological Society of London, London, pp 29–46

Bristow CS (2009) Ground penetrating radar in aeolian dune sands. In: Jol HM (ed) Ground penetrating radar: theory and applications. Elsevier, Amsterdam, pp 273–297

Bristow CS, Lancaster N, Duller GAT (2005) Combining ground penetrating radar surveys and optical dating to determine dune migration in Namibia. J Geol Soc 162:315–322

Bristow CS, Jol HM, Augustinus P, Wallis I (2009) Slipfaceless 'whaleback' dunes in a polar desert. Victoria Valley Antarctica: insights from ground penetrating radar. Geomorphology. doi:10.1016/j.geomorph.2009.08.001

Bučas J (2001) Curonian Spit National Park. Savastis, Vilnius (in Lithuanian with English summary)

Buynevich IV (2009) Taphonomic applications of georadar. J Taphonomy 7:47–52

Buynevich IV, Bitinas A, Pupienis D (2007a) Lithological anomalies in a relict coastal dune: geophysical and paleoenvironmental markers. Geophys Res Lett 34, L09707. doi:10.1029/2007GL029767

Buynevich IV, Bitinas A, Pupienis D (2007b) Reactivation of coastal dunes documented by subsurface imaging of the Great Dune Ridge, Lithuania. J Coast Res Spec Issue 50:226–230

Buynevich IV, Jol HM, FitzGerald DM (2009) Coastal environments. In: Jol HM (ed) Ground penetrating radar: theory and applications. Elsevier, Amsterdam, pp 99–322

Buynevich IV, Souza Filho PWM, Asp NE (2010a) Dune advance into a coastal forest, equatorial Brazil: a subsurface perspective. Aeolian Res 2:27–32

Buynevich IV, Bitinas A, Damušytė A, Pupienis D (2010b) Unique Baltic outcrops reveal millennia of ecological changes. EOS Trans Am Geophys Union 91:101–102

Buynevich IV, Souza Filho PWM, Bitinas A, Asp NE, Pupienis D (2010c) Aeolian sand invasion in coastal regions: geomorphological and geophysical aspects. In: Proceedings of the 6th Schukin Conference: geomorphological processes and their applied aspects – extended abstracts, Moscow State University, Moscow, pp 281–283

Buynevich IV, Bitinas A, Souza Filho PWM, Pupienis D, Asp NE, Goble RJ, Kerber LE (2011) Rapid coastal dune migration into temperate and equatorial forests: optical chronology of imaged upper slipface strata. J Coast Res Spec Issue 64:726–730

Carter RWG (1991) Near-future sea level impacts on coastal dune landscapes. Landsc Ecol 6:29–39

Clarke ML, Rendell HM (2006) Effects of storminess, sand supply and the North Atlantic Oscillation on sand invasion and coastal dune accretion in western Portugal. The Holocene 16:341–355

Clarke ML, Rendell HM, Tastet J-P, Clavé B, Massé L (2002) Late-Holocene sand invasion and North Atlantic storminess along the Aquitaine Coast, southwest France. The Holocene 12:231–238

Clemmensen LB, Andreasen F, Heinemeier J, Murray A (2001) A Holocene coastal aeolian system, Vejers, Denmark: landscape evolution and sequence stratigraphy. Terra Nova 13:129–134

Clemmensen LB, Bjørnsen M, Murray A, Pedersen K (2007) Formation of aeolian dunes on Anholt, Denmark since AD 1560: a record of deforestation and increased storminess. Sediment Geol 199:171–187

Dobrotin N, Bitinas A, Michelevičius D, Damušytė A, Mažeika J (2013) Reconstruction of the Dead (Grey) Dune evolution along the Curonian Spit, Southeastern Baltic. Bull Geol Soc Finl 85:53–64

Fillon L (1984) A relationship between dunes, fire and climate recorded in the Holocene deposits of Québec. Nature 309:543–546

Forman SL, Sagintayev Z, Sultan M, Smith S, Becker R, Kendall M, Marin L (2008) The twentieth-century migration of parabolic dunes and wetland formation at cape Cod National Sea Shore, Massachusetts, USA; landscape response to a legacy of environmental disturbance. The Holocene 18:765–774

Gaigalas A, Pazdur A (2008) Chronology of buried soils, forest fires and extreme migration of dunes on the Kursiu nerija spit (Lithuanian coast). Landf Anal 9:187–191

Girardi JD, Davis DM (2010) Parabolic dune reactivation and migration at Napeague, NY, USA: insights from aerial and GPR imagery. Geomorphology 114:530–541

Gudelis V (1998) Lithuanian coastal region. Lithuanian Academy of Sciences Monograph, Vilnius (in Lithuanian)

Gudelis V, Michaliukaitė E (1976) Ancient parabolic dunes on the spit of Kuršių Nerija. Geographia Lithuanica 5:59–63 (In Russian with English summary)

Halfen AF, Johnson WC (2013) A review of Great Plains dune field chronologies. Aeolian Res 10:135–160

Hart R, Peterson C (2007) Late-Holocene buried forests on the Oregon coast. Earth Surf Process Landf 31:210–229

Havholm KG, Ames DV, Whittecar GR, Wenell BA, Riggs SR, Jol HM, Berger GW, Holmes MA (2004) Stratigraphy of back-barrier coastal dunes, northern North Carolina and southern Virginia. J Coast Res 20:980–999

Hesp PA (2002) Foredunes and blowouts: initiation, geomorphology and dynamics. Geomorphology 48:245–268

Hesp PA, Thom BG (1990) Geomorphology and evolution of transgressive dunefields. In: Nortdstrom K, Psuty N, Carter RWG (eds) Coastal Dunes: processes and morphology. Wiley, Chichester, pp 253–288

Huntley DJ, Godfrey-Smith DI, Thewalt MLW (1985) Optical dating of sediments. Nature 313:105–107

Hurrell JW, Kushnir Y, Ottersen G, Visbeck M (2003) An overview of the North Atlantic Oscillation. Geophys Monogr Am Geophys Union 134:1–35

Jimenez JA, Maia LP, Serra J, Morais J (1999) Aeolian dune migration along the Ceará coast, NE Brazil. Sedimentology 46:689–701
Jol HM, Bristow CS (2003) GPR in sediments: advice on data collection, basic processing and interpretation, a good practice guide. In: Bristow CS, Jol HM (eds) Ground penetrating radar in sediments, Special Publication 11. Geological Society of London, London, pp 9–27
Junger H, Korjonen K, Heikkinen O, Wiedemann AM (2000) Luminescence and radiocarbon dating of a dune series at Cape Kiwanda, Oregon, USA. Quat Sci Rev 20:811–814
Klijn JA (1990) The younger dunes in the Netherlands: chronology and causation. In: Bakker TWM, Jungerius PD, Klijn JA (eds) Dunes of the European coasts, Catena Supplement 18. Catena Verlag, Heidelberg, pp 89–100
Lancaster N (1997) Response of aeolian geomorphic systems to minor climate change: examples from the southern Californian deserts. Geomorphology 19:333–347
Loope WL, Fisher TG, Jol HM, Goble RJ, Anderton JB, Blewett WL (2004) A Holocene history of dune-mediated landscape change along the southeastern shore of Lake Superior. Geomorphology 61:303–322
Mager F (1938) Die Landschaftsentwicklung auf der Kurischen Nehrung. [The landscape development of the Curonian Spit]. Graeffe und Unser, Koenigsberg, i. Pr., 140 pp. (in German)
Maia LP, Freire GSS, Lacerda LD (2005) Accelerated dune migration and aeolian transport during El Niño events along the NE Brazilian coast. J Coast Res 21:1121–1126
Moe D, Savukynienė N, Stančikaitė M (2005) A new ^{14}C (AMS) date from former heathland soil horizons at Kuršių Nerija, Lithuania. Baltica 18:23–28
Murray-Wallace CV, Banerjee D, Bourman RP, Olley JM, Broke BP (2002) Optically stimulated luminescence dating of Holocene relict foredunes, Guilchen Bay, South Australia. Quat Sci Rev 21:1077–1086
Neal A, Roberts CL (2000) Applications of ground penetrating radar (GPR) to sedimentological, geomorphological and geoarchaeological studies in coastal environments. In: Pye K, Allen JRL (eds) Coastal and Estuarine Environments: sedimentology, geomorphology and geoarchaeology, Special Publications 175. Geological Society of London, London, pp 139–171
Neal A, Roberts CL (2001) Internal structure of a trough blowout, determined from migrated ground-penetrating radar profiles. Sedimentology 48:791–810
Pedersen K, Clemmensen LB (2005) Unveiling past aeolian landscapes; a ground-penetrating radar survey of a Holocene coastal dunefield system, Thy, Denmark. Sediment Geol 177:57–86
Povilanskas R (2009) Spatial diversity of modern geomorphological processes on a Holocene dune ridge on the Curonian Spit in the South–East Baltic. Baltica 22:77–88
Povilanskas R, Chubarenko BV (2000) Interaction between the drifting dunes of the Curonian Barrier Spit and the Curonian Lagoon. Baltica 13:8–14
Povilanskas R, Baghdasarian H, Arakelyan S, Satkūnas J, Taminskas J (2009) Secular morphodynamic trends of the Holocene dune ridge on the Curonian Spit (Lithuania/Russia). J Coast Res 25:209–215
Pye K (1993) Late quaternary development of coastal parabolic megadune complexes in north-eastern Australia. In: Pye K, Lancaster N (eds) Eolian sediments: ancient and modern, International Association of Sedimentologists, Special Publication 16. Blackwell Publishing Ltd, Oxford, pp 23–44
Rick TC (2002) Eolian processes, ground cover, and archaeology of coastal dunes: a taphonomic case study from san Miguel Island, California, USA. Geoarchaeology 17:811–833
Schenk CJ, Gautier DL, Olhoeft GR, Lucius JE (1993) Internal structure of an eolian dune using ground-penetrating radar. In: Pye K, Lancaster N (eds) Eolian sediments: ancient and modern, International Association of Sedimentologists, Special Publication 16. Blackwell Publishing Ltd, Oxford, pp 61–69. ISBN 16
Seppälä M (1995) Deflation and redeposition of sand dunes in Finnish Lapland. Quat Sci Rev 14:799–809
Sevink J (1991) Soil development in the coastal dunes and its relation to climate. Landsc Ecol 6:49–56

Sherman DJ, Nordstrom KF (1994) Hazards of wind blown sand and sand drifts. J Coast Res Spec Issue 12:263–275

Sridhar V, Loope DB, Swinehart JB, Mason JA, Oglesby RJ, Rowe CM (2006) Large wind shift on the great plains during the medieval warm period. Science 313:345–347

van Dam RL, Nichol SL, Augustinus PC, Parnell KE, Hosking PL, McLean RF (2003) GPR stratigraphy of a large active dune on Parengarenga Sandspit, vol 22. The Leading Edge, New Zealand, pp 865–870

van der Meulen F (1990) European dunes: consequences of climate change and sea-level rise. In: Bakker TWM, Jungerius PD, Klijn JA (eds) Dunes of the European coasts, Catena Supplement 18. Catena Verlag, Heidelberg, pp 209–223

van Heteren S, FitzGerald DM, McKinlay PA, Buynevich IV (1998) Radar facies of paraglacial barrier systems: coastal New England, USA. Sedimentology 45:181–200

Wiles GC, McAllister RP, Davi NK, Jacoby GC (2003) Eolian response to Little Ice Age climate change, Tana Dunes, Chugach Mountains, Alaska, USA. Arct Alp Res 35:67–63

Wilson P, Orford JD, Knight J, Braley SM, Wintle AG (2001) Late-Holocene (post-4000 years BP) coastal dune development in Northumberland, northeast England. The Holocene 11:215–229

Žilinskas G, Jarmalavičius D, Minkevičius V (2001) Eolian processes on the shore. Institute of Geography, Vilnius, p 283 (in Lithuanian)

Chapter 6
The Joint History of Tróia Peninsula and Sado Ebb-Delta

Susana Costas, Luís Rebêlo, Pedro Brito, Christopher I. Burbidge, Maria Isabel Prudêncio, and Duncan FitzGerald

Abstract Traditionally, the study of coastal evolution has focused on emergent barriers or stratigraphic sequences on the adjacent shelf, but seldom are these two systems studied holistically or the information combined into a single model. Here, we combine data sets from the emerged and submerged sectors of a prograding coast, from the coastal dune to the innermost continental shelf, to reconstruct the long-term history of shelf reworking and spit elongation of Tróia Peninsula in Portugal. This analysis involves synthesizing high-resolution reflection seismic profiles from the shoreface, Ground Penetrating Radar images from the emerged sand barrier, high resolution digital terrain models, and Optically Stimulated Luminescence and radiocarbon dating of sediment samples from the emergent sand barrier and backbarrier. The results document the growth of the sandy peninsula in five major phases of progradation represented by massive foredunes separated by hiatuses of sedimentation and periods of shoreline stability. Formation of the peninsula began circa 6,500 years ago by spit elongation from the south as documented by the oldest beach sediments within the spit. The spit enlarged until a tidal inlet was formed around 3,300 years ago, which caused the construction of the ebb-tidal delta. The latter sequestrated the sand supplied to the spit inhibiting spit progradation until the ebb delta reached an equilibrium volume, allowing shoreline

S. Costas (✉)
Laboratório Nacional de Energía e Geología, Apartado 7586-Alfragide, 2610-999 Amadora, Portugal

Centro de Investigação Marinha e Ambiental (CIMA), Universidade do Algarve, Campus de Gambelas, Edifício 7, 8005-139 Faro, Portugal
e-mail: scotero@ualg.pt

L. Rebêlo • P. Brito
Laboratório Nacional de Energía e Geología, Apartado 7586-Alfragide, 2610-999 Amadora, Portugal

C.I. Burbidge • M.I. Prudêncio
C²TN, Campus Tecnológico e Nuclear, Instituto Superior Técnico, Universidade Técnica de Lisboa, 2695-066 Bobadela LRS, Portugal

D. FitzGerald
Boston University Earth & Environment, 685 Commonwealth Avenue, Boston, MA 02215, USA

G. Randazzo et al. (eds.), *Sand and Gravel Spits*, Coastal Research Library 12,
DOI 10.1007/978-3-319-13716-2_6

progradation out of phase with delta enlargement. Hiatuses in spit progradation are tentatively related to the onset of erosive conditions caused by enhanced storminess around 4,000 and 1,800 years ago and to a tsunami 250 years ago. These results suggest that the growth of the spit is largely controlled by self-adjustment processes and/or major climate shifts.

6.1 Introduction

Spits represent a subaerial projection of sediment forming on top of a larger and thicker submarine structure known as the spit-platform (Meistrell 1972). These features are generated by vigorous alongshore currents transporting sediment that ultimately accumulates along the distal position of the platform where there is an abrupt change in the direction of the shoreline; a coastal reentrant or a bay that drastically reduces wave energy and current velocity, forcing sediment deposition. A second mechanism for spit elongation is associated with the migration of a tidal inlet (Aubrey and Gaines 1982). Based on wave tank experiments and fieldwork, Meistrell (1972) found that the growth of spits and platforms are inversely related and occur in alternating cycles. Spit platform growth is controlled by sediment supply and water depth; as the spit platform builds into deeper water it requires a greater volume of sediment to fill the greater accommodation space, forcing the spit to elongate at lower rates.

Spits are relatively common features along the Portuguese coast, because of the strong and persistent longshore transport of sediment. However, the present sediment-starved condition of the Portuguese coast and the highly dynamic nature of these coastal features, limit the preservation of the sedimentary and historical record within these sedimentary deposits inhibiting our capacity to reconstruct and understand the major factors controlling their formation and evolution. The age of coastal barriers in Portugal has been inferred from the record of lagoonal sedimentation (Freitas et al. 2002, 2003), which is largely controlled by the presence and stability of the enclosing barriers. In this regard, the onset of lagoonal sedimentation about 5,300 years BP was interpreted as indicator of barrier formation (Freitas et al. 2002). Additionally, lagoonal sedimentation documents complex histories of barrier evolution dominated by changes in barrier permeability (Bao et al. 1999; Freitas et al. 2003). Similar to most barriers and deltas worldwide, Holocene formation of coastal barriers in Portugal has likely been influenced by a deceleration in the rate of sea-level rise (Freitas et al. 2002).

The Tróia peninsula is a very large coastal system located at the entrance to the Sado estuary. Because of its dimensions, presence of several dune ridges, and a long period of human occupation, beginning with the Romans (Etienne et al. 1994), there is a long and detailed history contained within its deposits. The Roman settlement is partially covered by a transgressive dunefield produced by northerly winds orographically enhanced by the Palmela valley (Fig. 6.1). The ruins provide information concerning a nucleus of Roman occupation at the downdrift spit end

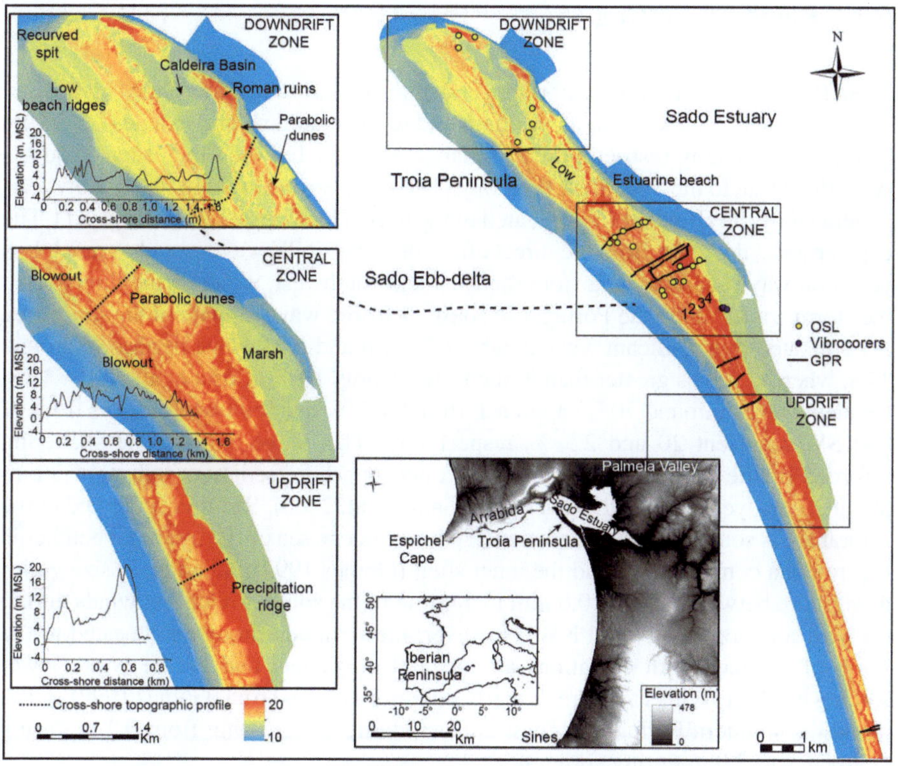

Fig. 6.1 Location of the study area within the Sado estuary. The subaerial morphology of the spit and its adjacent nearshore region are shown in the DTM obtained from the LiDAR dataset provided by the Direção-Geral do Território and the Agência Portuguesa do Ambiente. Surveyed GPR lines and collected sediment samples for OSL and vibrocorers are shown. Detailed images of the zones identified within the spit are shown together with a typical elevation crosshore profile in the *left* of the figure

associated with the development of important fish-salting factories between the end of the first and the first half of the fifth century AD (Etienne et al. 1994; Pinto et al. 2011).

Its size, age, and progradational history, indicate that the Tróia spit is an excellent candidate to study the onset and evolution of these coastal elements in relation to sea-level oscillations and climate and sediment supply variability. In addition, the presence of an ebb-tidal delta at the downdrift end of the spit represents an opportunity to explore the relationship between the submerged sand platform and emergent ridges. An integrated dataset from different sources (i.e. Ground Penetrating Radar, seismics, LiDAR, luminescence dating) allows us to image the evolution of the emerged spit and submerged embankment and obtain a holistic reconstruction of the coastal system.

6.2 Regional Setting

Tróia Peninsula represents the northern end of the 62 km-long Tróia-Sines littoral arc. This sandy spit is attached at its southern end to the mainland and extends for 24 km to the north thereby restricting the exchange of water between the estuary and the Atlantic Ocean to the Sado Inlet (Fig. 6.1). The spit is located within the shadow zone generated by the Espichel Cape located to the north (Jacob et al. 2009) (Fig. 6.1). The cape protects the coast from the direct effect of NW-WNW waves, but has much less effect on waves approaching from the west and southwest, which is the direction of the storm waves along the Portuguese coast. Offshore wave climate in the study area is dominated by significant wave heights of 1.7 m and peak wave periods of around 11 s, whereas waves greater than 3 m constitute only 10 % (Costa et al. 2001). Most frequent waves (around 70 %) approach from the NW quadrant while waves from W and SW represent 20 and 2.5 %, respectively. This wave climate and sheltering afforded by the cape are responsible for a net longshore sediment transport rate of about 10^5 m^3/year, directed to the north (Gama et al. 2006). Sediment supplied to the littoral arc is sourced from the erosion of Mio-Pliocene sea cliffs, intertidal benthonic macrofauna communities, and the inner shelf (Gomes 1992). Mean grain size varies alongshore between 0.3 and 0.6 mm increasing to the south. Prevailing winds during the year are from the N and NW. However, the Arrábida mountains located to the north of the bay, with elevations around 500 m, deflect and reduce wind power forcing northerly winds to more westerly directions along Tróia (Gomes 1992). Tides along the southern Portuguese coast are semidiurnal, increasing from 0.2 m during neap tides to 3.9 m during spring tides.

Gomes (1992) provided a detailed description of dune building along the entire littoral arc, documenting significant differences alongshore and demonstrating seaward progradation as the major mechanism of spit growth. Conversely, Psuty (1992) related the formation of offset foredune crests to the downdrift progradation of the spit and wave refraction on the ebb-tide delta. They hypothesized that decrease in dune elevation downdrift was a consequence of a longshore gradient from an updrift negative beach and dune sand budget to a downdrift positive budget producing beach ridges. Rebêlo et al. (2009) proposed episodic formation of spits alternating with erosive episodes to explain the growth of the Peninsula.

An important element of the coastal system at the mouth of the Sado estuary is the submerged ebb-tide delta. The delta has a maximum sediment thickness of approximately 40 m and has been classified as a high-angle half-delta by Brito (2010). It comprises a large reservoir of sediment (960×10^6 m^3) and its relatively shallow nature reduces wave energy reaching the northern part of the spit (Fig. 6.1). It has been speculated that the delta developed 7,500–5,000 years ago when the rate of sea-level rise rate slowed (Brito 2010). The internal stratigraphy of the delta suggests two major episodes of delta growth; a lower unit characterized by an aggrading stratigraphy and an upper prograding unit (Brito 2010). This shift in delta accretionary style has tentatively been related to a slowing of sea-level rise 3,000 years ago documented by Psuty and Moreira (2000) in the adjacent Sado estuary.

6.3 Methods

Reconstructing historical spit growth at the Tróia peninsula was studied as means of understanding the evolution of coastal barriers since the last transgression. To accomplish this, we must first explain the spatial and temporal evolution of the spit.

6.3.1 Subaerial Morphology

To determine the subaerial morphology and distribution of major depositional elements of the spit, a high-resolution (2 m) digital terrain model (DTM) was produced from detailed LiDAR datasets that were collected during 2011 by the Direção-Geral do Território and the Agência Portuguesa do Ambiente, to determine the number, size and morphology of the ridges along the spit. A series of cross-shore transects were subtracted from the DTM to describe and classify aeolian and beach elements within the spit.

6.3.2 Internal Stratigraphy

6.3.2.1 The Emerged Spit

Ground penetrating radar (GPR) has successfully resolved the distribution of sedimentary environments within coastal barriers (Costas and FitzGerald 2011; Bristow and Pucillo 2006; Hein et al. 2012) and reconstructed sea-level oscillations from coastal barriers by mapping the foreshore-shoreface (Tamura et al. 2008) or the dune-beach boundaries (Bristow and Pucillo 2006; Otvos 2000).

Images of the subsurface were acquired using an Ingegneria Dei Sistemi-Ground Penetrating Radar (IDS-GPR) system RIS MF Hi-Mod #1 equipped with a dual frequency antenna (200 and 600 MHz). Most GPR transects were run across the spit resolving the estuarine and seaward shorelines (Fig. 6.1). Maximum penetration (depth around 12 m) was achieved using the 200 MHz antenna. The high dielectric constant of water molecules within the freshwater aquifer and/or the presence of salt water attenuates electromagnetic waves and limited the penetration of some GPR lines (Daniels 2005). The GPR was synchronized to a RTK-GPS system in order to obtain the topographical information for static correction during the processing of the radargrams. Raw data were processed using the program package Reflex-Win Version 5.0.5 by Sandmeier Software. Processing included application of filters and gains, velocity profile estimate, migration and static corrections. An average subsurface velocity of 0.12 m/ns was estimated using the interactive hyperbola-adaptation method and is a typical velocity for dry sand.

6.3.2.2 The Submerged Spit

The analysis of the internal stratigraphy was extended to the submerged part of the spit and ebb-tidal delta using a very high-resolution seismic reflection system. The Seistec profiler system was expressly created for shallow water seismic profiling by IKB Technologies Limited (Simpkin and Davis 1993) and consists of a boomer source (1–10 kHz) associated with a line-in-cone receiver mounted in a catamaran. Signal penetration reached up to 15 ms with an estimated resolution of 0.3 m in the vertical and 1.0 m in the horizontal for a central frequency of 5 kHz. Data positioning used a R6 Trimble GPS receiver mounted in the system catamaran, at an equidistant position from the seismic source and receiver.

A total of 260 km of 2D seismic profiles were surveyed following parallel and normal transects to the coastline in three grids with a spacing ranging from 200 to 1,500 m. Additionally, seismic profiles (chirp, boomer and sparker) surveyed in previous cruises during 2003 and 2007, were revisited to support the present work. Seismic data processing and interpretation were carried out using Landmark software from the Halliburton, and included band-pass filter, tide correction, swell filter and deep scan stack or trace mixing.

6.3.3 Age Determination

Results from GPR and DTM were integrated in order to select the best locations for dating the growth of the Peninsula. Samples were taken in opaque tubes, from homogeneous layers above the base of 1 m test pits in each location. Optically stimulated luminescence (OSL) dating analyses were made by ETN, C^2TN, Instituto Superior Técnico.

Field gamma spectrometry measurements were conducted in situ to obtain concentrations of K, Th, and U. In the laboratory, water content as a fraction of dry sample mass was measured as received, saturated, and following free drainage. For instrumental neutron activation analysis (INAA), 50 g of dried sample was milled and 200 mg subsamples in HDPE phials were irradiated in the Portuguese Research Reactor at CTN/IST. Samples and standards were analyzed using high-resolution gamma spectrometry to quantify 27 elements including K, Rb, Th and U (Dias and Prudêncio 2007). Dose rates from alpha, beta and gamma radiation were calculated from elemental concentrations and corrected for time averaged water content (for further details see Burbidge et al. (2014)), which was constrained based on the measured values and the elevation of each sampling position relative to the water table. The dose rate from cosmic radiation was calculated for the site location and an estimate of the time averaged burial depth of each sample. For OSL measurements light protected material from each sample was wet sieved, acid washed, density separated, HF etched and re-sieved to obtain 160–200 μm quartz grains. Twenty-four cups each were measured using a single aliquot regenerative

(SAR-OSL) protocol with six calibration doses up to 10 Gy, and repeat points at 0 and 2–3 Gy, including following infrared exposure. These yielded on average 1,800 cts Gy^{-1} and repeat dose-response points within 6 %. Zero dose response was on average equivalent to 0.2 Gy, however Tróia samples with absorbed doses as low as 0.1 Gy yielded ages in agreement with historic records (Rebêlo et al. 2013). On average three results per sample were rejected, where systematic effects on absorbed dose as a function of preheat temperature (180–280 °C) and/or repeat points were significant relative to statistical uncertainties. The remaining results were analyzed using a weighted maximum likelihood estimate for a single population. Ages were calculated as the quotient of absorbed dose and dose rate, and converted to calendar years.

Additionally, three sediment cores were collected at the backbarrier to obtain samples for radiocarbon dating in order to determine the age of low-energy backbarrier sedimentation (Fig. 6.1). The vibra-cores were opened and described in the laboratory separating the sediments indicative of marsh deposition. The samples were treated at Leibniz Labor für Altersbestimmung und Isotopenforschung, Christian-Albrechts-Universität, Kiel where plant materials were selected for AMS radiocarbon dating. Calibrated or calendar ages were calculated using CALIB rev 5.01 (dataset: IntCal04, Reimer et al. (2004)).

6.4 Results

6.4.1 Subaerial Morphology

The DTM allowed the identification and mapping of the different morphological elements along the spit: dune ridges (foredunes and precipitation ridges), blowouts, parabolic dunes, beach ridges and interdune depressions. In this work we adopted the terminology proposed by Hesp (1999), which integrates the processes responsible for dune development.

Based on the spatial distribution and the dimensions of these elements, the spit was divided in three zones alongshore according to the morphology of the dune ridge sets: updrift, central and downdrift zone (Fig. 6.1). The distribution of these zones is related to the two axis of spit elongation displayed by the spit; the updrift axis is directed to the north and to the interior of the estuary whereas, the second or downdrift axis represents a westward or seaward rotation in the updrift axis downdrift of the central zone. The rotation of the former axis of elongation produced the realignment of the spit that evolved to a swash-aligned spit with a pronounced concave curvature. The spit curvature defines the transitional or central zone of the spit (Fig. 6.1).

The updrift zone extends from the anchor point of the spit to the central curving zone. This zone is the narrowest area of the spit (around 500 m wide) and is characterized by two to three dune ridges that increase in dimensions inland; larger ridges reach 25 m above MSL and 300 m wide (Fig. 6.1). The landward most and larger ridges are precipitation ridges that advanced onto the backbarrier marsh, which in some areas was transformed into rice fields. The seaward ridges are smaller and are represented by regular or fragmented foredunes. Interdune depressions are frequently around 4 and 6 m above MSL though they may reach 2 m above MSL (Fig. 6.1). Depressions present evidence of aeolian deflation producing a concentration of shell fragments at the surface suggesting the landward transfer of sand from a former shoreline or foredune to the precipitation ridge. The cross-shore present-day beach profile is characterized as an intermediate-barred coast (Masselink and Short 1993). The beachface is steep (~7.4°) and commonly exhibits two inflexion points at 2.5 m and 4.5 m above MSL. Both levels are related to the vertical migration of the berm crest associated with typical water oscillations generated fortnightly tidal cycles tide and effects of wave and wind climate.

The number of dune ridges increases to the central zone of the spit. This zone is characterized by a high diversity of dune forms ranging from the typical narrow, evenly spaced, and fragmented foredunes to parabolic dunes (Fig. 6.1). This zone reaches 1,500 m wide and comprises four to six massive foredune ridges with irregular spacings and crescentic morphologies. Highest elevations of dune ridges in the central zone are between 14 and 18 m above MSL. These forms correspond to parabolic dunes advancing to the south and to large blowouts moving landward (west to east) cutting through dune ridges. Interdune depressions in this zone are 5 and 6 m above MSL. The backbarrier region is characterized by the presence of a backbarrier marsh or estuarine beaches. The latter developed in the estuarine side of the spit due to the erosive action of the ebb channel over the backbarrier. Former wave-built beach-ridges can be observed at the estuarine side of the spit where they have not been covered by aeolian deposits (Fig. 6.1). Cross-shore beach profiles have a typical intermediate-barred profile with beachface slopes around 6°. Inflexions within the beachface are found at 2.3 and 3.6 m above MSL.

The downdrift part of the spit exhibits a high diversity of morphological elements with lower elevations and abundant evidence of post-depositional foredune reworking within the estuarine side. Massive foredunes are rare in this zone and are replaced by lower landforms that may, in turn, suggest faster shoreline progradation following sigmoidal planform dune crests (Fig. 6.1). These patterns repeat the complex morphology displayed by the present-day shoreline, which is dominated by cuspate features, entrances and the recurved downdrift spit end (Fig. 6.1). The estuarine side of this zone is characterized by the presence of parabolic dunes migrating to the southwest from the estuarine beaches. Seaward, these dunes are replaced by a very low area with elevations around 2 m above MSL (Fig. 6.1). The latter resembles the inter-ridge depressions found within the spit end or the deflation surfaces left by the advance of the parabolic dunes. This area becomes lower to the north where it contacts the Caldeira basin, a depressed area with average elevations below MSL (Fig. 6.1). The cross-shore beach profile in this zone is represented by reflective morphologies. Foreshore slope decreases towards the spit end where it terminates onto the adjacent ebb-delta with a slope around 5°. Beach berm in this zone is around 3 m above MSL.

6.4.2 *Internal Stratigraphy*

6.4.2.1 The Emerged Spit

Classification of the internal reflectors is primarily related to their continuity, attitude, and interval, using the terminology developed by Mitchum et al. (1977). The internal variability identified within the GPR images was simplified into three facies associations. Interpretation of the radar facies was aided by the morphology of present day coastal features resolved from the high-resolution DTM and previous works carried out in similar environments (van Heteren et al. 1998; Bristow et al. 2000).

A horizontal-paired, strong reflector running across the spit at 1.2–2.1 m above MSL is interpreted as the water table associated with a subsurface freshwater aquifer (Figs. 6.2 and 6.3). Reflection-free areas within the radargrams are a

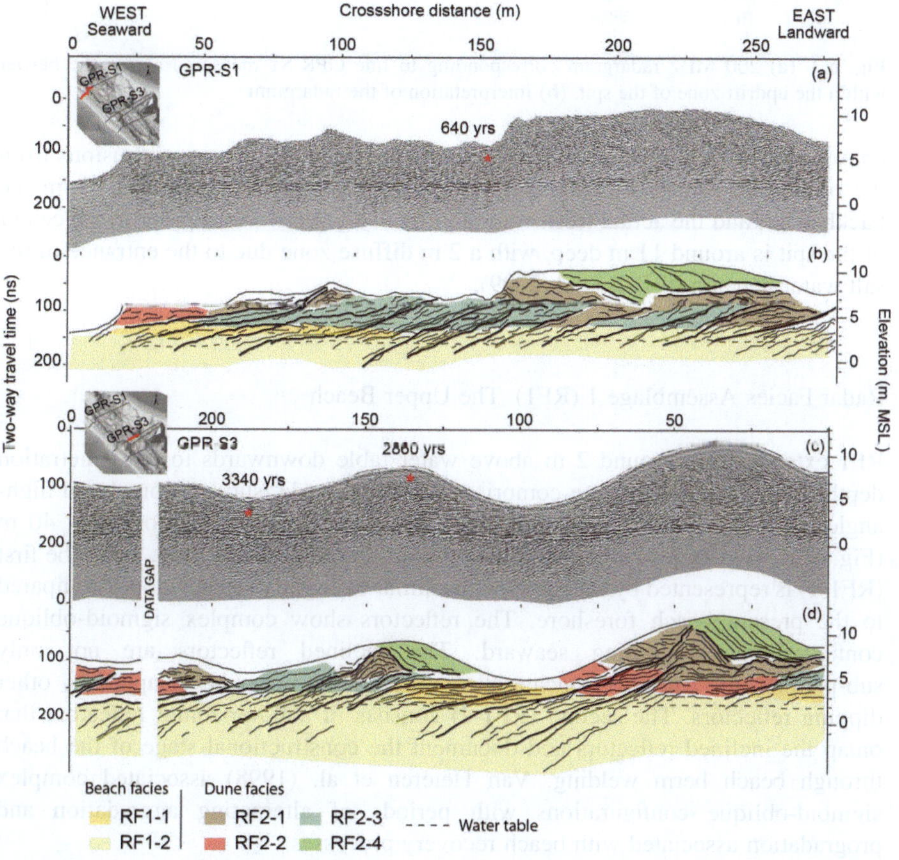

Fig. 6.2 (**a**) and (**c**) Segments of the 200 MHz radargrams GPR-S1 and GPR-S3 measured across the barrier spit. (**b**) and (**d**) Interpretation of the radargrams and location of the OSL samples (*asterisks*)

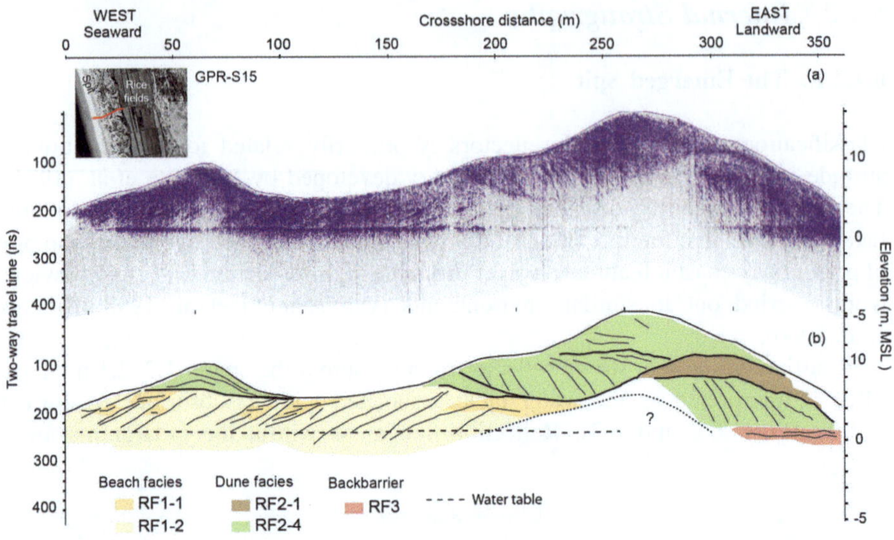

Fig. 6.3 (**a**) 200 MHz radargram corresponding to line GPR-S1 measured across the barrier within the updrift zone of the spit. (**b**) Interpretation of the radargram

consequence of signal attenuation due to the presence of salt water intrusions from the ocean and estuarine beach, the presence of fine-grained sediments from the backbarrier, and the actual freshwater aquifer. The freshwater aquifer in the center of the spit is around 11 m deep, with a 2 m diffuse zone due to the entrance of the salt water beneath it (Dill et al. 2009).

Radar Facies Assemblage 1 (RF1). The Upper Beach

RF1 extends from around 2 m above water table downwards to the penetration depth limit. This assemblage comprises high-amplitude, sub-horizontal and high-angle (4–7°) seaward-dipping reflectors that can be traced for more than 40 m (Fig. 6.2). Two facies can be identified within this assemblage (Fig. 6.2). The first (RF1-1) is represented by reflectors with similar to slightly steeper slopes compared to the present beach foreshore. The reflectors show complex sigmoid-oblique configurations prograding seaward. The inclined reflectors are not only sub-parallel, but can also truncate other reflectors and may be onlapped by other dipping reflectors. The second (RF1-2) consists of sub-horizontal reflectors that onlap the inclined reflectors and document the constructional stage of the beach through beach berm welding. Van Heteren et al. (1998) associated complex sigmoid-oblique configurations with periods of alternating aggradation and progradation associated with beach recovery processes.

Radar Facies Assemblage 2 (RF2). The Aeolian Cover

RF2 dominates most of the GPR radargrams, as aeolian sedimentation commonly forms the top 2 m above MSL. Internal configurations within this assemblage are highly complex and diverse mostly because of the presence of the vegetation cover (Figs. 6.2 and 6.3). The most common facies within RF2 are represented by hummocky discontinuous reflectors (RF2-1) represented by concave down reflectors that are frequently found at the termination of RF1. RF2-1 uses to concentrate within the internal architecture of large dune buildings extending upwards in pace with the vertical aggradation of the ridges (Fig. 6.2). This facies is interpreted as the vertical accretion of blown sand from the adjacent beach trapped by the vegetation, which is in turn the responsible for the discontinuous and chaotic configuration of this facies.

High-amplitude sub-horizontal and tangential-oblique reflectors (RF2-2) dipping seaward may downlap onto RF1-2 (Fig. 6.2). RF2-2 are parallel to the underlying RF1-2 and are interpreted to represent the vertical aggradation of the backshore through the accumulation of blown sand with reduced vegetation cover and frequent deflation surfaces. These surfaces resemble the inter-ridge depressions that can be found at 6 m above MSL. High-angle seaward dipping reflectors cutting through the water table may reach the dune crest and represent bounding surfaces erosively truncating RF1 and RF2 (Fig. 6.2). These surfaces represent the retreat of the shoreline and the eventual erosion of the frontal dune and are recorded as erosive scarps that use to affect the entire foredune. Tangential-oblique reflectors (RF2-3) dipping seaward onlap these bounding surfaces promoting the recovery and progradation of the dune seaward coincident with beach progradation (Buynevich et al. 2007; Costas et al. 2006).

Finally, a series of well-defined radar packages can be identified within the radargrams represented by oblique parallel reflectors (RF2-4) with erosional upper toplap terminations (Figs. 6.2 and 6.3). These facies were identified in the sectors of the GPR lines running across blowouts, precipitation ridges or parabolic dunes, which in turn explain their origin. RF2-4 reflectors downlap onto erosive or non-depositional bounding surfaces (Fig. 6.3).

Radar Facies Assemblage 3 (RF3). The Backbarrier

RF3 is represented by discontinuous and sub-horizontal reflectors and are found within the estuarine side of the spit (Fig. 6.3). These reflections constitute the onlapping surface of RF1-2 and in some cases the downlapping surface of RF2-4. Stratigraphy interpreted from vibracores collected in the estuarine margin of the spit support the interpretation of the RF3 facies. The core logs show a vertical succession of lithological facies represented by oxidized coarse sands at the base overlaid by well-sorted medium sands and organic-rich very fine sand to silt at the top of the cores. Oxidized coarse sands were also documented by Psuty and Moreira (2000) and interpreted as Plio-Pleistocene deposits. This sequence suggests the

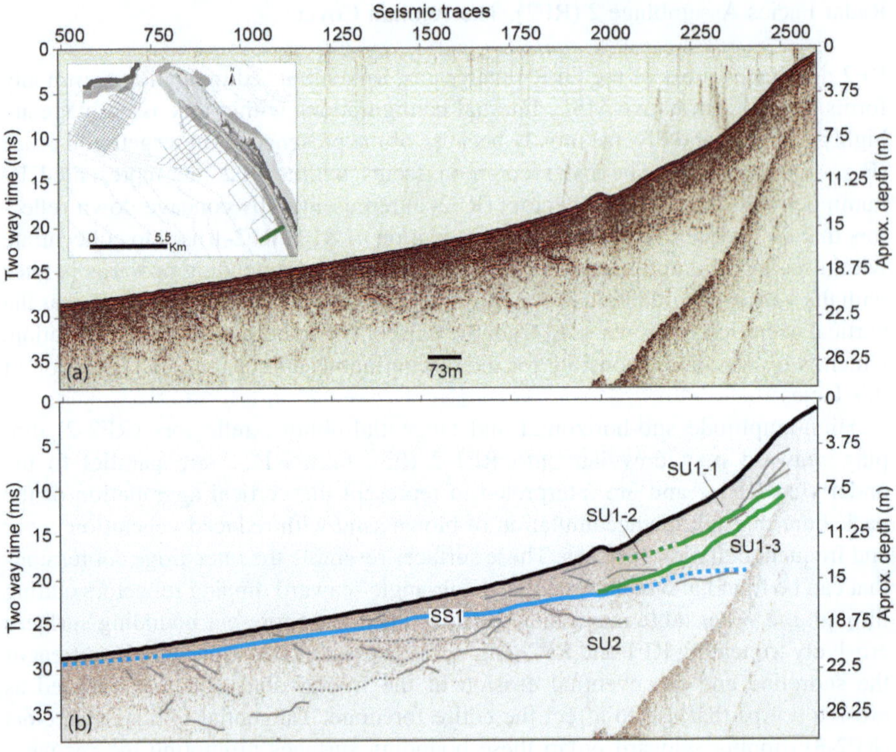

Fig. 6.4 (**a**) Seismic line SCARPS-69 running across the nearshore area of the updrift spit zone. (**b**) Interpretation of the seismic line showing major identified units and surfaces including surface SS1 interpreted here as the marine ravinement surface

onset of spit deposition on top of previous Plio-Pleistocene sediments and the subsequent onset of low-energy backbarrier environments that could explain the sedimentation of fine-grained deposits at the top of the sequence.

6.4.2.2 The Submerged Spit

As in the GPR data, the interpretation of the seismic reflection data was based on the principles developed by Mitchum et al. (1977). The interpretation of the seismic reflection data documents one main seismic surface (SS1) that individualizes two seismic units, SU1 and SU2, respectively above and below SS1. SS1 is marked by an offshore slopping reflection with high amplitude and continuity, whose erosive character is denoted by the truncation of the underlying reflections (Fig. 6.4). The interpolated morphology of this surface appears very similar to the present day sea bottom, reproducing both the general trend of the spit platform (Fig. 6.5). This

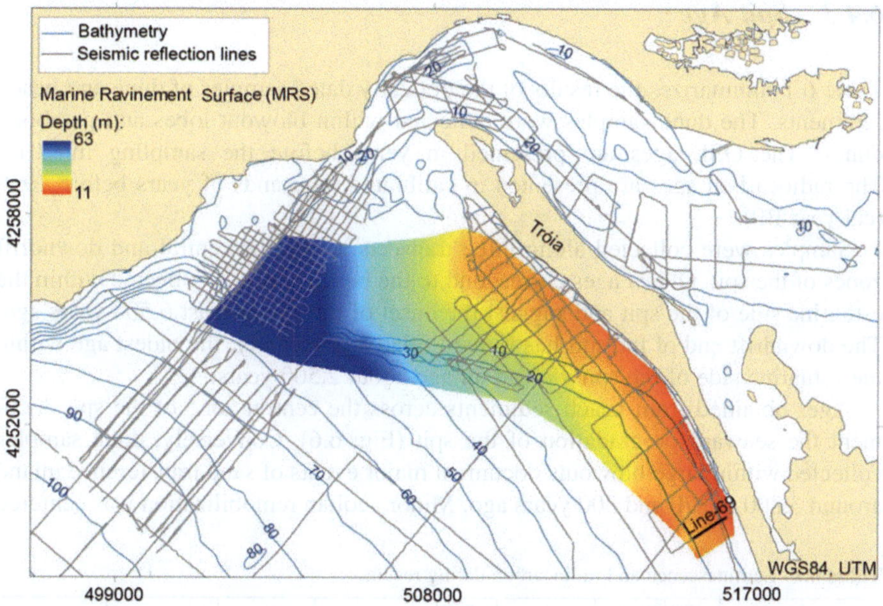

Fig. 6.5 Interpolation of the bounding surface SS1 identified within the surveyed seismic lines. This seismic surface was interpreted as the marine ravinement surface associated with the last marine transgression

seismic surface was interpreted as the marine ravinement surface associated with the last marine transgression.

SS1 is overlain by the seismic unit SU1, which is topped by the seafloor. SS1 represents a prograding unit with two medium-amplitude and medium- to low-continuity reflections that downlap onto SS1 and divide SU1 in three sub-units, SU1-1, SU1-2 and SU1-3 (Fig. 6.4). The two lower sub-units (SU1-2 and SU1 3) are thin (less than 5 and 2 ms TWT in thickness, respectively) and almost reflection free. These sub-units are bounded by two unconformity surfaces. The upper sub-unit (SU1-1) is characterized by high amplitude chaotic point reflections. Yet, it was possible to identify low-continuity reflections dipping offshore and downlapping onto each other or onto the top of SU1-2. The unit SU1 is interpreted as post-transgressive and it is here speculated that the two surfaces that limit its sub-units can eventually correspond to old shoreface erosive surfaces associated to coastline retreat while the subunits would represent the subsequent beach recovery. The lower unit SU2 underlies SS1 and its base is beyond the seismic signal penetration. SU2 was characterized by oblique to sigmoidal, onshore dipping, high-amplitude and medium-continuity reflections truncated by SS1 (Fig. 6.4). It was interpreted as the pre-transgression unit.

6.4.3 Spit Age

Table 6.1 summarizes the results of the optically dated samples of dune and beach sediments. The dune samples were collected within blowout lobes and parabolic dunes. The OSL ages are presented in years before the sampling in 2010. The radiocarbon ages are presented in calibrated thousands of years before 1950 (cal year BP).

Samples were collected along three transects across the central and downdrift zones of the spit. Oldest ages correspond to the beach deposits collected within the estuarine side of the spit and suggest the onset of the spit at least 6,500 years ago. The downdrift end of the spit has formed relatively recently; the oldest age within the estuarine side of the spit yielded an age about 2,500 years.

Ages obtained from beach sediments across the central zone of the spit document the seaward progradation of the spit (Fig. 6.6). Conversely, dune samples collected within large blowouts document major events of sand transference inland around 3,800, 1,870 and 300 years ago. Minor aeolian remobilization was centered

Table 6.1 Luminescence and radiocarbon dating results

Sample	ITNLUM#	Distance to present shoreline (m)	Elevation (m, MSL)	Burial depth (m)	H_2O (g/g)	Dose rate (mGy/year)	Absorbed dose (Gy)	Age (years)
TR-10-01	663	791	11	1.00	0.06	1.09±0.03	3.98±0.09	3,660±140
TR-10-02	664	620	11	0.90	0.04	1.19±0.03	3.42±0.08	2,860±110
TR-10-03	665	544	6	0.90	0.06	1.13±0.04	3.78±0.06	3,340±120
TR-10-04	666	221	16	0.85	0.06	1.20±0.04	0.30±0.03	250±20
TR-10-05	667	980	16	0.85	0.05	1.05±0.03	1.87±0.07	1,780±90
TR-10-06	668	1,181	18	0.75	0.07	1.02±0.04	3.87±0.21	3,870±210
TR-10-07	669	1,440	3	1.00	0.09	0.95±0.04	6.47±0.31	6,470±310
TR-10-08	670	152	5	0.75	0.05	0.91±0.03	0.76±0.05	760±50
TR-10-09	671	954	5	1.03	0.07	1.06±0.04	4.00±0.21	4,000±210
TR-10-10	672	1,051	11	0.90	0.03	1.12±0.04	1.72±0.09	1,720±90
TR-10-11	673	1,127	4	1.20	0.05	0.98±0.03	5.84±0.28	5,840±280
TR-10-12	674	582	3	1.10	0.09	1.22±0.05	3.24±0.16	3,240±160
TR-10-13	675	362	5	1.70	0.07	1.03±0.04	1.11±0.06	1,110±60
TR-10-14	676	337	10	1.05	0.06	1.20±0.05	0.95±0.05	790±60
TR-10-15	677	533	12	1.10	0.03	1.21±0.05	1.87±0.09	1,870±90
TR-10-16	678	220	4	1.05	0.03	1.26±0.04	0.81±0.03	640±30
TR-10-17	679	897	2	0.60	0.08	0.94±0.04	1.98±0.15	1,980±150
TR-10-18	680	1,116	3	0.90	0.14	1.04±0.04	2.46±0.12	2,460±120
TR-10-19	681	700	2	0.92	0.26	0.93±0.03	1.98±0.11	1,980±110
TR-10-20	682	349	4	1.30	0.02	1.52±0.05	2.43±0.04	1,600±60
TR-10-21	683	–	5	1.65	0.04	1.64±0.05	0.62±0.04	380±30
TR-10-22	684	890	2	1.25	0.06	0.93±0.04	1.00±0.04	1,080±60
TR-10-24	686	1,100	3	1.15	0.04	1.36±0.05	1.42±0.04	1,040±40
Sample	**Lab code**	**Depth (m)**	**$\delta^{13}C$ (%)**		**Conventional ^{14}C age BP**		**2σ calendar calibrated age BP (cal year BP)**	
D1-195	KIA36122	1.95	−26.37±0.14		2,405±25		2,492−2,350	
D4-227	KIA36123	2.27	−25.86±0.19		390±20		507−429	
D2-133	KIA36124	1.33	−22.76±0.33		3,135±35		3,445−3,319	

OSL ages are represented in years before 2010 while radiocarbon ages are presented in cal year BP. *Shadow rows* correspond to samples collected within the dunes

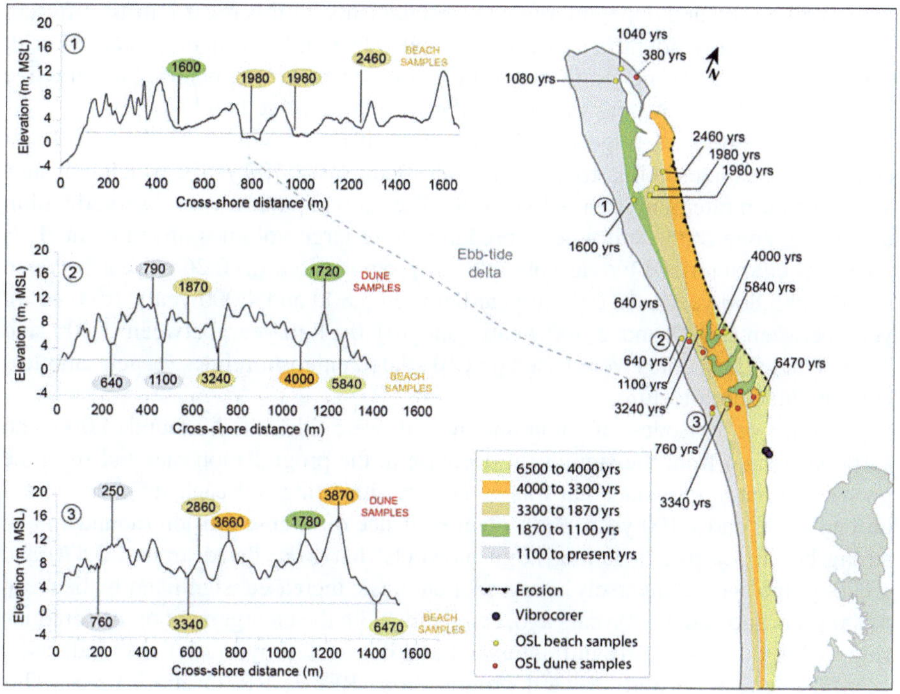

Fig. 6.6 Results of the OSL ages for the beach samples and major phases of spit progradation and elongation. (*1*), (*2*) and (*3*) show the crosshore elevation profiles with the location and age of the OSL samples for the dune and beach sediments

around 2,800 and 790 years. In contrast, the aeolian activity that centered around 1,700 years was related to the formation and southward advance of parabolic dunes from the estuarine beaches.

The age of the backbarrier deposits determined by radiocarbon dating yielded ages between 3,400 and 420 cal year BP suggesting the onset of the marsh after 3,300 cal year BP. The sample from the transition between the sandy deposits and organic-rich silty sediments was collected 2 m below present MSL, which could correspond to low marsh sedimentation.

6.4.4 Progradation Rates

Figure 6.6 shows the age relationship between sample locations and the present shoreline. The occurrence of beach deposits located 1 km from the present shoreline with ages between 2,000 and 4,000 years alongshore (Fig. 6.6) supports the model of downdrift spit elongation. The major morphological differences observed along and across the barrier are related to the volume of sediment stored within the foredune ridges, which in turn yields information concerning differences in the progradation rate and dune enlargement mechanisms. In this regard, three different patterns have been identified within the spit: (1) rapid beach progradation with small foredunes, (2) moderate beach progradation with fragmented and massive foredunes, and (3) low beach progradation with large ridge heights.

The first pattern of ridge morphology was observed within the downdrift spit end. Most foredunes in this zone formed during the last 1,100 years, which indicates a progradation rate of at least 0.30 m/year. The second pattern was observed within the central zone of the spit where foredunes store large volumes of sediment. This zone was characterized by variable rates of progradation: (a) 0.26 m/year between 6,500 and 5,800 years, (b) 0.09 m/year between 5,800 and 4,000 years, (c) 0.45 m/year between 4,000 and 3,300 years, and (d) 0.14 m/year between 3,300 and 1,100 years. These rates were based on OSL-dated paleoshorelines (beach samples) and foredune architecture.

Two major episodes of shoreline retreat were identified around 4,000 and 1,100 years ago from the significant decrease in the progradation rate and from the presence of large blowouts that fragmented foredune ridges. Beach sediments dated around 4,000 and 1,100 years show clear evidence of intense erosion (abundant and strong beach scarps), including large blowouts that were dated around 3,870 and 1,870 years ago. Conversely, progradation rates increased significantly between 4,000 years ago and 3,300 years ago, coinciding with the enlargement of the foredune ridge. This ridge is mostly unfragmented and dominated by radar facies indicating vertical aggradation and seaward progradation (RF2-1, RF2-2 and RF2-3). This architecture is interpreted as a temporary stabilization of the shoreline around 3,300 years ago. Shoreline progradation continued around 1,100 years ago before being interrupted by the last major phase of instability and hiatus in deposition along the spit around 250 years ago, coinciding with the Lisbon tsunami (Rebêlo et al. 2013).

Finally, the third evolutionary pattern was observed in the updrift zone of the spit. This narrower zone should record all of the identified phases, because it is the oldest part of the spit. In fact, if we look into detail within the internal architecture of the barrier, these phases can be isolated even though they merged into a single very large landform, the precipitation ridge. The latter is the result of the landward transfer of sand from a former shoreline during several sediment pulses. In this regard, the transgressive character of the aeolian landforms may be related to the retrograding character of the shoreline in this zone. In fact, this zone has shown lower architecture than the northern and central zones probably due to its higher exposure to wave action, lower sediment input and greater accommodation space.

6.5 Discussion

6.5.1 Initiation and Elongation of the Spit

A model of spit evolution was reconstructed by integrating information from diverse data sources: high-resolution subsurface stratigraphy revealed by GPR and seismic profiler system, OSL dating of dunes and beaches, and high resolution surface morphology (DTM). Dated beach deposits from the estuarine margin of the central zone of the spit suggest its onset at least 6,500 years ago, coincident with a significant reduction on the rate of sea-level rise ca 7,000 cal year BP (Leorri et al. 2012). This suggests spit formation approximately 1,000 years before previously published dates for SW Portugal (Alday et al. 2006; Freitas et al. 2002; Cearreta et al. 2007), but simultaneous with similar systems along the northern coast of Portugal (Naughton et al. 2007) and southern coast of Spain (Goy et al. 2003). Differences in timing could be explained by the fact that the age of barriers has been previously inferred from depositional changes within lagoons or estuaries whereas the present work is based on the actual age of the coastal barrier.

The analysis of barrier architecture and mapping of the spatial (vertical and horizontal) distribution of the transition between the upper beach and the foredune, suggest that sea-level during the first stages of barrier formation was about 2 m below present MSL supporting the sea-level curve proposed by Leorri et al. (2012) for the SW of Portugal. Conversely, previous reconstructions of sea-level in the Sado estuary suggest a sea-level rise of about 6 m during the last 7,000 years (Psuty and Moreira 2000), which conflicts with documented beach deposits in a position close to present MSL and is 4 m below the estimates presented here.

Coincident with spit formation was a reduction in the rate of sea-level rise, which is interpreted as an unequivocal signal of causality (Thom and Roy 1985; Stapor et al. 1991). However, other factors such as sediment supply, wave exposure and accommodation were important in determining the actual shape and location of the spit. In this regard, the Cape Espichel diminished the impact of energetic waves that approach from N and NW thereby faciliting the formation and preservation of the spit as well as promoting its northward progradation and enclosure of the estuary. On the other hand, the morphology of the ravinement surface documented here suggests the presence of inherited topography that may have had a significant role on the construction and growth of the spit and the delta by reducing accommodation space. The latter would promote rapid barrier enlargement by reducing the volume of sediment required for barrier growth while also providing an anchor point for spit accretion. In addition, the presence of this topographic high could have acted as part of the spit platform encouraging the accumulation of sand that ultimately was transferred to the spit system promoting spit growth.

As the spit elongated northward (Fig. 6.6), it restricted the circulation within the embayment eventually transforming it into an estuary. The transformation to a more restricted environment may have had consequences on the hydrodynamics within the bay promoting the onset of low energy sedimentation. This may have contributed to the formation of the backbarrier marsh documented in the work around 3,500 years ago and to the expansion of fringing marshes in the estuary around 2,600 years ago observed by Psuty and Moreira (2000) and previously as attributed to a reduction in the rate of sea-level rise.

Spit elongation is also expected to have influenced ebb delta formation. As the spit elongated, the cross-sectional area of the opening of the embayment gradually decreased while the tidal prism remained approximately constant. The large tidal prism of the estuary should ensure a permanent communication with the open sea for the foreseeable future. As the entrance narrowed and tidal currents increased, sand would have been transported offshore causing the progradation and vertical accretion of the ebb-tidal delta (Brito 2010). The change in the facies architecture of the delta had previously been related to a deceleration in the rate of sea-level rise as documented by Psuty and Moreira (2000) for the Sado estuary ca 3,000 years ago (Brito 2010). Alternatively, and considering the magnitude of sea-level change inferred from the present work, we propose a process of self-organization related to local hydrodynamic conditions (tides and waves) as the responsible of delta stratigraphy. The general shape of ebb-tidal deltas is determined by a balance between tidal currents, waves and the amount of supplied sediment (Hayes 1980). The aggradational phase identified at Sado ebb-tidal delta represents the formation of the delta once tidal flow became constricted at the inlet. This process could only occur after spit elongation narrowing the opening to the estuary and forming the tidal inlet.

Evidence points to the development of the ebb delta around 3,300 years ago when the downdrift end of the spit reached a position close to the present terminus. This resulted in backbarrier mash sedimentation and a significant change in the progradation rate from rapid progradation to stabilization. Despite a relatively stable sea-level, the delta aggraded to attain an equilibrium volume with the tidal prism and the dominant hydrodynamic conditions. The shift from aggradation to progradation may mark the point that the former delta attained its maximum elevation and further enlargement was accommodated by delta progradation. This shift could be indicative of the period when the estuary achieved its approximate present circulation pattern. This circulation regime within the estuary also likely coincided with the development of the flood delta. This shoal development could have forced the channel towards the estuarine margin of the spit, thereby promoting erosion of the shoreline. The eroded backbarrier shoreline became exposed to northerly winds that provoked the formation of blowouts and parabolic dunes around 1,700 years ago. In turn, spit elongation provided new land for habitation by future human communities during Roman occupation. Comparing the age of the parabolic dunes that advanced from the north covering the Roman ruins, we speculate that sand invasion from the estuarine beaches could have forced the temporal abandonment of the spit by the end of the second century.

6.5.2 Barrier Progradation

Foredune ridges are the most prominent morphological elements within spit systems, and understanding the processes responsible for ridge formation is critical to determine spit evolution. To accomplish this, we have integrated information from the surficial morphology of the dunes and the internal stratigraphy. We have documented the presence of foredunes exhibiting different degrees of complexity, parabolic dunes and precipitation ridges. In spite of the increasing number of studies (Mauz et al. 2013; Bristow and Pucillo 2006; Tamura 2012) and definitions of beach ridges (Otvos 2000), we have adopted the nomenclature defended by Hesp et al. (2005) as it integrates the mechanisms responsible for the formation and growth of coastal ridges and maintains a clear relationship with the ridge and adjacent former shoreline. One of our goals was to understand the landforms in terms of growth mechanisms; whether or not are they wind- or wave-built, and in situ or transgressive landforms with significant landward migration. Indeed, we have found the latter process particularly important to understand and reconstruct the history of the system in terms of sediment budget and progradation rates. Therefore, we have adopted the term foredune to describe the seaward-most vegetated sand dune formed on the backshore zone of beaches formed by aeolian sand deposition within vegetation (Hesp 1999). In turn, blowouts, parabolic and precipitation ridges were included into the erosive and transgressive dune types and were excluded from the backshore dunes (Hesp 1999). These latter types cannot be used as indicators of sea-level elevation or shoreline position as they have been deposited independently of the position of sea-level.

Most blowouts were associated with erosion of a former foredune, which in many cases had already been eroded by wave action while the depositional lobe advanced landward, merging into an antecedent foredune (Fig. 6.2). Conversely, blowout advancement may destroy the original topography or foredune, and may create a deflation surface contributing sand to enlarge of inland foredunes or precipitation ridges (Fig. 6.3). Therefore, the frequency and type of dune may indicate the dynamic state of the spit.

The subaerial morphology of the spit documents four massive foredune ridges that can be traced alongshore and are ubiquitously represented within the central zone (Fig. 6.1). The large ridges document the progradation of the spit as they result from cumulative processes of beach progradation, beach and dune scarping, foredune aggradation and aeolian sediment transport inland as informed by the GPR images. The age of the ridges was estimated by dating the beaches within the interdune depressions, which separate major ridges or progradation packages (Fig. 6.6). Interdune depressions are interpreted as the result of a change in the progradation usually associated with shoreline retreat and landward aeolian sediment transfer due to the onset of a temporary negative sediment budget in a context of shoreline progradation. Interestingly, interdune depressions could be related to the bounding surfaces separating progradational subunits in the shoreface beach profile. If so, this would confirm the erosive character of the depressions separating successive ridges that would provoke the retreat of the entire beach profile.

The occurrence of erosive events or periods prevents an estimate of the actual temporal boundaries separating progradation packages. The upper limit of these packages is represented by the age at which the beach starts recovering and not the age of the onset of the shoreline instability. Conversely, aeolian sediments that have been dated from blowouts are supportive evidence of shoreline instability, since they originate from foredune fragmentation, erosion and a slight negative sediment budget in the beach (Psuty 1992).

The first phase of spit progradation (6,500–4,000 years) corresponds to the initial stages of elongation and progradation. The end of this stage is marked by a significant retreat of the barrier prior to 4,000 years ago. Shoreline retreat between 1,870 and 1,100 could be related to an increase on the frequency and/or intensity of storms that could promote shoreline scarping and enhanced longshore sediment transport to the north. Storm winds in this region approach from the west and southwest, explaining the occurrence of very large blowouts around this age. Shoreline instability and inland aeolian sediment transference were also documented in other regions along the coasts of Portugal (Costas et al. 2013), France (Billeaud et al. 2009; Sabatier et al. 2012), England (Spencer et al. 1998) or Denmark (Clemmensen et al. 2009) and have been related to the onset of storminess conditions in the North Atlantic and the southward displacement of the westerlies (Sorrel et al. 2012).

Following this erosive event, the spit continued elongating and prograding until it stabilized approximately 3,300 years ago. It is speculated that this elongation of the spit and concomitant constriction of the estuary could have initiated the formation of the ebb delta. During this time (between 3,300 and 1,100 years) the longshore transport of sediment was intercepted by the ebb delta substantially reducing the sand supply to the spit and diminishing spit progradation. The aggradation of the delta likely lasted between 3,300 and 1,700 years. Once the delta attained its equilibrium volume, progradation commenced. This evolution suggests shoreline progradation was out of step with delta progradation resembling the spit elongation mechanism proposed by Meistrell (1972). Seaward rotation and progradation of the spit is expected to become more pronounced once growth of the ebb delta equilibrates. Moreover, shoreline retreat, barrier scarping and blowout formation around 250 years ago were related to the erosive effect of the tsunami of Lisbon in 1755 AD (Rebêlo et al. 2013).

6.6 Conclusions

In this study we have presented the evolution of a barrier spit since its formation and discussed the major drivers involved in its elongation and progradation. Although elongation processes dominate most spits worldwide, at Tróia seaward progradation may also play a significant role. This multi-disciplinary approach allowed a reconstruction of the entire system by integrating geophysical data sets from the emerged and submerged barrier, high-resolution digital terrain models, and OSL and radiocarbon dating.

Results suggest the initiation of the spit occurred at least 6,500 years ago while elongation continued until around 3,300 years ago. Spit elongation provoked the confinement of the estuary entrance and the subsequent formation of a single tidal inlet, which in turn determined the end of spit elongation and caused the construction of the ebb-tidal delta. The process of delta construction involved sequestration of part of the sand supplied to the spit, inhibiting its progradation until the ebb delta reached an equilibrium volume. Foredunes have recorded complex histories of barrier progradation and retreat or instability, the larger the foredune, the more complex will be the growth history of the coastal barrier. In this regard, we have demonstrated that the internal architecture and morphology of dune ridges may provide significant information regarding the progradational history of a spit based on a conceptual model proposed by Psuty (1992).

It is notable that changing progradation rates do not necessarily require changes in sediment supply but rather, changes in accommodation or self-organization within the spit embankment. In this regard, we have invoked changes in accommodation space, storminess regime, tsunamis and processes of self-adjustment within the spit system, the delta or even the estuarine circulation to explain the occurrence of major shifts in the progradation of the spit.

Acknowledgments The authors would like to thank Gabriel Menezes, Marco Ferraz and Rita González for their participation during fieldwork. This research was founded by the Portuguese Science Foundation (FCT) through the projects SCARPS (PTDC/CTE-GIX/101466/2008). Susana Costas thanks the FCT for financial support through the program Ciência-2007.

References

Alday M, Cearreta A, Cachão M, Freitas MC, Andrade C, Gama C (2006) Micropalaeontological record of Holocene estuarine and marine stages in the Corgo do Porto rivulet (Mira River, SW Portugal). Estuar Coast Shelf Sci 66(3–4):532–543

Aubrey DG, Gaines AG Jr (1982) Rapid formation and degradation of barrier spits in areas with low rates of littoral drift. Mar Geol 49(3–4):257–277, http://dx.doi.org/10.1016/0025-3227(82)90043-3

Bao R, Freitas MC, Andrade C (1999) Separating eustatic from local environmental effects: a late-Holocene record of coastal change in Albufeira Lagoon, Portugal. The Holocene 9:341–352

Billeaud I, Tessier B, Lesueur P (2009) Impacts of late Holocene rapid climate changes as recorded in a macrotidal coastal setting (Mont-Saint-Michel Bay, France). Geology 37(11):1031–1034. doi:10.1130/g30310a.1

Bristow CS, Pucillo K (2006) Quantifying rates of coastal progradation from sediment volume using GPR and OSL: the Holocene fill of Guichen Bay, south-east South Australia. Sedimentology 53:769–788

Bristow CS, Chroston PN, Bailey SD (2000) The structure and development of foredunes on a locally prograding coast: insights from ground-penetrating radar surveys, Norfolk, UK. Sedimentology 47(5):923–944. doi:10.1046/j.1365-3091.2000.00330.x

Brito P (2010) Na Fronteira do Mar – Evolução Geológica do Estuário do Sado e da Plataforma Continental entre Sesimbra e o Canhão de Setúbal nos Últimos ~ 50000 anos). Câmara Municipal de Sesimbra, Sesimbra, Portugal

Burbidge CI, Trindade MJ, Dias MI, Oosterbeek L, Scarre C, Rosina P, Cruz A, Cura S, Cura P, Caron L, Prudêncio MI, Cardoso GJO, Franco D, Marques R, Gomes H (2014) Luminescence dating and associated analyses in transition landscapes of the Alto Ribatejo, central Portugal. Quat Geochronol 20(0):65–77, http://dx.doi.org/10.1016/j.quageo.2013.11.002

Buynevich IV, FitzGerald DM, Goble RJ (2007) A 1500 yr record of North Atlantic storm activity based on optically dated relict beach scarps. Geology 35(6):543–546

Cearreta A, Alday M, Freitas MC, Andrade C (2007) Postglacial foraminifera and paleoenvironments of the Melides lagoon (SW Portugal): towards a regional model of coastal evolution. J Foraminifer Res 37(2):125–135. doi:10.2113/gsjfr.37.2.125

Clemmensen LB, Murray A, Heinemeier J, de Jong R (2009) The evolution of Holocene coastal dunefields, Jutland, Denmark: a record of climate change over the past 5,000 years. Geomorphology 105(3–4):303–313. doi:10.1016/j.geomorph.2008.10.003

Costa M, Silva R, Vitorino J (2001) Contribuição para o estudo do clima de agitação marítima na costa Portuguesa. In: II Jornadas Portuguesas de Engenharia Costeira e Portuária. Sines, Portugal

Costas S, FitzGerald D (2011) Sedimentary architecture of a spit-end (Salisbury Beach, Massachusetts): the imprints of sea-level rise and inlet dynamics. Mar Geol 284(1–4):203–216. doi:10.1016/j.margeo.2011.04.002

Costas S, Alejo I, Rial F, Lorenzo H, Nombela MA (2006) Cyclical evolution of a modern transgressive sand barrier in NW-Spain elucidated by GPR and aerial photo. J Sediment Res 76:1077–1092

Costas S, Brito P, Fitzgerald D, Goble R (2013) Climate-driven episodes of dune mobilization and barrier growth along the central coast of Portugal, Special Publications 388. Geological Society, London. doi:10.1144/sp388.6

Daniels DJ (2005) Ground penetrating radar. In: Encyclopedia of RF and microwave engineering. John Wiley & Sons, Inc., Hoboken. doi:10.1002/0471654507.eme152

Dias MI, Prudêncio MI (2007) Neutron activation analysis of archaeological materials: an overview of the ITN NAA laboratory, Portugal. Archaeometry 49:383–393

Dill AC, Turberg P, Müller I, Parriaux A (2009) The combined use of radio-frequency electromagnetics and radiomagnetotellurics methods in non-ideal field conditions for delineating hydrogeological boundaries and for environmental problems. Environ Geol 56(6):1071–1091. doi:10.1007/s00254-008-1208-1

Etienne R, Makaroun Y, Mayet F (1994) Un grand compexe industriel à Tróia (Portugal). de Boccard, Paris

Freitas MC, Andrade C, Cruces A (2002) The geological record of environmental changes in southwestern Portuguese coastal lagoons since the Lateglacial. Quat Int 93–94:161–170

Freitas MC, Andrade C, Rocha F, Tassinari C, Munhá JM, Cruces A, Vidinha J, Da Silva CM (2003) Lateglacial and Holocene environmental changes in Portuguese coastal lagoons 1: the sedimentological and geochemical records of the Santo André coastal area. The Holocene 13:433–446

Gama C, Taborda R, Andrade C (2006) Longshore sediment transport in the Tróia-Sines Littoral Ribbon (SW Portugal). Paper presented at the VII Congresso Nacional de Geologia, Estremoz

Gomes NMN (1992) Dinâmica dunar do arco litoral Tróia-Sines (Portugal). MSc, University of Lisbon, Lisboa

Goy JL, Zazo C, Dabrio CJ (2003) A beach-ridge progradation complex reflecting periodical sea-level and climate variability during the Holocene (Gulf of Almería, Western Mediterranean). Geomorphology 50(1–3):251–268

Hayes MO (1980) General morphology and sediment patterns in tidal inlets. Sediment Geol 26(1–3):139–156. doi:10.1016/0037-0738(80)90009-3

Hein CJ, FitzGerald DM, Carruthers EA, Stone BD, Barnhardt WA, Gontz AM (2012) Refining the model of barrier island formation along a paraglacial coast in the Gulf of Maine. Mar Geol 307–310(0):40–57, http://dx.doi.org/10.1016/j.margeo.2012.03.001

Hesp PA (1999) The beach backshore and beyond. In: Short AD (ed) Handbook of beach and shoreface morphodynamics. Wiley, New York, pp 145–169

Hesp PA, Dillenburg SR, Barboza EG, Tomazelli LJ, Ayup-Zouain RN, Esteves LS, Gruber NLS, Toldo-Jr EE, Tabajara LLCA, Clerot LCP (2005) Beach ridges, foredunes or transgressive dunefields? Definitions and an examination of the Torres to Tramandaí barrier system, Southern Brazil. An Acad Bras Cienc 77:493–508

Jacob J, Gama C, Salgado R, Liu JT, Silva A (2009) Shadowing effects on beach morphodynamics during storm events on Tróia-Sines embayed coast, southwest Portugal. J Coast Res Spec Issue 56 (Proceedings of the 10th International Coastal Symposium):73–77

Leorri E, Cearreta A, Milne G (2012) Field observations and modelling of Holocene sea-level changes in the southern Bay of Biscay: implication for understanding current rates of relative sea-level change and vertical land motion along the Atlantic coast of SW Europe. Quat Sci Rev 42(0):59–73, http://dx.doi.org/10.1016/j.quascirev.2012.03.014

Masselink G, Short AD (1993) The effect of tide range on beach morphodynamics and morphology: a conceptual beach model. J Coast Res 9(3):785–800

Mauz B, Hijma MP, Amorosi A, Porat N, Galili E, Bloemendal J (2013) Aeolian beach ridges and their significance for climate and sea level: concept and insight from the Levant coast (East Mediterranean). Earth Sci Rev 121(0):31–54, http://dx.doi.org/10.1016/j.earscirev.2013.03.003

Meistrell FJ (1972) The spit-platform concept: laboratory observation of spit development. In: Schwartz ML (ed) Spits and bars. Dowden, Hutchinson & Ross, Stroudsburg, pp 225–283

Mitchum RMJ, Vail PR, Thompson S (1977) Seismic stratigraphy and global changes of sea level, Part 2: The depositional sequence as a basic unit for stratigraphic analysis. In: Payton CE (ed) Seismic stratigraphy – applications to hydrocarbon exploration, vol AAPG Memoir 26, pp 53–62

Naughton F, Sanchez Goni MF, Drago T, Freitas MC, Oliveira A (2007) Holocene changes in the Douro Estuary (Northwestern Iberia). J Coast Res 23:711–720

Otvos EG (2000) Beach ridges – definitions and significance. Geomorphology 32(1–2):83–108, http://dx.doi.org/10.1016/S0169-555X(99)00075-6

Pinto IV, Magalhães AP, Brum P (2011) O complexo industrial de Tróia desde os tempos dos Cornelii Bocchi. Escritor Lusitano da Idade de Prata da Literatura Latina. In: Cardoso JL, Almagro-Gorbea M (eds) Lucius Cornelius Bocchus. Academia Portuguesa da História. Real Academia de la Historia, Tróia, pp 133–167

Psuty NP (1992) Spatial variation in coastal foredune development. In: Carter RWG, Curtis TGF, Sheehy-Skeffington MJ (eds) Coastal dunes: geomorphology, ecology, and management for conservation. Balkema, Rotterdam, pp 3–13

Psuty NP, Moreira ME (2000) Holocene sedimentation and sea level rise in the Sado Estuary, Portugal. J Coast Res 16(1):125–138

Rebêlo L, Ferraz M, Brito P (2009) Tróia Peninsula evolution: the dune morphology record. J Coast Res SI56:352–355

Rebêlo L, Costas S, Brito P, Ferraz M, Prudêncio MI, Burbidge C (2013) Imprints of the 1755 tsunami in the Tróia Peninsula shoreline, Portugal. In: Conley DC, Masselink G, Russell PE, O'Hare TJ (eds) 12th international coastal symposium, Plymouth, England, J Coast Res, pp 814–819

Reimer PJ, Baillie MGL, Bard E, Bayliss A, Beck JW, Bertrand CJH, Blackwell PG, Buck CE, Burr GS, Cutler KB, Damon PE, Edwards RL, Fairbanks RG, Friedrich M, Guilderson TP, Hogg AG, Hughen KA, Kromer B, McCormac FG, Manning SW, Ramsey CB, Reimer RW, Remmele S, Southon JR, Stuiver M, Talamo S, Taylor FW, van der Plicht J, Weyhenmeyer CE (2004) IntCal04 terrestrial radiocarbon age calibration, 26–0 ka BP. Radiocarbon 46:1029–1058

Sabatier P, Dezileau L, Colin C, Briqueu L, Bouchette F, Martinez P, Siani G, Raynal O, Von Grafenstein U (2012) 7000 years of paleostorm activity in the NW Mediterranean Sea in response to Holocene climate events. Quat Res 77(1):1–11. doi:10.1016/j.yqres.2011.09.002

Simpkin PG, Davis A (1993) For seismic profiling in very shallow water, a novel receiver. Sea Technol 34(9):21–28

Sorrel P, Debret M, Billeaud I, Jaccard SL, McManus JF, Tessier B (2012) Persistent non-solar forcing of Holocene storm dynamics in coastal sedimentary archives. Nat Geosci 5(12):892–896, http://www.nature.com/ngeo/journal/v5/n12/abs/ngeo1619.html#supplementary-information

Spencer CD, Plater AJ, Long AJ (1998) Rapid coastal change during the mid- to late Holocene: the record of barrier estuary sedimentation in the Romney Marsh region, southeast England. The Holocene 8(2):143–163. doi:10.1191/095968398673197622

Stapor FJW, Mathews TD, Lindfors-Kearns FE (1991) Barrier-island progradation and Holocene sea-level history in Southwest Florida. J Coast Res 7(3):815–838

Tamura T (2012) Beach ridges and prograded beach deposits as palaeoenvironment records. Earth Sci Rev 114(3–4):279–297, http://dx.doi.org/10.1016/j.earscirev.2012.06.004

Tamura T, Murakami F, Nanayama F, Watanabe K, Saito Y (2008) Ground-penetrating radar profiles of Holocene raised-beach deposits in the Kujukuri strand plain, Pacific coast of eastern Japan. Mar Geol 248(1–2):11–27. doi:10.1016/j.margeo.2007.10.002

Thom BG, Roy PS (1985) Relative sea levels and coastal sedimentation in Southeast Australia in the Holocene. J Sediment Res 55(2):257–264. doi:10.1306/212f8693-2b24-11d7-8648000102c1865d

van Heteren S, Fitzgerald DM, McKinlay PA, Buynevich IV (1998) Radar facies of paraglacial barrier systems: coastal New England, USA. Sedimentology 45(1):181–200

Chapter 7
The Historical Evolution of the Tindari-Marinello Spit (Patti, Messina, Italy)

Antonino Crisà, Stefania Lanza, and Giovanni Randazzo

Abstract The aim of this paper is to investigate the evolution of Tindari-Marinello spit, which comprises a natural reserve and small lakes between Tindari (west) and Oliveri (east) towns of Messina province (Sicily, Italy). In particular, the spit is certainly worthy of analysis because it represents a significant case study of interaction between human activities and geological evolution of the lakes and coastline. The paper will also provide data on antiquarian studies on Tindari and information on cartography, which are beneficial to understand the spit evolution through the past and present history.

7.1 Introduction

This paper focuses on the geological evolution of the Tindari-Marinello spit over the last 2,000 years, examining the coastline of Tindari (Fig. 7.1) and the history of the site in relation to the sea and its exploitation by humans.

This chapter provides a reconstruction of the Tindari-Marinello lagoon system (Fig. 7.2), according to historical and recent cartographic evidence, demonstrating the progressive changes and evolution of the site. This methodological approach is significant for two main reasons. Firstly, it enables us to analyse the system chronologically in relation to the historical and cartographic representations. Secondly, it provides an outline of the on-going transformations of the coastal territory. Natural and anthropic events have constantly occurred due to a mutual cross-adaptation. Historical and recent maps show how the Tindari-Marinello spit has been altered its shape over the past few centuries, presenting different stages of growth, consolidation, stagnation, and ultimately a progressive decline and closure.

A. Crisà (✉)
School of Archaeology and Ancient History, University of Leicester, Leicester, UK
e-mail: ac472@le.ac.uk

S. Lanza • G. Randazzo
Dipartimento di Fisica e Scienze della Terra, Università degli Studi di Messina, Messina, Italy
e-mail: lanzas@unime.it; grandazzo@unime.it

G. Randazzo et al. (eds.), *Sand and Gravel Spits*, Coastal Research Library 12,
DOI 10.1007/978-3-319-13716-2_7

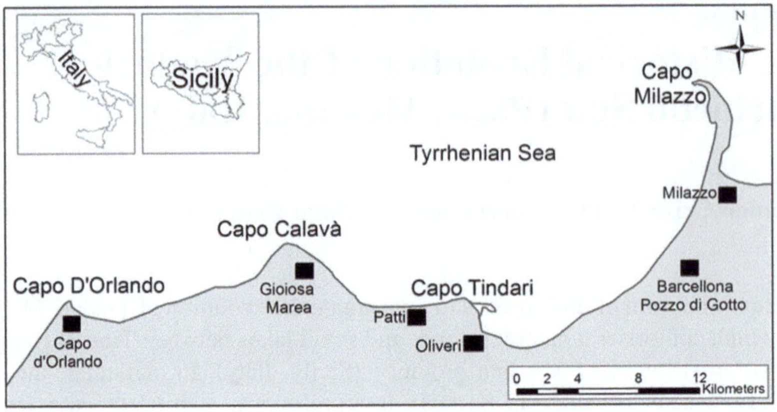

Fig. 7.1 Map of Sicily showing Tindari and the Gulf of Patti

Fig. 7.2 View of the lakes of Marinello (From the Sanctuary of Tindari)

Tindari, now a religious site in the province of Messina (northern Sicily) about 60 Km from the provincial capital and under the jurisdiction of municipality of Patti, stands on the promontory of Capo Tindari (38°13′ N-15°05′ E) along the Tyrrhenian coastal strip. This area encompasses the large Gulf of Patti, which includes Capo d'Orlando (Western limit), Capo Calavà and Capo Tindari (central seaboards) and Capo Milazzo (Eastern limit).

Tindari is also important for two reasons. First, it corresponds to the ancient Greco-Roman centre of *Tyndaris*, which antiquarians and archaeologists have investigated since eighteenth century. Now, visitors and scholars can visit many ancient remains (e.g. the theatre, the Roman *Basilica*, the *insula* IV private quarter, Cercadenari, etc.). Secondly, the Sanctuary of the Holy Black Madonna, represented by a medieval wooden sculpture, is still visited by pilgrims (Reitano 1961; Mollica 2000; Ross Holloway 2000; Spigo 2005; Leone and Spigo 2008; Crisà 2012b; Fasolo 2013).

Archaeological investigations, performed by Luigi Bernabò Brea and M. Cavalier in the *insula* IV (1950–1952), have revealed finds of an Early Bronze

Age settlement (*facies* Rodì-Tindari, twentieth to seventeenth century BC). The discovery reveals that the site was occupied during the Bronze Age (Cavalier 1970; Ross Holloway 2000; Spigo 2005).

Ancient *Tyndaris* was founded in 396 BC by Dionysius I, tyrant of Syracuse, who occupied the promontory of Tindari, probably subjugating territories of the indigenous centre of *Abakainon*-Tripi (Messina). *Tyndaris* became a notable city, due to the strategic and advantageous position for commerce in the Tyrrhenian Sea between northern Sicily and the Aeolian Islands. For instance, historical sources report that naval battles occurred in this area. In the Gulf of Patti commerce was prosperous from antiquity to the modern age (Diod. 15.5–6; Pettignano and Riccobono 1992; Spigo 2005; Crisà 2012a).

Soil exploitation and human activity along the Gulf of Patti coastal strip have constantly altered its complex environment. Recently, the coastal wetland system of Marinello-Tindari has been declared by the Government of Sicilian Region a natural reserve (Riserva Naturale Orientata – Natura 2000) (Decree n. 745/44, December 1988).

7.2 Geological and Geomorphological Aspects of Tindari-Marinello Spit

In historical times, the coastal system of the Gulf of Patti has undergone major geomorphic change. This change has favoured the deposition of a wide alluvial plain in the western area of the gulf and the formation of the Tindari Marinello spit on the slopes of the promontory. Furthermore, the progressive alteration has left geological evidence in the area, which have been reported in historical maps and documents as well.

The entire area of the Gulf of Patti from Capo d'Orlando to Capo Tindari represents an extended portion of land showing most of the main characteristics of a typical Sicilian landscape. The most significant is certainly the historical stratification, linked to the different past land uses. Geology, land morphology, biology, and ecology, have persistently interacted with the development/exploitation of land, sometimes influencing the historical evolution of the area making the Gulf of Patti and the Tindari-Marinello spit of ongoing scholarly interest (Crisafi et al. 1981; Giacobbe and Leonardi 1986; Giacobbe et al. 1990; Amore et al. 1991; Di Natale and La Loggia 1991; Bonfiglio et al. 2003a, b; Randazzo et al. 2004, 2005a, b; Nigro and Renda 2005; Leonardi et al. 2013; Privitera and Torre 2013).

There are three main natural physiographic units or coastal tracts (Fig. 7.3: see Units A-C) along the shorelines of the Gulf of Patti. These can be defined as a spatially contiguous set of morphological units, which represent a "sediment-sharing coastal cell" (Cowell et al. 2003).

These units include large headland bays and narrow pocket beaches with sand, pebble, and large boulder deposits (Ø 10–12 m), detached by steep and high

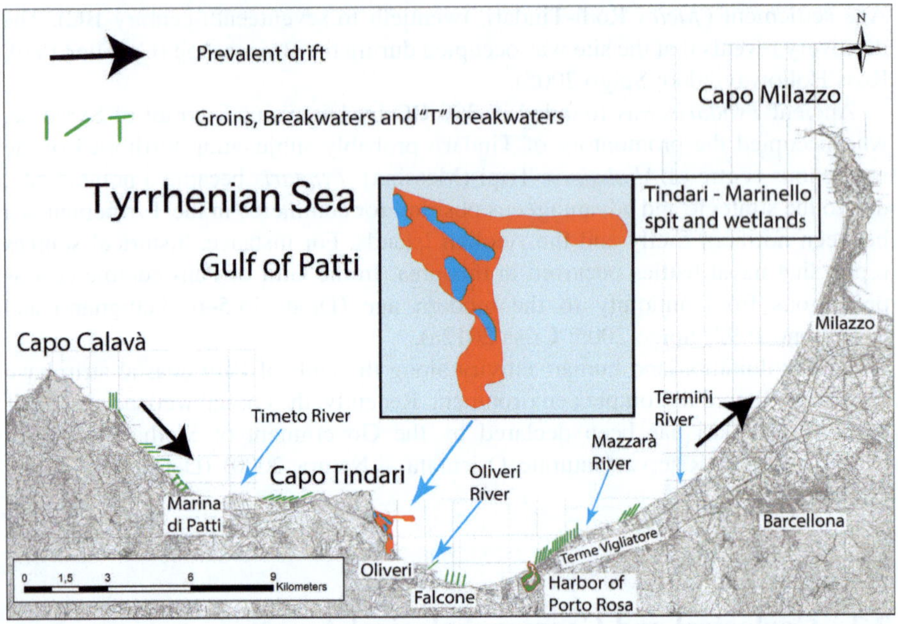

Fig. 7.3 The Gulf of Patti between Capo Calavà and Capo Milazzo headlands with two main coastal cells divided by Capo Tindari

rocky cliffs. Along the rocky coast of the main four headlands (from W to E: Capo d'Orlando, Calavà, Tindari, and Milazzo), erosional processes are predominant mainly where the slope angle is dependent on the jointing, bedding, and hardness of the cliff's geological materials. Several sea-stacks and caves represent a mechanism for slowing down the erosion of beaches and act as efficient reflector of high wave energy levels at predominate here (Venneri and Randazzo 1999; Randazzo et al. 2005b).

In this coastal portion, several waterways have gradually brought substantial sediments during flooding, with torrential flow in the western (Timeto, Montagnareale and Fetente) and east areas (Elicona, Mazzarà, Termini and Mela) of Capo Tindari.

The promontory of Capo Tindari, located in the northern area of the Peloritani Mountains, is oriented NW-SE, following the main tectonic paths and shows a high overhanging cliff (290 m a.s.l.). Capo Tindari is subjected to intensive tectonic activities, being crossed by a system of strike-slip faults (NW-SE orientation) which intersect with direct faults (NE-SW orientation). Recent tectonic activity of the main fault, called "Tindari-Novara di Sicilia" (Ghisetti 1979; Catalano and Di Stefano 1997; Lanzafame and Bousquet 1997), has revealed and raised up the metamorphic substrate, giving rise to steep cliffs and escarpments that surround the promontory (Barbano et al. 1979; Lentini et al. 2000; Nigro and Renda 2005; Billi et al. 2006).

Fig. 7.4 The faulted cliff of Cape Tindari

The ridge of Capo Tindari was quickly uplifted during the Late Pleistocene, as the elevation of the coeval deposits suggests. Today the deposits emerge for several hundred metres above the sea level (Bonfiglio et al. 2003b).

The recent discovery at Capo Tindari headland of an emerged marine notch (2–4 m a.s.l.), partially coated by sterile coarse clastic beachrock and not laterally continuous, supports a new interpretation of the Late Quaternary evolution of Capo Tindari region. In fact, it can be inferred that Capo Tindari has uplifted all together with the entire south-eastern sector following the same uplifting rate after one or more post-Tyrrhenian tectonic phases. Then, the northern margin increasingly receded (Bonfiglio et al. 2003a).

In Capo Tindari, a bounded block fault, the on-going morphological processes structurally affect the steep cliffs (Fig. 7.4), represented by very young tectonics which have not yet been dated (Bonfiglio et al. 2010). Moreover, ancient and present coastal features show evidence of erosional forms cut into carbonate rocky cliffs of Capo Tindari headland (e.g. caves and marine notches). Thus, they testify that the site area has been subjected to a recent neo-tectonic vertical motion and catastrophic landslides.

At the base of the promontory the Tindari-Marinello spit is visible, extending to the 30 m isobath, the emerged portion of which is about 3 km long (NW-SE direction). Since their formation, due to the currents that regulate the coastal drift, the emerged sections have taken the shape of elongate barriers extending south-easterly. These portions are continuously moved, cut and anastomosed by the wind.

In fact, the bar has undergone intense changes in terms of its size and shape, and recently researchers have observed a directional variation from E to the coast, showing a type of closure of the system. According to this complex development, the lakes of Marinello initially formed on the land surface (maximum of six lakes) and then subsequently, the lakes decreased because of an evident involution of the system.

Recently, some sand barriers have developed along the north-eastern side with alongshore sediment transport developing a microtidial, wave-dominated coarse sandy-gravelly spit platform. It is connected to a wetland system, formed by four permanently flooded littoral lacustrine habitats. The spit and the steep cliffs have an unstable equilibrium, characterized by a curvature of advanced beach ridge ends and a long backshore and embryo dune zone.

The lagoon of Marinello probably formed in the seventeenth to eighteenth centuries, due to the first intensive local deforestation. This circumstance plausibly increased the erosion processes in the hydrographic basins and accumulation of Timeto river sandy debris to the mouth and coastal areas. Particularly, deforestation of substantial woods of western Peloritani mountains intensified during that period.

In summary, the sandy spit and brackish lakes of Marinello represent a natural environment of high scientific interest because of its rapid and continuous evolution. Above all, these bodies of water are currently the only coastal lagoons still active along the Tyrrhenian Sicilian coast. The spit is also a valuable case study for understanding the relationship between human activity and the surrounding environment. Its genesis and recent evolution depend not only on currents, waves, wind, sediment budget, and tidal range (Bastos et al. 2012), but also on past human activities.

7.3 Historical Aspects and Geomorphological Evolution

Archestratus of Gela, author of the second half of the fourth century BC, mentioned the importance of tuna fishing at *Tyndaris* (Ath. 7.63.302). Two naval battles were fought between the Romans and the Carthaginians on the coast of *Tyndaris* during the third century BC (Zon. 8:12; Poly. 1.25.1–5). Pliny the Elder (first century AD) even indicated the good taste of the pearly razorfish (*Xyrichtys novacula*), caught in the sea of *Tyndaris* (Plin. *Nat. Hist.* 22.53.11).

Following the creation of the province of Sicily (241 BC), the substantial grain exports towards Rome also involved *Tyndaris*. The grain, which derived from the *decima* tax (1/10 of the harvest), was transported by sea to *Halaesa Archonidea*-Tusa (north-western Sicilian coast) or other crop-collecting centres. Then, it was moved to Rome (Facella 2006; Crisà 2012a).

During the Civil Wars (42–36 BC) Tindari became the cornerstone of Pompey's troops. Octavius, who became *Augustus* a few years later, was encamped quite close *Tyndaris* commanding a huge army (21 legions and 25,000 soldiers). It is not so clear where the fleet took refuge. However, we can speculate that at that time the bay of Patti from Capo Calavà and Tindari was much deeper than now. Thus, the Timeto stream was plausibly navigable in the section near its estuary (Lo Iacono 1997).

Pliny the Elder writes of a devastating earthquake that hit Tindari during the first century AD. This event destroyed half of the town, which collapsed into the sea

(Plin. *Nat. Hist.* 2.94). The earthquake, which might have damaged *Tyndaris*'s economy, also activated the progressive formation of the Tindari-Marinello spit.

The mass of rubble that slumped into the sea from the cliff likely changed the bathymetry across the seabed. The collapse of the city resulted in the formation of a substantial deposit, partially below sea level, which eased the settling of materials coming from the nearby Timeto river. Therefore, the river contributed to the sandbar over the next 1,500 years.

These deposits, which must have been quite visible during the long Imperial period (first to fifth century AD), created a convenient landing area, mostly used for navigation, commerce, and import of goods, as shown by archaeological finds. Above all, scholars consider *Tyndaris* to have been a stopover on the routes to Africa. The centre was probably an intermediate step in navigation along the coast of northern Sicily, often traversed by ships sailing in the open Tyrrhenian Sea. Of course, Tindari was also the departure centre en route to the Aeolian Islands (Uggeri 1997–1998; Crisà 2012a).

The area of *Tyndaris* has also been affected by numerous high-intensity earthquakes in late antiquity, such as the violent earthquake in AD 365, which caused serious damage to most of the Roman centres on the Tyrrhenian coast (Spigo 2005; Bottari 2008; Leone and Spigo 2008; Fasolo 2013). These events and the constant marine erosion may have caused the collapsing of rock blocks still visible today. More importantly, it is significant that the whole cliff, which borders the promontory of Capo Tindari, is still subject to continuous detachment of blocks or sometimes rock falls.

A few years after the Arab invasion (AD 836), *Tyndaris* was still occupied by its inhabitants, but it lost much of its importance in comparison with the past centuries. Meanwhile, Patti played a leading role in that territorial district, at least in the last years of the eleventh century (Cadili 2000; Irato 2004; Fasolo 2011).

During the medieval period, Tindari became a centre of pilgrimage. At that time pilgrims probably venerated the Black Madonna. According to historical sources, since twelfth century, there was monastery of St. Elias de Scala. It was built in 1110, "not so far from the ruins of the ancient Tindari close to climbing passage (*La Scala*), which started at the mouth of river Oliveri and went towards Patti". Scala is now a hamlet (less than 300 inhabitants) under the jurisdiction of the municipality of Patti (Cadili 2000; Crisà 2012a).

Muhammad al-Idrisi, a distinguished scholar born in Morocco in 1099, who spent 20 years in Sicily at the court of King Roger II, offers insights for the reconstruction of the territory of Patti in the twelfth century. Al-Idrisi's work documents the continuity of this advantageous exploitation of the sea from antiquity to the middle ages, carried out by the inhabitants along the coast between Patti and Oliveri-Tindari and already attested to by Archestratus Gela (Edrisi 2004; Crisà 2012a).

According to historical sources, archival records and building remains, the coastal area of the province of Messina (section from Falcone to San Giorgio di Gioiosa Marea), had many *tonnare* (factories for tuna fishing), intensely used from the modern age until the twentieth century. In particular, the *tonnara* of Oliveri (east of Tindari) was one of the most remarkable plants: built in the fourteenth century, it reached peak production in the early twentieth century. San Giorgio's *tonnara*, closed after World War II, has been recently transformed into a residence (Consolo 2008; Crisà 2012a).

During the modern era the production and trade of ceramics has been vigorous. Most of ceramics were massively produced in Patti Marina, especially between the fifteenth and twentieth centuries. The beach of Patti Marina was very close to the production workshops and represented a key place of these activities. In fact the products, moved from the workshops to the beach, were loaded on numerous ships. They could arrive already full of goods to be traded, then emptied and filled again with ceramics. According to records, it can be also inferred that the seaport of Tindari-Oliveri offered additional support. It is clear that ships could wait to be loaded in Patti Marina, on the occasion that the landing was too hard here due to bad weather and sea condition. Unfortunately, only scant remains of this substantial manufacturing system still survive in Patti, because of past demolition of workshops and structures in favour of local urban development (Pettignano and Riccobono 1992; Crisà 2012a).

7.4 Antiquarian Studies and Cartography on the Tindari-Marinello

Tiburzio Spannocchi (1543–1606), military engineer at the court of King Philip II and Philip III of Spain, carried out the earliest surveys of the Sicilian coasts. Spannocchi visited Sicily from 1577 to 1578 on the orders of Viceroy Marco Antonio Colonna. The main aim of Spannocchi's survey was to make detailed maps of the coastlines for defensive purposes. Regarding the coast between Mongiove and Capo Tindari, Spannocchi reported no natural harbours, sand banks or lagoons, although he detailed the above-mentioned *tonnara* of Oliveri (Mazzamuto 1986; Polto 2001).

Even the Belgian cartographer Gerardus Mercator (1512–1594), who accurately mapped the *Regum Siciliae* (1830) (scale: approximately 1:700,000) did not report lakes or sand banks on the slopes of Capo Tindari. On the contrary, he drew the coastal wetlands at Ganzirri (Messina) and Vendicari (Syracuse). This is important information indeed, as Mercator's maps testify that the formation of the spit (at least only its emerged and visible part) began between the second half of the seventeenth century and early eighteenth century (Fig. 7.5).

The deserted ruins of the ancient *Tyndaris*, especially the theatre and the Roman basilica, aroused interests among travellers and antiquarian scholars from the second half of the eighteenth century. First of all, Ignazio Paternò Castello (1719–1786), Royal Custodian of Antiquities in Sicily, mentioned the presence of underwater archaeological remains ("*sommerse rovine*"), which he linked to the first-century AD earthquake. Jean Houel (1735–1813), French painter and author of

Fig. 7.5 Mercator's map of the *Regum Siciliae*

Fig. 7.6 View of *Tyndaris (Houel 1782–1787, I: pl. 53)*

the *Voyage pittoresque des isles de Sicilie, de Malte et de Lipari* (1782) (Fig. 7.6), reported similar information. Tindari's inhabitants informed him that they had sometimes observed ruins in the sea below the promontory. Nevertheless, Houel could not trace or see these remains (Houel 1782; Paternò Castello 1817; Pagnano 2001; Crisà 2012a).

Early nineteenth-century historical sources detail the existence of a natural harbour and offshore sand bar. There were also three small lakes at Marinello, which appeared only in a later map (1842), discussed below. In particular, in 1806 Colonel Errico Sanches, who was Director of the Hydraulic Unit of the Royal Navy ("Direttore del Corpo Idraulico della Real Marina"), ordered an inspection of the Tindari-Marinello harbour to assess its potential for sailing and military use. The Library of the City Council of Palermo ("Biblioteca Comunale di Palermo") still keeps a detailed report on Tindari harbour, written by Camillo Manganaro on 15 September 1808 ("Porto detto della Marinella del Tindaro"). This report described the harbour as an ancient sand deposit from marine currents and winds. Unfortunately, this caused many problems to ships. Even two small vessels coming from Cefalù (Palermo) ("due Schifazzi Cefalutani") remained beached in the harbour's mouth for months, waiting to be emptied and moved with difficulty. Therefore, we know that sailors could use the harbour only occasionally, being aware of headwinds and dangerous sandbanks (Columba 1991; Crisà 2012a).

As stated above, the report provides remarkable details on the potential military uses of Tindari's harbour. In particular, Manganaro asserts that the area was naturally protected because of the substantial sandbanks. Therefore, the Bourbon Royal Navy did not need to build military structures (e.g. fortified docks) to protect the area. Nevertheless, ships and trade vessels could struggle to enter or exit from the harbour due to currents, tides and adverse winds (Crisà 2012a).

Francesco Ferrara (1767–1850), Royal Custodian of Antiquities, described the beach in the western area of Capo Tindari (1822), where he made many scattered finds (e.g. fragmentary pottery and bricks) evidently from the rubble corresponding to the first-century AD earthquake. Stunningly, Ferrara argued that they were the same fragments that could be seen on top of *Tyndaris*' ruins. Nicola Giardina, priest at the Sanctuary of Tindari and author of *L'antica Tindari* (1882), confirmed what Ferrara had reported about 60 years before. Thus, we can argue that those remains were still on the beach, uneffaced by coastal erosion, tides or sea storms (Ferrara 1822; Giardina 1882; Crisà 2012a).

William Henry Smyth (1788–1865), astronomer and admiral, visited Tindari between 1815 and 1818, when he was charged by the British Navy with drawing maps of Sicily. After arriving at Marinello and carrying out deep dredging, he discovered fragments of bricks and mortar ("sand mixed with triturated brick"). Then, Smyth argued that they could be part of the ancient port of *Tyndaris* (called Port Madonna). This information testifies to the presence of archaeological remains at Marinello. Above all, local fishermen often find fragmentary bricks in that area (Crisà 2012a).

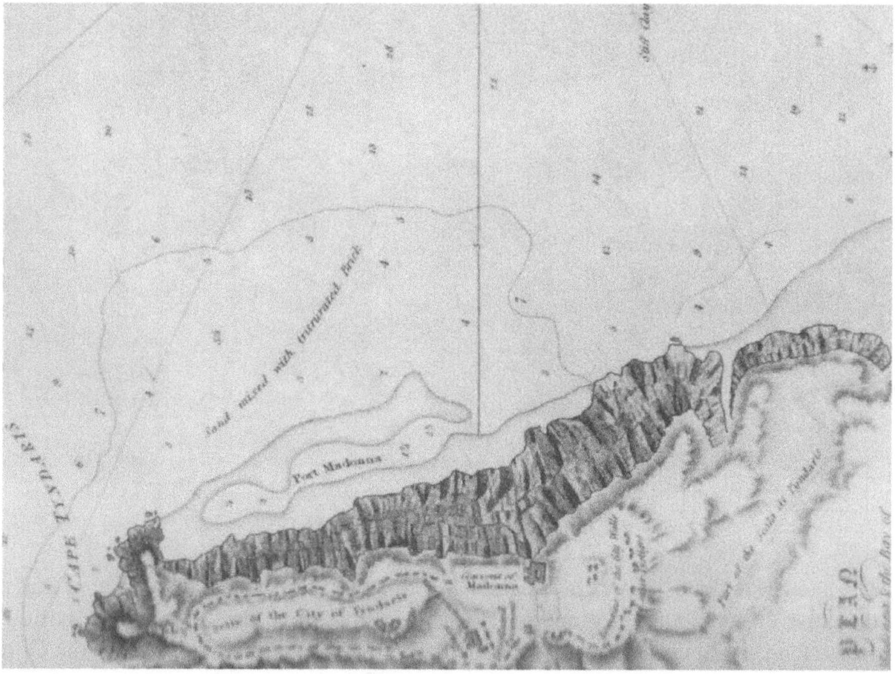

Fig. 7.7 Plan of Port Madonna and the Bay of Olivieri (Smyth 1824)

> Part of *Tyndaris* is said to have been precipitated into the sea by an earthquake (the same, add the good monks of the convent, that took place at the crucifixion of our Saviour); but as I traced the walls, and found them continuous, I should imagine to have been a suburb that fell, or rather a necropolis, as the whole of that side of the rock abounds with fragments of vases, lachrymatories, lamps, and idols. The cliff that was separated, no doubt, damaged the port beneath, as I found not only the dry sand, but also that which I dredged up in four fathoms' water, on the bank, mixed with numerous pieces of brick and cement triturated into small pebbles (Smyth 1824).

The *Plan of Port Madonna and the Bay of Olivieri* reveals a long sandbank in parallel to the coast and a small deep creek corresponding to a natural harbour (Fig. 7.7):

> Point *Tyndaris* is bold, and easily distinguishable from afar, by the monastery on its highest summit. On the east side is a curious little anchorage, called Port Madonna, where small vessels find secure shelter against all winds; the circumference is not above half a mile, and its depth is fourteen feet in the middle; and this is all that remains of the ancient harbour. A sand bank extends from it to nearly half a mile in the offing, which ships bound into the bay must avoid by giving it a wide birth, stearing well to the eastward, and bringing the castle Scalaproto to bear S.S.W., before standing to southward (Smyth 1824, *Appendix* 1, *Chapter* I: VII).

Smyth's accurate description clearly adds to our knowledge of the conformation of Marinello's lakes in the early nineteenth century. In addition, the author pointed

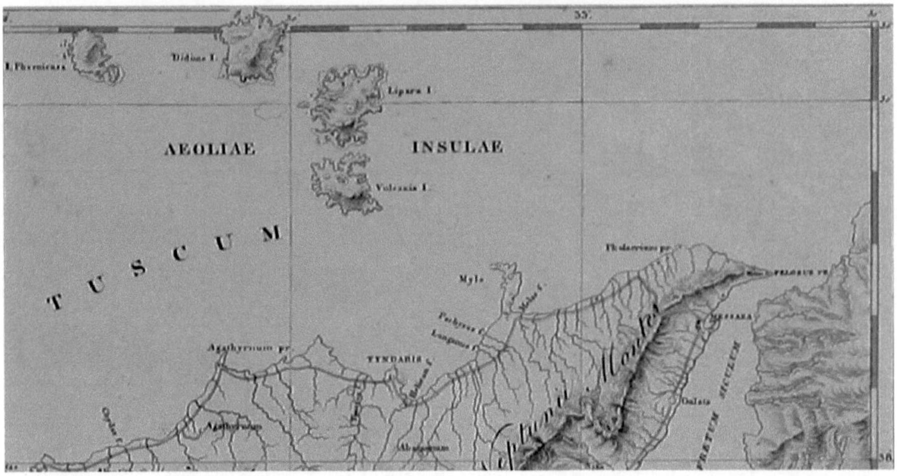

Fig. 7.8 Sicilia antiqua (Lo Faso 1834)

out a landslide deposit, which basically gave rise to the spit. It can be argued that a particular circumstance, formed between the Roman age and the modern period, altered the flow of currents, favouring a changing of deposition. Above all, the Gulf of Patti was shallower and streams more navigable than now.

A few years later, the Bourbon Topographic Office (*Officio Topografico*), based in Naples, published a new map of Sicily (*Carta generale della isola di Sicilia*, 1826; scale: approximately 1:270,000). This map showed a long sandy offshore bar (S-E direction) at the base of the rocky cliff and a lagoon behind it. Subsequently, the architect Francesco Saverio Cavallari (1809–1896) drew a map of the ancient *Tyndaris*, published in the 4-volume work *Le Antichità di Sicilia* (1835–1842) by Domenico Lo Faso (1783–1863), Duke of Serradifalco. The new map by Cavallari reported details of the archaeological ruins and displayed a sandbank at Marinello (direction NW-SE) forming a natural harbour (Figs. 7.8 and 7.9) (Lo Faso 1834–1842, V, pl. XXX).

Regarding the cartography during the Kingdom of the Two Sicilies, we finally mention the *Carte des Còtes de la Sicilie et de la Regence de Tunis* (1840) (scale: 1:1,500,000), published by the admiral Albin Reine Roussin (1781–1854) on the behalf of King Ferdinand I. Again, the map showed a sandbank at Marinello (direction S-E).

After the Unification of Italy (1861), the Geographic Military Institute (*Istituto Geografico Militare*, IGM) published a map (1865) (scale: 1:25,000) (Fig. 7.10). It reported a vast beach on the slopes of Tindari's cliff (178,620 m^2) and a small harbour in the eastern area, close to the escarpment (*Porto di Tindaro*). The latter is formed by two inlets and protected by two sandy strips (direction NE-SW).

The nautical map, produced by the Hydrographic Institute of Navy (*Istituto Idrografico della Marina*) (1881) and based on data previously collected by the Royal Navy of Washington in 1877, presents a coastal sandbank (c. 1,500 m long

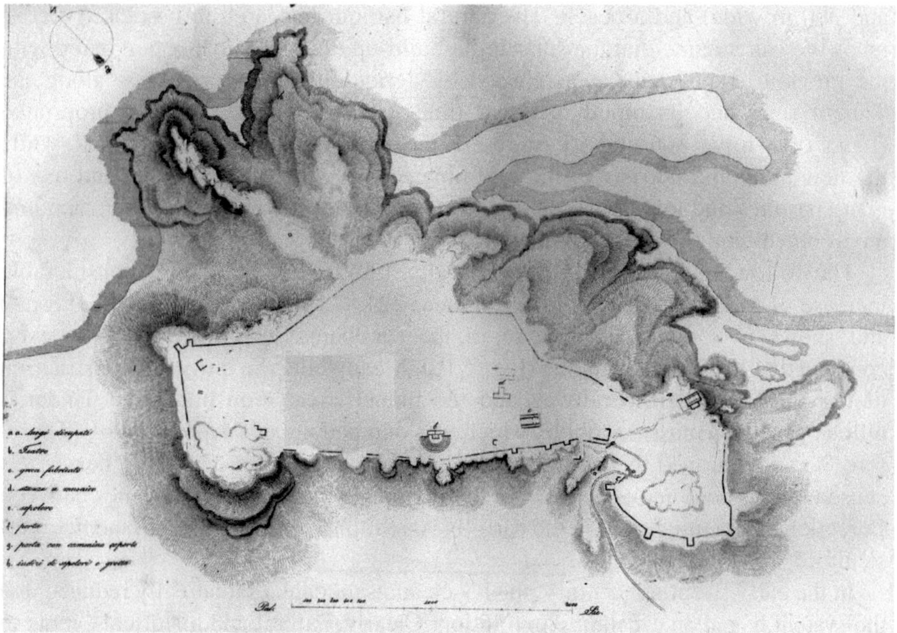

Fig. 7.9 Map of Tindari (Lo Faso 1834–1842)

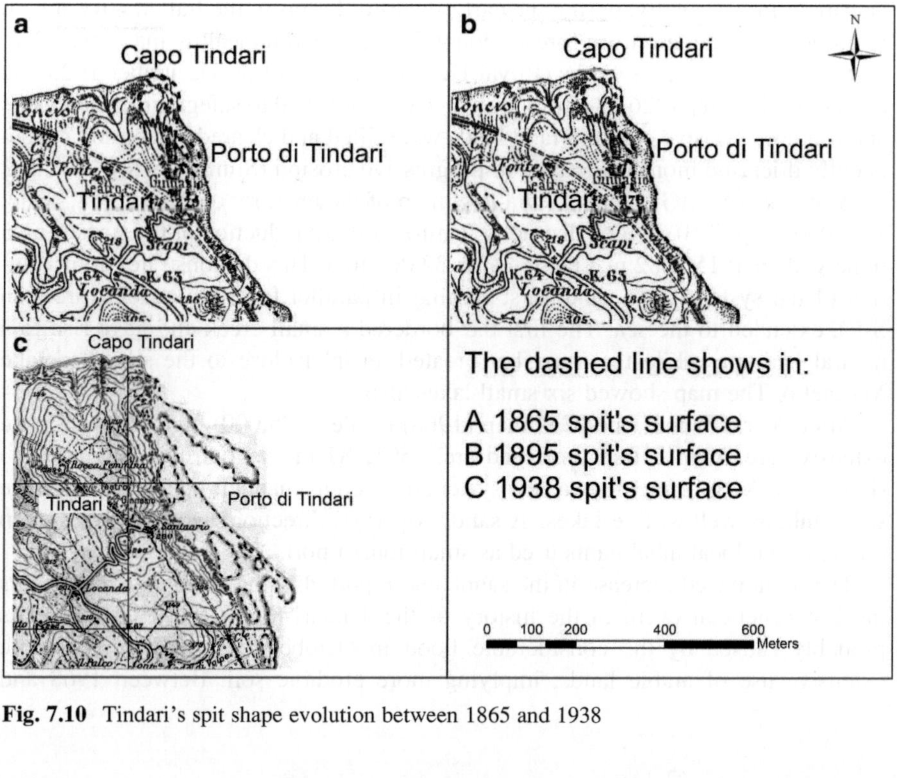

Fig. 7.10 Tindari's spit shape evolution between 1865 and 1938

and 300 m wide) and a beach. The natural harbour appeared to be closed on the map. We can assume that the substantial build-up of the sandbank, compared with the previous 1865 shape, was caused by agricultural transformations along the Timeto river and its mouth. At that time, Church's estates were expropriated (Laws of 7 July 1866 and 15 August 1867) and divided in smaller lands, while the new agricultural crisis of the mid-nineteenth century changed the end use of many pastures and forests. These circumstances evidently changed the surrounding environment and the sandbank.

The following IGM map (1895) (scale: 1:25,000) highlights a considerable increase of Marinello beach (emerged area: 224,878 m^2; lagoon area: 14,198 m^2) and shows two small lakes close to the sandbank, which are not, however, connected to the Tyrrhenian Sea (Fig. 7.10). Meanwhile, the on-going construction of the Messina-Palermo railway and the tunnel excavation under the Tindari's hillock produced massive rubble, which was dumped under the promontory. Therefore, it can be argued that the substantial surface increase of Marinello beach was caused by this continuous disposal of debris. Finally, the construction sites between Barcellona and Patti closed in the early 1890s (http://www.ferrovieitaliane.net/rete-sicula).

In the early twentieth century, the beach and sandbank dramatically reduced and the system began an escalating involution. Clearly, natural and historical events in the province of Messina stimulated this phenomenon along the Tindari-Oliveri coast. First of all, the catastrophic earthquake and tsunami of Messina (1908) substantially moved back the coastline and softened the bathymetric external isobath (5 m). Secondly, during the early stages of Fascism, the Italian government imposed new regulations on forest, mountain areas, and unhealthy marshes, subsidizing land reclamation works (Royal Decrees n. 3267 of 30 December 1923 and n. 23 of 03 January 1926). A strict project was established to safeguard coastal belts and promote massive works. Many lands were tilled and cleared, becoming consequently drier and more vulnerable to progressive erosion (Sturani 1984).

Afterwards the IGM published a new map of the territory of Patti (1938, scale: 1:25,000) (Fig. 7.10), which distinctly features a sharp reduction of the surface area (emerged area: 150,682 m^2; lagoon area 32,060 m^2). This demonstrates an involution of the system. Two sand bars, starting in parallel from the beach (direction S-E), extended to the sea. The first bar bordered a small creek and gave rise to a natural harbour, while the inner bar created an inlet close to the so-called lake Marinello. The map showed six small lakes also.

After World War II, the IGM map (1968) (scale: 1:25,000) highlighted a more extensive area than before (emerged area: 595,287 m^2; lagoon area: 40,743 m^2) (Fig. 7.11). At the same time, a new system of internal channels appeared inside the sandbank, as well as five lakes. A sand strip (S-E direction) gave rise to a deep creek, which local inhabitants used as small tourist port.

The pronounced increase in the sandbank, reported in the 1968 map, represents the last beneficial event in the history of the Tindari-Marinello. The event was probably caused by the considerable flood in October 1951 and by the more extensive use of arable lands, implying more erodible soil. Between 1968 and

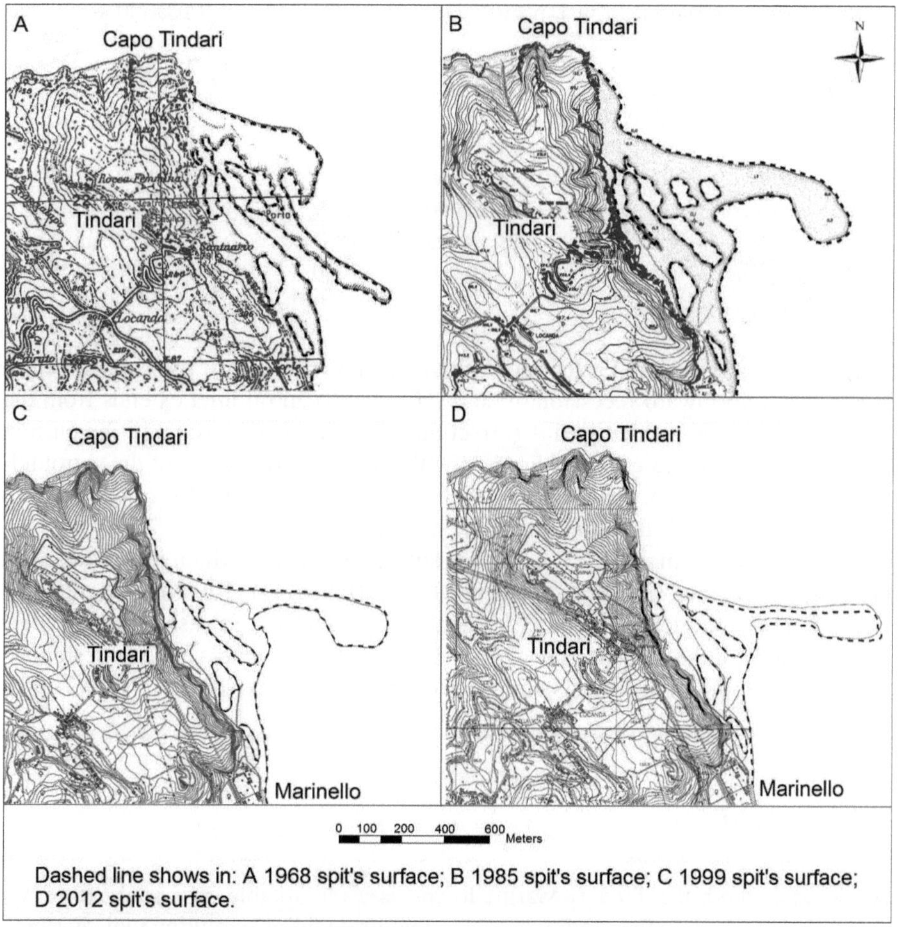

Fig. 7.11 Tindari's spit shape evolution between 1985 and 2012

1978 the system appeared to be quite stable, although the sandbank's tip started to continuously change its shape and conformation. The cause could be the construction of the new port of Capo d'Orlando (1972), which partially deviated the coastal currents (Venneri and Randazzo 1999; Zanghì and Randazzo 2000).

The 1985 map (scale: 1:10,000) demonstrates how the emerged area remains constant (623,783 m^2), while the lagoon increased in surface area (110,313 m^2). At the time, the final part of the sandbank followed an easterly direction and the natural harbour was shallower than before (Fig. 7.11). Thus the inlet was partially closing. In addition to the above-mentioned construction of the new port of Capo d'Orlando, the expansion of the local area from 49 ha (1968) ha to 139 ha (1989) provoked this change in Marinello beach (Di Natale and La Loggia 1991).

The Regional Technical Map (*Carta Tecnica Regionale*) (1999) (scale 1:10,000) reflects an increase of both surface areas (emerged area: 565,380 m^2; lagoon area:

Table 7.1 Scheme showing a time progression of the surface area of Tindari-Marinello spit (1865–1999)

Year	Author	Scale	Emerged area (m^2)	Lagoon area (m^2)
1865	*Istituto Geografico Militare*	1:25,000	178,620	Absent
1895	*Istituto Geografico Militare*	1:25,000	224,878	14,198
1938	*Istituto Geografico Militare*	1:25,000	150,682	32,060
1968	*Istituto Geografico Militare*	1:25,000	595,287	40,743
1985	*Regione Siciliana*	1:10,000	623,783 m^2	110,313 m^2
1999	*Regione Siciliana*	1:10,000	565,380	125,500

125,500 m^2) (Fig. 7.11). Furthermore, the map shows a shift of the spit (direction E-SE) and a northward recession of coast. Thus, the central inlet extends from the sea to form two expanded basins (direction NW-SE). This is roughly the current shape of Tindari-Marinello spit. Evidently, the massive concreting of the surrounding seaboard, especially during the 1990s, increased the coastal erosion, causing those changes in the lakes and sandbank shape.

The most recent maps (survey 2012) testify to the way in which the sandbank is becoming thinner and shifting towards the E-SE (Fig. 7.11). Therefore, the involution of the geological system is evident. More recently, construction of new embankments of the Timeto river and installation of breakwaters have both stopped the natural supply of sediments coming from the western territories (Table 7.1).

7.5 Conclusion

As demonstrated, the Tindari-Marinello spit is a remarkable case study for environmental research in the province of Messina. Its complex evolution indicates how the lagoon and lakes have been changed in the past centuries, due to on-going human activity, modifying the surrounding environment and landscape.

We reiterate that the early origins of the Tindari-Marinello sandbank came from a sudden catastrophic event. As historical sources report, an extensive landslide occurred in the ancient period, creating an underwater structure on which the lagoon slowly developed. Immediately after the landslide, sediments carried by local rivers were quite scant. In fact, Sicily was almost entirely covered by forests at that time and soil eroded less than today because of this substantial woodland coverage. Thus, it is possible that a very small beach arose in the area between the cliff and the collapsed debris.

For many centuries, the situation remained largely unchanged until the deforestation of the seventeenth and eighteenth centuries drove a significant morphological change. In fact, deforestation facilitated the accumulation of river sediments, which began to accumulate on the undersea step, gradually giving rise to the sandbank. It seems that Manganaro was the first scholar to describe the sandbank in 1808 and to detail the natural harbour at Marinello.

Finally, the analysis of historical and recent cartography has clarified how the geological system underwent a rapid evolution, which was strictly connected to the human activity. Today, many transitory small lakes appear and disappear. On-going massive constructions on the province of Messina's seashore (e.g. breakwaters) are still transforming the coastal conformation and altering the marine currents, thereby dispersing the sediments and augmenting erosion.

References

Amore C, Giuffrida E, Zanini A (1991) Evoluzione temporale e dinamica litorale dell'area lagunare di Oliveri-Tindari (Messina). Bollettino Accademia Gioenia di Scienze Naturali 24(337):117–131

Barbano MS, Bottari A, Carveni P, Cosentino M, Federico B, Fonte G, Lo Giudice E, Lombardo G, Patanè G (1979) Macro-seismic study of the Gulf of Patti earthquake in the geostructural frame of north-eastern Sicily. Boll Soc Geol Ital 98:155–174

Bastos L, Bioa A, Pinho JLS, Granja H, Jorge da Silva A (2012) Dynamics of the Douro estuary sand spit before and after breakwater construction. Estuar Coast Shelf Sci 109:53–69

Billi A, Barberi G, Faccenna C, Neri G, Pepe F, Sulli F (2006) Tectonics and seismicity of the Tindari Fault System, southern Italy: crustal deformations at the transition between ongoing contractional and extensional domains located above the edge of a subducting slab. Tectonics 25:1–20

Bonfiglio L, Di Maggio C, Marra AC, Masini F, Petruso D (2003a) Bio-chronology of Pleistocene vertebrate faunas of Sicily and correlation of vertebrate bearing deposits with marine deposits. Il Quaternario 16 (bis):107–114

Bonfiglio L, Formica S, Geremia F, Lanza S, Mangano G, Randazzo G (2003b) Evoluzione morfotettonica tardoquaternaria di Capo Tindari. In: Antonioli F, Monaco C (eds) Il contributo dello studio delle antiche linee di riva alla comprensione della dinamica recente. Escursioni nello Stretto di Messina. Atti del Convegno di Studi, Messina, 5–8 Maggio 2003, Messina, pp 29–30

Bonfiglio L, Mangano G, Pino P (2010) The contribution of mammal-bearing deposits to timing Late Pleistocene tectonics of Cape Tindari (North Eastern Sicily). Riv Ital Paleontol Stratigr 116(1):103–118

Bottari C (2008) Evidence of seismic deformation of the paved floor of the *decumanus* at Tindari (NE, Sicily). Geophys J Int 174:213–222

Cadili N (2000) Il Castello di Patti dal mille al duemila. Edizione Altavilla, Patti

Catalano S, Di Stefano A (1997) Sollevamento e tettogenesi Pleistocenica lungo il margine tirrenico dei Monti Peloritani: integrazione dei dati geomorfologici, strutturali e biostratigrafici. Il Quaternario 10:337–342

Cavalier M (1970) La stazione preistorica di Tindari. Bullettino di Paletnologia Italiana 21, 79, 1:61–94

Columba GM (1991) I porti della Sicilia. Accademia Nazionale di Scienze Lettere e Arti, Palermo

Consolo V (2008) La pesca del tonno in Sicilia. Sellerio, Palermo

Cowell PJ, Stive MJ, Niedoroda AW, de Vriend HJ, Swift DJ, Kaminsky GM, Capobianco M (2003) The coastal-tract (Part I): a conceptual approach to aggregated modelling of low-order coastal change. J Coast Res 19(4):812–827

Crisà A (2012a) La fascia costiera di Tindari e Patti dall'antichità agli inizi dell'Ottocento. Rassegna Storica dei Comuni 36(158–159):8–34

Crisà A (2012b) Numismatic and archaeological collecting in northern Sicily during the first half of the nineteenth century (BAR international series 2411). Archaeopress, Oxford

Crisafi E, Giacobbe S, Leonardi M (1981) Nuove ricerche idrobiologiche nell'area lagunare di Oliveri-Tindari (Messina). Morfologia dei bacini e caratteristiche fisico- chimiche delle acque e dei sedimenti. Memorie di Biologia Marina e Oceanografia 4:139–186

Di Natale R, La Loggia G (1991) Analisi dell'evoluzione della costa prospiciente Capo Tindari ai fini della conservazione dei laghetti di Marinello. In: VV.AA., Problemi ambientali e tecnologie avanzate. Atti del Convegno, Catania 1991, Ente Autonomo Fiera del Mediterraneo, Palermo, p 21

Edrisi A (2004) La Sicilia nel Libro di Ruggiero. Edi.bi.si, Palermo

Facella A (2006) Per una storia di Alesa Archonidea. Ricerche su un'antica città della Sicilia tirrenica. Edizioni della Normale, Pisa

Fasolo M (2011) L'assetto del territorio ad ovest di Tindari in età normanna. In: VV.AA., Contributi alla conoscenza del territori dei Nebrodi. *Da* Halaesa *ad* Agathyrnum. Studi in memoria di Giacomo Scibona. Edizioni del Rotary Club, Sant'Agata di Militello, pp 161–184

Fasolo M (2013) Tyndaris e il suo territorio. Volume I: Introduzione alla carta archeologica del territorio di Tindari. MediaGeo, Ciampino

Ferrara F (1822) Guida dei viaggiatori agli oggetti più interessanti a vedersi della Sicilia. Tipografia di Francesco Abbate, Palermo

Ghisetti F (1979) Relazione tra strutture e fasi trascorrenti e distensive lungo i sistemi Messina Fiumefreddo, Tindari-Letojanni e Alì-Malvagna (Sicilia nord-orientale): uno studio microtettonico. Geologia Romana 18:23–58

Giacobbe S, Leonardi M (1986) L'area lagunare di Oliveri-Tindari: sue variazioni morfologiche recenti ed evoluzione del popolamento a molluschi. In: Bregant D, Fonda Umani S (eds) Atti del VII Congresso della Associazione Italiana di Oceanologia e Limnologia, Consiglio Nazionale delle Ricerche, 11–14 Giugno 1986, pp 355–366

Giacobbe S, Leonardi M, Azzaro F, Rinelli P (1990) Descrizione di un esempio di "confinamento": l'area lagunare di Oliveri-Tindari (Messina). Oebalia 16(2):675–678

Giardina N (1882) L'antica Tindari: cenni storici. Tipografia S. Bernardino, Siena

Houel J (1782) Voyage pittoresque des isles de Sicilie, de Malte et de Lipari. De l'Imprimerie de Monsieur, Paris

Irato F (2004) Patti nella storia. Pungitopo, Messina

Lanzafame G, Bousquet JC (1997) The Maltese escarpment and its extension from Mt. Etna to the Aeolian Islands (Sicily): importance and evolution of a lithosphere discontinuity. Acta Vulcanol 9:113–120

Lentini F, Catalano S, Carbone S (2000) Carta geologica della provincia di Messina: scala 1:50.000 e nota illustrativa. S.E.L.C.A., Firenze

Leonardi M, Azzaro F, Bergamasco A, Lanza S (2013) L'ecosistema litorale di Marinello (Messina): dinamica geomorfologica e sedimentaria degli ultimi cinquant'anni. In: Il monitoraggio costiero mediterraneo: Problematiche e tecniche di misura. Atti del IV Simposio, Fondazione LEM, Livorno, 12–14 giugno 2012, pp 277–284

Leone R, Spigo U (eds) (2008) Tyndaris 1. Ricerche nel settore occidentale: campagne di scavo 1993–2004. Regione Siciliana, Assessorato dei Beni Culturali, Ambientali e della Pubblica Istruzione, Ambientali e della, Palermo

Lo Faso D (1834–1842) Le Antichità di Sicilia, esposte ed illustrate da Domenico Lo Faso Duca di Serradifalco. 5 vols. Presso la Reale Stamperia, Palermo

Lo Iacono N (1997) Nauloco e Diana Facellina. Un'ipotesi sul territorio di Patti fra Mitologia, Storia e Archeologia. Armando Siciliano Editore, Messina

Mazzamuto A (1986) Architettura e Stato nella Sicilia del '500. I progetti di Tiburzio Spannocchi e di Camillo Camilliani del sistema delle torri di difesa dell'isola. S. V. Flaccovio, Palermo

Mollica M (2000) Tindari dalla città greca al culto della Madonna Nera. A. Siciliano, Messina

Nigro F, Renda P (2005) Plio-Pleistocene strike-slip deformation in NE Sicily: the example of the area between Capo Calavà and Capo Tindari. Boll Soc Geol Ital 124(2):377–394

Pagnano G (2001) Le Antichità del Regno di Sicilia. 1779. I plani di Biscari e Torremuzza per la Regia Custodia. A. Lombardi, Palermo

Paternò Castello I (1817) Viaggio per tutte le antichità della Sicilia, descritte da Ignazio Paternò principe di Biscari. Dalla Tipografia di Francesco Abbate, Palermo

Pettignano A, Riccobono F (1992) Antiche ceramiche di Patti. Marina di Patti, Pungitopo

Polto C (2001) La Sicilia di Tiburzio Spannocchi. Una cartografia per la conoscenza e il dominio del territorio nel secolo XVI. Istituto Geografico Militare, Firenze

Privitera S, Torre S (2013) Le lagune di Oliveri-Tindari: un caso di interazione tra azione antropica ed evoluzione ambientale. Annali della Facoltà di Scienze della Formazione Università degli Studi di Catania 12:111–157

Randazzo G, Geremia F, Lanza S (2004) Temporal evolution of Tindari headland spit and Marinello coastal wetland system (NE Sicily). In: Viaroli P, Lassere P, Campostrini P (eds) International conference lagoons and coastal wetlands in the global change context: impact and management issues. International conference, UNESCO-CORILA, Venice, 26–28 Apr 2004, pp 17–18 (abstract)

Randazzo G, Geremia F, Lanza S (2005a) Negative response to remedial measures of shore protection in Sicily: the case of the Tindari headland spit (Northern Sicily). In: ICCCM05 international conference on coastal conservation management in the Atlantic and Mediterranean, Tavira, 17–20 Apr 2005, pp 199–200

Randazzo G, Geremia F, Lanza S (2005b) The headland spit of Tindari (NE Sicily): past, present and prospective. In: Proceedings of the international conference "Lagoons and Coastal Wetlands in the Global Change Context". Dossier n. 3 UNESCO, pp 100–105

Reitano A (1961) Memorie sul santuario del Tindari. Edizione Rondine, Roma

Ross Holloway R (2000) Archaeology of ancient Sicily. Routledge, London/New York

Smyth WH (1824) Memoir descriptive of the resources, inhabitants, and hydrography, of Sicily and its islands, interspersed with antiquarian and other notices. John Murray, Albermarle-Street, London

Spigo U (ed) (2005) Tindari. L'area archeologica e l'Antiquarium. Rebus Edizioni, Milazzo

Sturani E (1984) Il grande libro della Sicilia. Arnoldo Mondadori Editore, Milano

Uggeri G (1997–1998). Itinerari, strade, rotte, porti e scali della Sicilia tardoantica. Kokalos, 43–44, 1, 1:299–365

Venneri S, Randazzo G (1999) Erosional processes in an urbanized area between Capo d'Orlando and Capo Tindari (NE Sicily). COASTline 199(3):20–21

Zanghì A, Randazzo G (2000) Relazione tecnica sullo stato di fatto della portualità siciliana propedeutica alla redazione della carta vocazionale dei porti. Centro di Ric Economica e Scientifica 4(2):7–41

Chapter 8
Anthropogenic Influence on Spit Dynamics at Various Timescales: Case Study in the Bay of Cadiz (Spain)

L. Del Río, J. Benavente, F.J. Gracia, C. Alonso, and S. Rodríguez-Polo

Abstract Human interventions are one of the main drivers of coastal change in many areas, often generating undesired impacts like shoreline retreat. Sandspits are especially sensitive to anthropogenically-induced changes, especially those related to sediment supply. This work presents a case study of Valdelagrana spit in SW Spain, a sandbody where anthropogenic influence has been evident since Roman times. A variety of methods were applied to assess geomorphological and morphodynamic changes in the area at various timescales. Historical interventions involve mainly river course diversion, which caused important changes in sediment supply. More recently, coastal engineering structures and land reclamation deeply modified wave and current patterns in the area, triggering massive coastal erosion. As a consequence of this, the system has evolved from a drift-aligned spit to a swash-aligned barrier. This study provides insights into the consequences of human interventions at similar coastal settings.

8.1 Introduction

The dynamic nature of coastal systems is related to a variety of drivers, of both natural and anthropogenic origin, often acting simultaneously at different spatial and temporal scales. In this respect, it has been widely stated that anthropogenic influence on coastal behaviour has greatly increased worldwide over recent decades due to the accelerated occupation of littoral zones which had remained pristine until recent times (Komar 2000). However, in some areas where development started in ancient times, anthropogenic transformations of coastal landscape and dynamics have been going on for centuries.

L. Del Río (✉) • J. Benavente • F.J. Gracia • S. Rodríguez-Polo
Earth Sciences Department, University of Cádiz, Av. República Saharaui s/n, 11510 Cádiz, Puerto Real, Spain
e-mail: laura.delrio@uca.es

C. Alonso
Centre for Maritime Archaeology (CAS), IAPH, Consejería de Cultura, Junta de Andalucía, Av. Duque de Nájera 3, 11002 Cádiz, Spain

G. Randazzo et al. (eds.), *Sand and Gravel Spits*, Coastal Research Library 12,
DOI 10.1007/978-3-319-13716-2_8

This human influence on coastal behaviour generally involves changes in the conditions that determine the geomorphological and morphodynamic character of each particular coastal area. The most common small-scale impacts are related to interventions altering local sediment budget on the coast, such as building shoreline defence structures, sand mining, artificial beach nourishment or destruction of coastal natural systems by urban development, among others (Komar 2000). The main medium- and large-scale impacts are associated with human interventions on river basins, such as dam construction (Syvitski et al. 2005), and those related to sea level rise and coastal erosion (FitzGerald et al. 2008).

The extent and nature of the effects of these activities will depend on the type of coastal system involved. Sandy coasts are extremely dependant on sediment supply, so their evolution and behaviour will be highly determined by human interventions that alter the sediment budget. In this regard, sandspits are particularly sensitive to changes in conditions such as wave approach direction or sediment input (Komar 1998).

This work presents a case study of Valdelagrana spit barrier in SW Spain, a sandbody where different types of human interventions at various timescales have been altering the natural behaviour of the system during the past 2,000 years. The analysis of this study case provides an understanding the response of a sandspit to anthropogenic modifications at diverse spatial and temporal scales. Thus, it can ultimately contribute to predict future consequences of planned interventions at similar settings, in order to avoid undesired impacts.

8.2 Study Area

Valdelagrana spit barrier is located in the Bay of Cádiz (SW Spain) (Fig. 8.1). It is a north-south oriented sandy body with a total length of 7.2 km and an average width of 1.5 km, running from the Guadalete River mouth to the outlet of Río San Pedro tidal channel. Guadalete River is the main fluvial course in Cádiz province with a total length of 170 km and a drainage basin of 3,677 km^2. The river mouth at the northern limit of the study area is a 200 m wide, tide-dominated estuary.

Valdelagrana shoreline shows a Zeta-bay shape developing downdrift of Santa Catalina headland (Martínez del Pozo et al. 2002), but due to wave refraction around Cádiz tombolo it is a mixed geomorphic system between a drift-aligned spit and a swash-aligned barrier (Benavente et al. 2006). It includes a broad range of morphosedimentary environments, both active and relict (Fig. 8.1): a sandy beach (Valdelagrana or Levante beach); embryo dunes and discontinuous ridges of foredunes less than 2 m high; four Holocene beach ridges indicating old growth stages of the spit (Zazo et al. 1994); mud flats; and wide areas of vegetated salt marshes. The area is crossed by two narrow tidal creeks. At the southern end of the spit barrier (Saboneses Point), a wide tidal delta composed of two main sandy lobes is developed at the mouth of Río San Pedro tidal channel (Fig. 8.1).

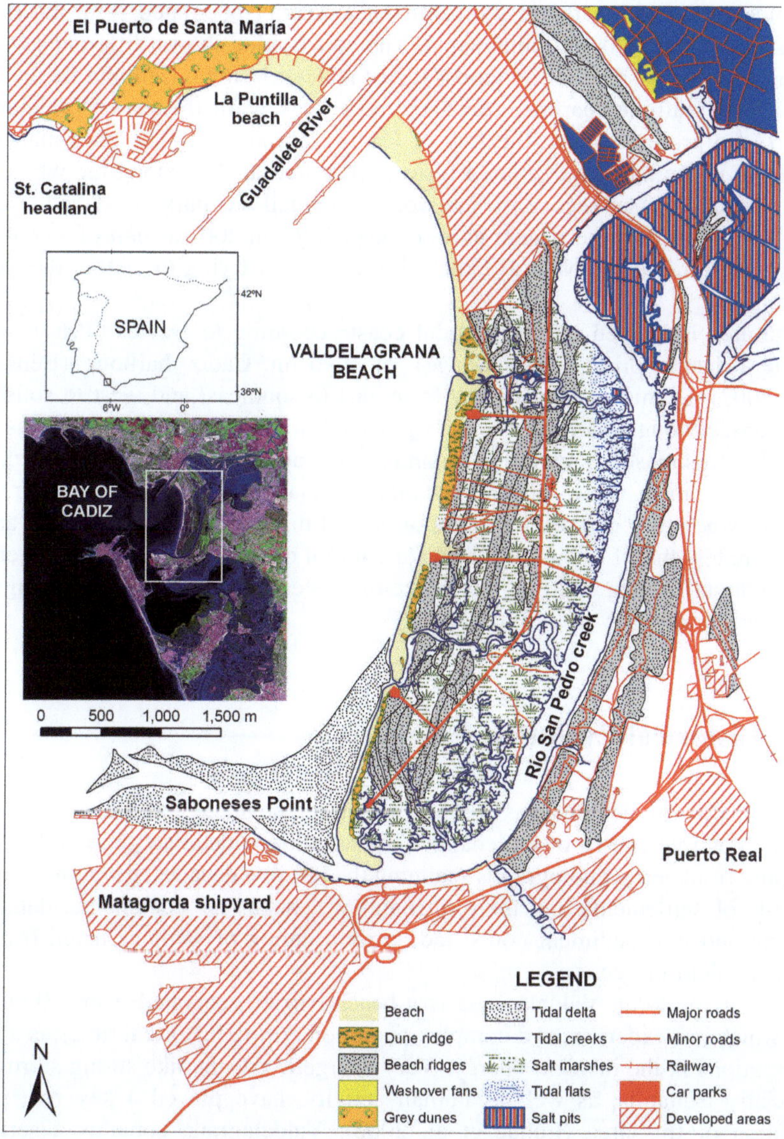

Fig. 8.1 Location and geomorphological sketch of Valdelagrana spit barrier

The beach is composed of fine quartz sand, and it shows a typically dissipative profile about 250 m wide in the northern and central sectors of the spit. Intertidal slope tends to decrease southwards, and at the southernmost end the profile turns ultradissipative and is nearly 500 m wide; here the beach almost transforms into a tidal flat as it connects to the tidal delta at Saboneses Point (Benavente et al. 2006).

Valdelagrana is located in the centre of a heavily populated area, the Bay of Cádiz, with over 400,000 inhabitants. The spit barrier shows different levels of human occupation: the northernmost sector is densely urbanized and the beach is backed by a seafront, apartments and summerhouses, while the rest of Valdelagrana is part of a protected area, the Bay of Cádiz Natural Park. The only man-made structures in the natural area are a former road that ran N-S along the whole spit, with four transversal secondary junctions and small car parks next to the beach (Fig. 8.1). Most of these structures were dismantled in 2002 as part of a conservation programme of the Natural Park, with the aim of trying to restore the natural environment.

The area is located on a mesotidal coast according to Davies (1964), with a MSTR (Mean Spring Tidal Range) of 2.96 in Cádiz harbour (Benavente et al. 2007). Dominant winds blow from east to southeast and west to southwest directions, although the former is not significant in wave generation due to its short fetch. For this reason, prevailing sea and swell waves approach the coast from the west. The highest waves occur in winter associated with Atlantic low pressure systems, when they can reach heights of up to 4 m. However, over 70 % of annual waves are less than 1 m high, so the Cádiz littoral can be classified as a low-energy coast. Coastal setting and the prevalence of westerly waves determines longshore drift to generally flow in a SE direction.

8.3 Historical Evolution

A combination of methods was used to reconstruct the main phases of spit evolution and the possible causes of the changes. Data included detailed geomorphological mapping from aerial photographs spanning the last five decades, geoarchaeological analysis of settlements and their surroundings, historical documents, data from existing and new sediment cores, radiocarbon dating of samples taken from the cores and field inspection.

Strictly speaking, Valdelagrana is a beach-barrier generated over 3,000 years ago, which has experienced a complex evolution during historical times as a result of variations in the incident energy. Very energetic events, like strong storms and especially tsunamis, as well as human activity, have played a key role in the evolution of this area (Luque et al. 2002). Valdelagrana contains a series of historical beach ridges, more or less parallel to the present coastline. In the southern area, many of the ridges form hooks characteristic of spit ends, indicative of the prevailing longshore drift and wave refraction processes. Detailed mapping of ridges and hooks allowed the identification of a set of over 20 different prograding episodes (Rodríguez-Polo et al. 2009), with complex inter-relationships.

Since the eustatic sea level highstand was reached in the zone, about 5,300–4,800 year BP (Gracia et al. 2006), sea level has remained in a more or less stable position until present; this favoured the development of coastal prograding sedimentary systems. The oldest beach ridges in Valdelagrana date to 3,500 year

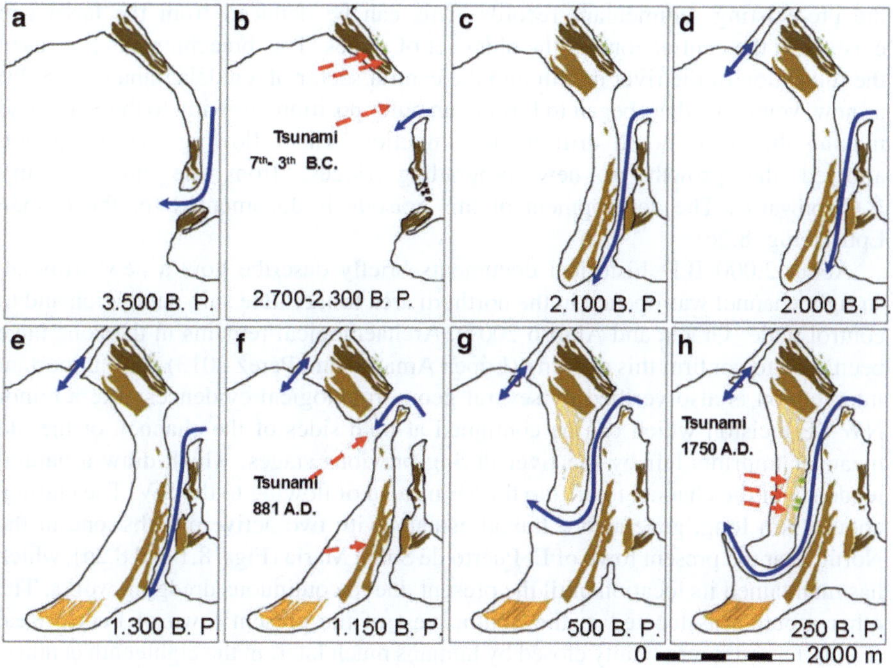

Fig. 8.2 Diagrammatical representation of historical evolution of Valdelagrana spit barrier. *Brown zones* indicate present remains of different sets of historical beach ridges

BP (Zazo et al. 1994), by which stage they were already occupied by humans (López Amador and Pérez 2013). This age is significantly younger than the oldest Holocene beach ridges in other nearby coastal prograding systems, such as Doñana (Goy et al. 1996). This apparent absence of older ridges in Valdelagrana has been interpreted as the consequence of an initial sedimentary infilling of the original Guadalete estuary, by which a slow aggradational process would have prevented the generation of coastal prograding bodies (Dabrio et al. 2000). Nevertheless, some new cores in the back-barrier marshlands seem to indicate that perhaps beach ridges formed earlier, approximately at the same time as in other nearby systems, but in Valdelagrana they were fossilized by subsidence and burial under marsh sediments. Other outcrops may have been eroded by lateral migration of the Guadalete River near its mouth, as revealed by the numerous abandoned channels and meanders found in its floodplain near the river mouth.

The southward growth of the first set of ridges caused the migration of the Guadalete River mouth in the same direction (Fig. 8.2a). During pre-Roman and early Roman times several very energetic events, probably tsunamis, affected the Gulf of Cadiz coast. Archaeological and sedimentary records point to events that likely occurred in the seventh and third centuries B.C. (Luque et al. 2002; Lario et al. 2010). Very probably, one or both of them broke the Valdelagrana ridges, opening a breach in the central-northern sector and destroying a significant part of the pre-existing sedimentary records. This can be deduced from the large gap existing in the central zone of the older set of ridges. This breaching surely caused the relocation of the river mouth into the central sector of Valdelagrana (Fig. 8.2b) as new, younger ridges began to form in an outer position, growing to the South and making the river mouth drift in this direction. The following stability period allowed the growth of new prograding ridges, from the third century B.C. onwards. The development of this episode is documented for the Roman Epoch (Fig. 8.2c).

About 2,000 B.P. historical documents briefly describe how a new artificial, straight channel was opened in the northern area to facilitate ship navigation and to control trade (Gracia and Alonso 2009). Archaeological remains in the zone have been used to confirm this opening (López Amador and Pérez 2013). This important intervention is also verified by several geomorphological evidences, like a minor NW-SE incision which can be continued at both sides of the channel, or the old meander imprints left by the river during previous stages, which draw a natural tendency of the channel to drift to the SE instead of flowing to the SW. The cutting, about 2 km long, generated a fluvial estuary with two active mouths, one in the North, near the present town of El Puerto de Santa María (Figs. 8.1 and 8.2d), which has maintained its location until the present due to continuous dredging works. The other mouth was located in the South, west of the present town of Puerto Real (Fig. 8.1), and it was finally closed by humans much later, in the eighteenth century, through works described in detail in different historical documents.

The intervention in 2,000 B.P. occurred at the time of maximum historical occupation in the Bay of Cadiz, during which several cities and villages developed, leading to the construction of roads, port facilities, coastal and fluvial hydraulic works (aqueducts, river and tidal channels, salt pits, etc.). From that moment on the sediment supply of the Guadalete River discharged directly into the coastal zone around Valdelagrana. This favoured a period of very rapid coastal progradation. In fact, radiocarbon dating of different ridges suggests a shoreline advance of over 1,000 m in only 800 years (Fig. 8.2d, e; Gracia and Martin 2009; Rodríguez-Polo et al. 2009). This progradation could have been affected by several historically documented energetic events, both of climatic and seismic origin, recorded in the Gulf of Cadiz (Campos 1992; Gutiérrez-Mas et al. 2009). According to historical records, a devastating tsunami affected the Gulf of Cadiz in 881 A.D. Dating of samples taken in Valdelagrana suggests that this event deeply eroded most part of the Roman-age ridges and opened a new breach (Fig. 8.2f), again in the central sector – the most vulnerable one according to its position, orientation and nearshore morphology. Sedimentary records of this event on the ridges indicate that they were directly exposed to the marine action and that the Guadalete River had never flowed to the west of this position (Gutiérrez Mas et al. 2009, Alonso et al. 2014). More recent historical maps (seventeenth century) show how the river flowed to the west of this point, as it does at present. Hence, it can be deduced that as a consequence of the 881 A.D. energetic event, a new capture of the southern river course occurred, which moved its mouth to the breaching point. Subsequent modal, low-energy conditions led to a rapid growth of new sedimentary ridges, which in this case developed as barrier-spits that migrated southwards; this can be deduced from the geometrical relationships between the new sets of ridges and the former one (Fig. 8.2g) As this new set of ridges grew, the southern river mouth also migrated in the same direction and progressively eroded part of the previous Roman ridges. Some remains of the ancient Roman ridges, not completely eroded by the southward drifting of the river mouth, can still be recognized in 1950s aerial photographs to the south of the present Río San Pedro creek (Fig. 8.1). These Roman ridges were destroyed in the 1970s and occupied by a shipyard.

In the eighteenth century a new human intervention deeply altered the fluvial dynamics of the Guadalete River. An artificial cut was excavated in 1721 in the inner marshlands that made the Guadalete river flow to the sea entirely through the Northern mouth, leaving the Southern one completely inactive (Baldera and Falcón 1987). This intervention had two purposes: increasing the water depth in the Puerto de Santa María harbour, of growing importance for its commercial activity with America at that time, and preventing the formation of wide intertidal sandbars in front of the Northern mouth (López Amador and Pérez 2013). As a result, the southern mouth transformed into a tidal channel, losing any significant sediment supply and the ability of active lateral migration, as it remains nowadays (Río San Pedro tidal creek).

In 1755, the devastating tsunami associated with the Lisbon earthquake, with waves about 8 m high in this zone (Campos 1992), flattened all the previous historical ridges (Fig. 8.2h). Two new transverse breaches, which abruptly cut all the previous ridges by forming small E-W microcliffs, could have been produced by this tsunami. These breaches affect recent ridges dated to 340 cal. year B.P. (Gracia and Martin 2009). Two minor tidal channels occupy the bottom of these breaches and at present they form small tidal deltas that partly block longshore sediment transport. From then on, a new system of beach ridges developed in front of the historical ones, although aerial photographs from the last six decades suggest that the rate of progradation has been much lower than in previous historical epochs (Martínez del Pozo et al. 2001).

8.4 Recent Behaviour

Recent changes occurred in Valdelagrana spit barrier were assessed using 14 sets of aerial photographs and orthophotographs spanning the period between 1956 and 2008 (Del Río et al. 2009, Rodríguez-Polo et al. 2010). GIS tools were employed to georectify the photographs and digitize the high-water line or HWL (Crowell et al. 1997) and the dune toe (Moore and Griggs 2002). Detailed rates of change were calculated by linear regression on shore-normal transects drawn every 20 m along the spit, by means of the DSAS extension for ArcGIS® developed by the USGS (Thieler et al. 2005).

Shoreline behaviour in Valdelagrana over the last five decades shows very clear patterns, with marked contrasts between both ends of the spit (Fig. 8.3) (Del Río

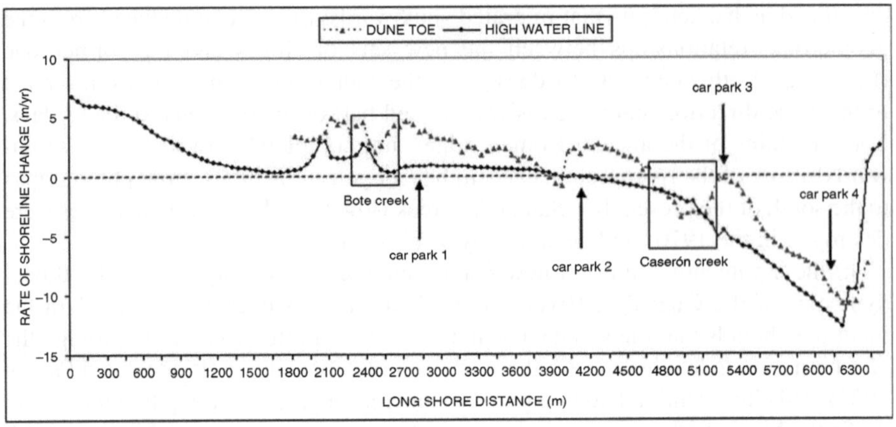

Fig. 8.3 Shoreline change rates along Valdelagrana spit between 1956 and 2008. The left section of the graph corresponds to the northern sector of the spit, i.e. the urban zone where dunes are absent. For reference purposes, the areas affected by tidal creek dynamics have been highlighted, as well as the location of the car parks

et al. 2009). The evolution of the HWL reveals a strongly accreting trend in the northern section, with average rates of up to 6 m/year in the shadow zone of the Guadalete river jetty that gradually diminish southwards. The accretion rate was maximum between 1977 and the late 1980s. The central portion of the spit exhibits generally low variations, except at points affected by local migration of the minor tidal creeks. In contrast, the southern spit end has recorded extreme erosion, with rates between −6 m/year immediately to the south of the second creek and up to −14 m/year in front of the southernmost (dismantled) car park at Saboneses Point (Del Río et al. 2009).

These trends are also reflected in the evolution of the dune toe (Fig. 8.3), with an average accretion of 3.2 m/year in the northern and central zones and important changes in the areas adjacent to tidal creeks. A new dune ridge developed immediately to the south of the first parking lot from 1977 onwards, revealing the intrinsically stable character of this sector (Benavente et al. 2006). The inflection point between the accretionary and erosive zones is located in the second creek. Here a change in the position of the inlet was recorded between 1977 and 1982, possibly related to the breaching of the dunes by a very energetic storm event; this could have occurred in February 1979, when wave height exceeded 7 m according to SIMAR-44 dataset from the HIPOCAS project (Del Río et al. 2012).

As with the HWL, the dune toe also recorded extreme recession at the southern spit end (Figs. 8.3 and 8.4). Here average retreat rate for the last decades was −7.7 m/year, increasing southwards up to a maximum of nearly −12 m/year in front of the last car park (Del Río et al. 2013). This erosion around Saboneses Point has involved the loss of a 200–450 m wide strip of beach, dune and salt marsh areas (Rodríguez-Polo et al. 2010) (Fig. 8.4). Modern dunes in this zone are low, narrow

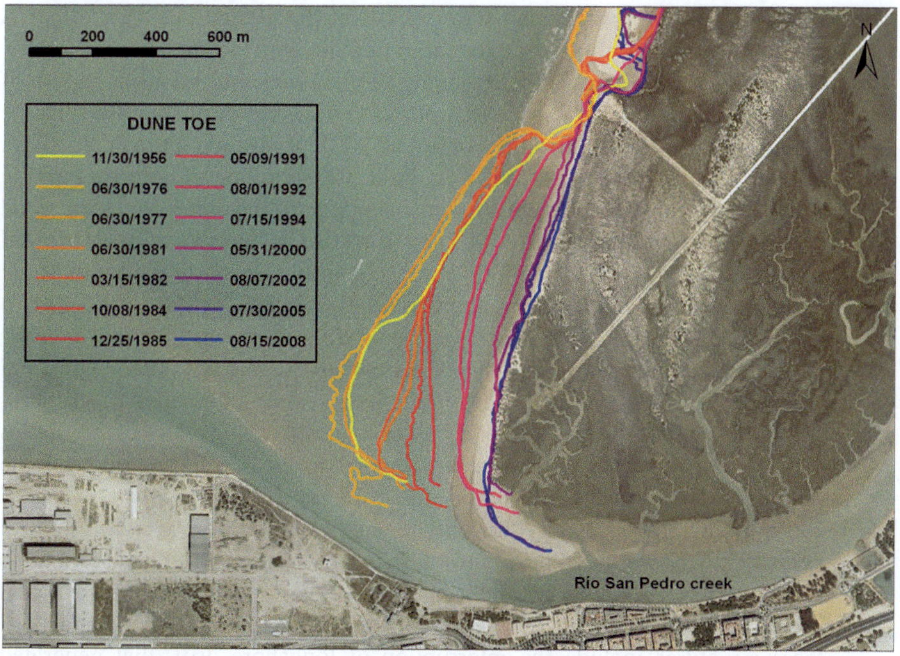

Fig. 8.4 Changes in the position of the dune toe at Saboneses Point along the study period (1956–2008). Note the initial stability of the shoreline until 1977, followed by very fast erosion until mid-1990s

and discontinuous, further increasing the vulnerability of this area to erosion, flooding and overwash processes in case of storms (Benavente et al. 2006).

Temporal analysis of shoreline changes reveals that stability and accretion dominated in the dunes between 1956 and 1977; by that time, however, the beach had already started to erode at Saboneses Point. Dunes turned then rapidly to erosion, and maximum dune retreat took place in the period 1977–1982 (up to 30 m/year), and at Saboneses Point also between 1982 and 1992 (Benavente et al. 2006) (Fig. 8.4). These periods of maximum dune toe erosion coincide with the maximum HWL accretion at the northern zone of Valdelagrana. Over the last years this accretion has stopped and dune erosion in the southern sector has greatly slowed down, although recession rates were between −1.5 and −6 m/year in the period 2002–2008.

From the above data it is clear that recent behaviour in Valdelagrana spit barrier represents an example of fast response to anthropogenic changes in sediment budget. In this way, the evolution of this zone over the last five decades has been determined by three main factors (Del Río et al. 2009):

(i) The building of three dams in Guadalete river basin between 1956 and 1977 probably contributed to the initiation of beach retreat at the southernmost end of the spit, due to their effect of fluvial sediment trapping (Rodríguez-Polo et al. 2010). Two bigger dams were built in the 1990s, but their effect is overlapped by that of other factors affecting the area (see below).

(ii) The timing of building and lengthening of the jetties located at the mouth of Guadalete river (Fig. 8.1) is clearly related to the patterns of shoreline change in the study zone (Martínez del Pozo et al. 2001; Rodríguez-Polo et al. 2010). In 1956, before these structures were built, the shoreline of both La Puntilla and Valdelagrana beaches constituted a single unit; the planform shape fitted a log-spiral function (Yasso 1965) or a parabolic bay equation (Hsu and Evans 1989), being adapted to wave conditions according to the static equilibrium model of Terpstra and Chrzastowski (1992) (Martínez del Pozo et al. 2002). The southern beach started eroding in the mid-1970s when the jetties were built, while the dunes started eroding a few years later and recorded the strongest erosion when the jetties were lengthened in the early 1980s (Fig. 8.4). In this headland-bay system, the main effect of the jetties was the shifting of the upcoast control point of the log-spiral, which is the point from which wave diffraction starts (Terpstra and Chrzastowski 1992). The building and especially the lengthening of the jetties moved this control point southwestwards, so the planform shape of the spit barrier had to rotate to adapt to these new wave conditions, by eroding in the southern end and accreting in the northern (Martínez del Pozo et al. 2002). Other impacts of these structures include the interruption of longshore drift, that generated massive sand accumulation at La Puntilla beach (Del Río et al. 2013), and the injection of fluvial sediments into the outer Bay of Cádiz, where transport by tidal ebb currents prevents sediments from reaching Valdelagrana spit end. Overall, this produced a transformation of the system from a sand spit controlled by longshore drift to a swash-aligned beach controlled by wave refraction.

(iii) The land reclamation performed in 1976 for the building of Matagorda shipyard, which is located south of Río San Pedro inlet, significantly affected the evolution of Saboneses Point (Martínez del Pozo et al. 2001). Tidal hydrodynamics in the area were completely changed by this reclamation, as the mouth of the Río San Pedro tidal channel was narrowed, thus increasing the velocity of the ebb flow currents. This triggered erosion of its southern bank, where defence works and artificial nourishments had to be performed (Herrera et al. 2010). As a consequence, the tidal delta, which is mainly formed by sediments eroded from Valdelagrana beach, started developing and growing westwards (Martínez del Pozo et al. 2001); at present it covers an area of about 2 km^2 and constitutes a problem for operations at the shipyard.

In the last few years it seems that Valdelagrana shoreline is reaching a new equilibrium with the surrounding conditions of sediment supply and forcing agents, especially wave patterns, as northern accretion has stopped and southern erosion has greatly slowed down. However, coastal recession on the southern section of the spit is still significant, especially after energetic storm seasons such as the winter of 2009–2010 (Rangel-Buitrago and Anfuso 2011).

8.5 Present Dynamics

A 3-year program of topographic beach monitoring was performed to evaluate short-term morphodynamic behaviour of Valdelagrana beach. The surveys were performed on a monthly basis on shore-normal topographic profiles distributed along the whole spit, extending from the dune toe or the seafront to a depth equivalent to the mean spring low water level (Gracia et al. 2005). Beach sediment samples were collected and analysed by dry sieving. Wave data used to calculate beach morphodynamic parameters were obtained from the offshore buoy "Cádiz" (National Ports Authority), located at a depth of 20 m about 10 km from Valdelagrana. Present dynamics of the tidal delta at Saboneses Point was assessed by detailed DGPS surveys and deployment of current meters and pressure transducers on the delta (Rodríguez-Polo 2009).

Beach monitoring reveals a clear alongshore (north-south) gradation in morphological and sedimentological parameters. Beach profile morphology is visually dissipative in the northern sector, with a gentle slope around 2.3 % and some intertidal bars (Anfuso et al. 2006). Beachface slope gradually decreases southwards, and in the southern sector of the spit the profile morphology is visually ultradissipative and resembles a tidal flat (slope 0.7 %) without bars, cusps or other features (Gracia et al. 2005). Beach profile variability also shows a north-south gradient, with moderate seasonal changes occurring in the northern sector, while the southern profiles show a continuously retreating trend (Fig. 8.5) without any significant seasonal variations. In fact, dune disappearance was recorded at Saboneses Point during the study period, as well as outcropping of fossil saltmarsh

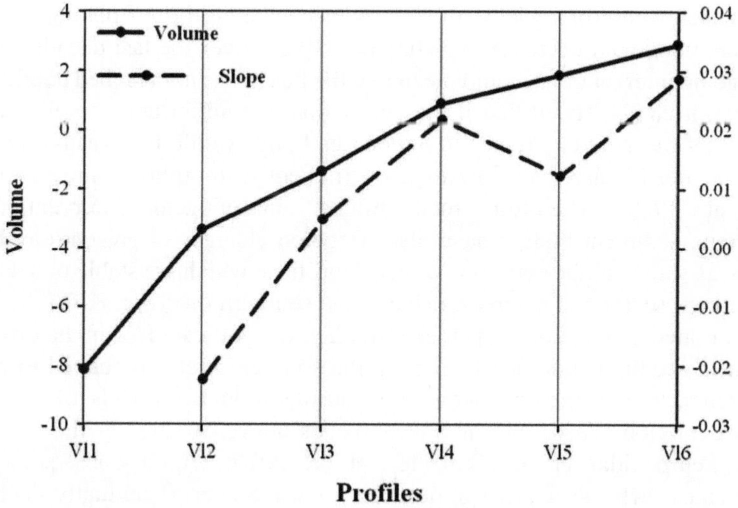

Fig. 8.5 Evolution of beach profile volume (monthly rate of volume change in cubic meters per linear meter of beach) and intertidal slope at Valdelagrana beach during the study period. Note that Vl1 is located at Saboneses Point and Vl6 is located at the northern end of the study area

sediment in the intertidal zone after storm periods, clearly indicating landward recession of the spit. Similarly, volumetric changes reveal a slightly accreting pattern on the northern profiles, while the southern area displays an erosive trend (Gracia et al. 2005) (Fig. 8.5). The areas close to the tidal creeks are greatly influenced by inlet-related processes and so they differ from this general behaviour.

Regarding sediment characteristics, the beach is composed of well-sorted fine sand; average grain size (D_{50}) is 0.16 mm, with a slight decrease in the southward direction. According to Benavente (2000), this can be related to the differential transport of finer sediments by longshore drift (Komar 1998) from the source area, i.e. the Guadalete River. Grain size patterns in the southern profiles show a distinct increase in mean grain size, indicating erosive processes.

As for beach morphodynamic state, Valdelagrana is clearly dissipative according to classic models of Wright and Short (1984) or Masselink and Short (1993). No changes in morphodynamic state were recorded during the monitoring program (Anfuso et al. 2006). However, and as mentioned above, gradients in beach slope and grain size indicate the northern area is closer to intermediate states while Saboneses Point shows some typical ultradissipative morphologies (Benavente 2000).

From the above information it can be stated that present morphodynamic behaviour in Valdelagrana is determined by three main factors: tidal influence, sediment supply and evolution of beach planform shape, the two latter being highly dependent on human activities. In this respect, strong tidal currents are responsible for generating beach morphologies more typical of macrotidal environments, despite the study area being mesotidal (Benavente et al. 2007). On the other hand, there is a gradient in beach characteristics and hence morphodynamic behaviour as the distance to the sediment source increases. Sediment supply is an important factor controlling how the beach reaches equilibrium planform shape, so the aforementioned decrease in sediment delivery over the last decades due to anthropogenic interventions would mean that the beach has not reached equilibrium yet. In fact, it can be stated that if the beach was in equilibrium, shoreline at the distal zone of the Zeta-bay (close to Saboneses Point) would be swash-aligned so there would not be any longshore drift or gradient in morphodynamic variables (Hsu et al. 1989). Therefore, medium-term anthropogenic interventions in Valdelagrana sediment budget have also triggered changes in present-day beach dynamics, as shown in the evolution of beach profiles, which are stable or accreting in the northern sector and clearly erosive in the southern one (Fig. 8.5).

As mentioned above, tidal influence also plays a significant role in the observed behaviour, since the gentle beach slope at the southern sector is related to strong tidal currents, which are not taken into consideration by models of Zeta bay planform evolution. These currents have speeds above 50 cm/s at the mouth of Río San Pedro tidal creek (González et al. 2010). As a consequence, in Valdelagrana beach the relative importance of wave energy gradually decreases southwards due to the presence of the ebb tidal delta, while the importance of tide energy increases, determining contrasting beachface morphologies and an ultradissipative profile in the southernmost sector (Benavente et al. 2007).

The development and growth of the tidal delta at Saboneses Point greatly influences beach behaviour. On one hand, wave dissipation on the delta determines the beach at the southern end of the spit being a very low-energy area, similar to the estuarine beaches described by Nordstrom and Jackson (1992), where the upper intertidal zone works as a proper beach but the lower intertidal shows a tidal flat behaviour. During high-energy storm conditions, erosion of the beach profile occurs by parallel retreat so that shoreline recession is recorded but beach slope remains unchanged (Rodríguez-Polo 2009).

On the other hand, water flow on the tidal delta is directed through two main channels: a main one running WNW-ESE close to Matagorda shipyard breakwater, and a secondary one running NNE-SSW in a shore-parallel direction at Saboneses Point. Current velocities in the main channel are around 60 cm/s and the highest velocities are recorded during ebb tide, while in the secondary channel they reach 80 cm/s during the flood tide, due to the coupling between the tide and prevailing wave direction and hence longshore drift. As a consequence of this, the secondary channel probably contributes to the scouring and erosion of the lower intertidal beach at this area (Rodríguez-Polo 2009).

8.6 Conclusions

Valdelagrana spit barrier is a clear example of multi-scale anthropogenic disruption of natural coastal behaviour. A variety of geomorphological and morphodynamic changes, including particular patterns of shoreline evolution, have been triggered by human activities in the area since Roman times. These are superimposed on other important natural sources of coastal change, such as energetic storms and tsunami events.

On the historical scale, major anthropogenic interventions occurred in Roman times and in the 18th century, and were focused on improving navigation in the Guadalete estuary. They generated massive progradation of beach ridges and the isolation of the Río San Pedro as a tidal channel, respectively. During the second half of the twentieth century, the building of two jetties at the river mouth and the reclamation works performed south of Río San Pedro produced significant beach accretion at the northern sector of Valdelagrana and extreme shoreline retreat at the southern sector, as well as the development of a wide tidal delta at Saboneses Point. The changes in spit dynamics generated by these interventions could still be noticed on beach profile behaviour over recent years.

Plans for future human interventions around Valdelagrana spit are lately being considered by national authorities. These include dredging the tidal delta, increasing harbour development both in Guadalete River mouth and Matagorda shipyard, and restoring the natural behaviour of old salt marshes which are currently dry. It is likely that dredging the delta would trigger erosion by waves due to the increase in shoreline exposure; nevertheless, present-day beach erosion at Saboneses Point is partly generated by the secondary channel of the delta during ebb tide, so the dredging could reduce this impact. Regarding the building of new harbour structures at Guadalete River mouth, they would produce a more efficient blockage of longshore sediment supply to Valdelagrana, as well as a further modification of the control point in the Z-bay planform. These processes would combine to intensify coastal erosion in the Southern end spit. As for the restoration of the reclaimed salt marshes, this would increase the tidal prism at Río San Pedro tidal channel, thus increasing the size of the tidal delta.

In any case, further research must be performed before any of the above interventions are executed, in order to avoid undesirable consequences in coastal stability. This research should be mainly focused on modelling wave propagation and longshore sediment transport in the coastal zone under different weather conditions, as well as assessing the volume of sediment supplied to the area by the Guadalete River.

In any case, it is important to note that besides human interventions, the long term evolution of Valdelagrana is also highly influenced by the short term erosion produced by storms during the most energetic winter seasons, so the storminess conditions in the future will be determinant in future shoreline evolution.

Acknowledgements This work is a contribution to the research group RNM-328 of the PAI and to the projects GERICO (CGL 2011-25438) and P10-RNM-6547.

References

Alonso C, Gracia FJ, Rodríguez-Polo S (2014) Modelo de evolución histórica de la flecha-barrera de Valdelagrana (Bahía de Cádiz). In: Proceedings XIII Reunión de la Sociedad Española de Geomorfología. S.E.G., Cáceres (accepted)

Anfuso G, Benavente J, Gracia FJ, Del Río L (2006) Morphodynamic characterization of Cadiz beaches (SW Spain). J Coast Res Spec Issue 48:8–15

Baldera J, Falcon MA (1987) Descoordinación de las grandes actuaciones y sus efectos en la desorganización del territorio. In: Evolución de los paisajes y ordenación del territorio en Andalucía Occidental – Bahía de Cádiz. Diputación de Cádiz and Casa Velázquez, pp 49–86

Benavente J (2000) Morfodinámica litoral de la Bahía externa de Cádiz. PhD thesis, University of Cádiz

Benavente J, Del Río L, Gracia FJ, Martínez JA (2006) Coastal flooding hazard related to storms in Valdelagrana spit (Cadiz Bay Natural Park, SW Spain). Cont Shelf Res 26:1061–1076

Benavente J, Del Río L, Anfuso G, Gracia FJ, Nachite D, Rodríguez-Ramírez A, Cáceres L (2007) Efecto de la marea en la clasificación morfodinámica de playas. In: Gómez-Pujol L, Fornós JJ (eds) Investigaciones recientes (2005–2007) en geomorfología litoral. Palma de Mallorca, Spain, pp 17–21

Campos ML (1992) El riesgo de tsunamis en España. Análisis y valoración geográfica. IGN, Monografías 9

Crowell M, Douglas BC, Leatherman S (1997) On forecasting future US shoreline positions: a test of algorithms. J Coast Res 13(4):1245–1255

Dabrio CJ, Zazo C, Goy JL, Sierro FJ, Borja F, Lario J, González JA, Flores JA (2000) Depositional history of estuarine infill during the last postglacial transgression (Gulf of Cadiz, Southern Spain). Mar Geol 162:381–404

Davies JL (1964) A morphogenic approach to world shorelines. Z Geomorphol 8:27–42

Del Río L, Rodríguez-Polo S, Gracia FJ, Benavente J (2009) Spatial and temporal patterns of human-related coastal changes in the Bay of Cadiz (SW Spain). In: Abstracts of the 7th international conference on geomorphology, Melbourne, Australia, p 769

Del Río L, Plomaritis TA, Benavente J, Valladares M, Ribera P (2012) Establishing storm thresholds for the Spanish Gulf of Cadiz coast. Geomorphology 143–144:13–23

Del Río L, Gracia FJ, Benavente J (2013) Shoreline change patterns in sandy coasts. A case study in SW Spain. Geomorphology 196:252–266

FitzGerald DM, Fenster MS, Argow BA, Buynevich IV (2008) Coastal impacts due to sea-level rise. Annu Rev Earth Planet Sci 36:601–647

González CJ, Álvarez O, Reyes J, Acevedo A (2010) Two-dimensional modeling of hydrodynamics and sediment transport in the San Pedro tidal creek (Cádiz Bay): morphodynamic implications. Cienc Mar 36(4):393–412

Goy JL, Zazo C, Dabrio CJ, Lario J, Borja F, Sierro FJ, Flores JA (1996) Global and regional factors controlling changes of coastlines in southern Iberia (Spain) during the Holocene. Quat Sci Rev 15:773–780

Gracia FJ, Alonso C (2009) El cambiante paisaje de la bahía gaditana. In: Fernández-Palacios JM (ed) Cádiz de la Constitución de 1812. Serie Agua, Territorio y Sociedad. Agencia Andaluza del Agua, Junta de Andalucía, pp 28–31

Gracia FJ, Martín C (2009) Tasas de sedimentación en las marismas del Parque Natural de la Bahía de Cádiz a partir de sondeos geotécnicos: Una aplicación para la reconstrucción paleoambiental. Demarcación de Costas de Andalucía Atlántico, Spanish Ministry of Environment. Cádiz, 71 pp. (unpublished)

Gracia FJ, Benavente J, Anfuso G, Reyes JL, Del Río L (2005) Velocidades y tendencias de cambio morfológico interanual en las playas del entorno de la Bahía de Cádiz. In: Sanjaume E, Mateu J (eds) Geomorfologia litoral i Quaternari. Universidad de Valencia, Spain, pp 181–193

Gracia FJ, Del Río L, Alonso C, Benavente J, Anfuso G (2006) Historical evolution and present state of the coastal dune systems in the Atlantic coast of Cádiz (SW Spain): palaeoclimatic and environmental implications. J Coast Res SI48:55–63

Gutiérrez-Mas JM, Juan C, Morales JA (2009) Evidence of high-energy events in shelly layers interbedded in coastal Holocene sands in Cadiz Bay (south-west Spain). Earth Surf Proc Land 34:810–823

Herrera A, Gómez-Pina G, Fages L, de la Casa A, Muñoz-Pérez JJ (2010) Environmental impact of beach nourishment: a case study of the Río San Pedro beach (SW Spain). Open Oceanogr J 4:32–41

Hsu JRC, Evans C (1989) Parabolic bay shapes and applications. Proc Inst Civ Eng 87(2):557–570

Hsu JRC, Silvester R, Xia YM (1989) Static equilibrium bays: new relationships. J Waterw Port C-ASCE 115(3):285–298

Komar PD (1998) Beach processes and sedimentation, 2nd edn. Prentice Hall, Englewood Cliffs

Komar PD (2000) Coastal erosion – underlying factors and human impacts. Shore Beach 68(1):3–16

Lario J, Luque L, Zazo C, Goy JL, Spencer C, Cabero A, Bardají T, Borja F, Dabrio CJ, Civis J, González-Delgado J, Borja C, Alonso-Azcárate J (2010) Tsunami vs. storm surge deposits: a review of the sedimentological and geomorphological records of extreme wave events (EWE) during the Holocene in the Gulf of Cadiz, Spain. Z Geomorph 54(Suppl 3):301–316

López Amador JJ, Pérez E (2013) El puerto gaditano de Balbo. Ed. El Boletín, Puerto de Santa María

Luque L, Lario J, Civis J, Silva PG, Zazo C, Goy JL, Dabrio CJ (2002) Sedimentary record of a tsunami during Roman times, Bay of Cádiz, Spain. J Quat Sci 17:623–631

Martínez del Pozo JA, Anfuso G, Gracia FJ (2001) Recent evolution of a tidal delta in Cádiz Bay (SW Spain) due to human interventions. In: Özhan E (ed) Proceedings of MEDCOAST '01. Hammamet, Tunisia, vol 3, pp 1425–1433

Martínez del Pozo JA, Benavente J, Gracia FJ, Anfuso G (2002) Evolución reciente de la forma de equilibrio en planta de la playa de Valdelagrana (Bahía de Cádiz). In: Pérez-González A, Vegas J, Machado M (eds) Aportaciones a la geomorfología de España en el inicio del tercer milenio. IGME and Ministerio de Ciencia y Tecnologia, Madrid, Spain, pp 355–361

Masselink G, Short AD (1993) The effect of tide range on beach morphodynamics and morphology: a conceptual beach model. J Coast Res 9(3):785–800

Moore LJ, Griggs GB (2002) Long-term cliff retreat and erosion hotspots along the central shores of the Monterey Bay National Marine Sanctuary. Mar Geol 181:265–283

Nordstrom KF, Jackson NL (1992) Two dimensional change on sandy beaches in estuaries. Z Geomorph 36(4):465–478

Rangel-Buitrago N, Anfuso G (2011) Morphological changes at Levante Beach (Cadiz, SW Spain) associated with storm events during the 2009–2010 winter season. J Coast Res SI64:1886–1890

Rodríguez-Polo S (2009) Estudio de geomorfología ambiental de la playa de Valdelagrana y Parque Metropolitano de Los Toruños. Implicaciones en la gestión. MSc Thesis, University of Cádiz

Rodríguez-Polo S, Gracia FJ, Benavente J, Del Río L (2009) Geometry and recent evolution of the Holocene beach ridges of the Valdelagrana littoral spit (Cádiz Bay, SW Spain). J Coast Res SI56:20–23

Rodríguez-Polo S, Gracia FJ, Del Río L (2010) Retroceso costero en la flecha de Valdelagrana, El Puerto de Santa María (Cádiz). In: Úbeda X, Vericat D, Batalla R (eds) Avances de la geomorfología en España 2008–2010. Solsona, Spain, pp 75–78

Syvitski JPM, Vörösmarty CJ, Kettner AJ, Green P (2005) Impact of humans on the flux of terrestrial sediment to the global coastal ocean. Science 308:376–380

Terpstra PD, Chrzastowski MJ (1992) Geometric trends in the evolution of a small log-spiral embayment on the Illinois shore of Lake Michigan. J Coast Res 8(3):603–617

Thieler ER, Himmelstoss EA, Zichichi JL, Miller TL (2005) Digital Shoreline Analysis System (DSAS) version 3.0: an ArcGIS extension for calculating shoreline change. US Geological Survey Open-File Report 2005–1304

Wright LD, Short AD (1984) Morphodynamic variability of surf zones and beaches: a synthesis. Mar Geol 56:93–118

Yasso W (1965) Plan geometry of headland-bay beaches. J Geol 73(5):702–714

Zazo C, Goy JL, Somoza L, Dabrio CJ, Belluomini G, Improta S, Lario J, Bardají T, Silva PG (1994) Holocene sequence of sea-level fluctuations in relation to climatic trends in the Atlantic-Mediterranean linkage coast. J Coast Res 10:933–945

Chapter 9
The Development and Management of the Dingle Bay Spit-Barriers of Southwest Ireland

Robert J.N. Devoy

Abstract The palaeoenvironmental and morphodynamic functioning of three closely linked spit-like structures are examined from the high energy Atlantic margin of Europe, at the head of Dingle Bay, southwest Ireland. The 'spits' are formed within a long (c. 40 km) and narrow (c. 10 km) sedimentary compartmentalised embayment and are controlled by a mixed wave and tidal dominated regime. The features demonstrate the quasi-unique, local (micro- to meso-scales) functioning of coastal systems. The 'spits' represent essentially composite beach- and dune-barriers, developed under Holocene sea-level rise (SLR) on N-S aligned glacial end-moraines formed at the end of the last glacial stage (MIS 2–4) and orientated normal to present onshore wave action. Minor drift aligned shoreline spits are found at the distal ends of these barriers. The two seaward fronting structures of Inch and Rossbehy are separated by an ebb-tidal delta and probably formed as a single barrier in the early- to mid-Holocene across Dingle Bay, before moving by *roll over* mechanisms to their present positions. Extensive back-barrier wetlands began to form behind this structure in the mid-Holocene. This earlier barrier was breached c. 3,000 years BP, leading to the formation of the three present spit-like morphologies. Whilst the Inch Spit appears to be relatively stable today, Rossbehy Spit was breached by a storm surge in 2008 and continues to erode at rates of 30–50 m/year along its core-length and at c. 25 m/year on the seaward shore face, with the breach doubling in size to 1,400 m wide from 2012 to 2014. Under future climate warming this structure is likely to disintegrate, with significant impacts on Inch and the morpho- and hydrodynamics of neighbouring coastal systems. The processes now in operation evidence the likely *coastal squeeze* that will occur on many World coasts under SLR in the twenty-first century.

R.J.N. Devoy (✉)
The Coastal and Marine Research Centre, Beaufort Research and the Department of Geography, University College Cork, Cork, Ireland
e-mail: r.devoy@ucc.ie

G. Randazzo et al. (eds.), *Sand and Gravel Spits*, Coastal Research Library 12,
DOI 10.1007/978-3-319-13716-2_9

9.1 Introduction: The Context of Spit-Barriers on European Atlantic Coasts

Understanding of the primary mass and energy drivers operating in coastal systems and in the creation of coastal morphological and process operation variability is well established (e.g., Hails and Carr 1976; Clifford et al. 1993; Carter and Woodroffe 1994; van Rijn 1998; Short 1999; Cooper et al. 2001; Aagaard et al. 2004; Cowell et al. 2006; Cooper and Pilkey 2007a, b, 2008; Cronin 2010). The outcomes of coastal geomorphological research over many decades (e.g., Carter 1988; Woodroffe 2002; Davidson-Arnott 2010; Masselink et al. 2011; French and Burningham 2011) show that coasts are comprised of almost unique sets of components. Further, that coastal functioning often operates at local time and spatial scales (micro- to meso-levels) of minutes – weeks and m^2 – km^2, rather than at regional (macro-scale) levels (Cooper et al. 2004, 2007); part of the still ongoing classic 'geomorphological processes and events operation debate' (e.g., Hart 1986; Thorn 1988).

Many generic controls influence all coastal systems, e.g., for mid latitude, clastic dominated; paraglacial and high latitude, glacial; carbonate and reefs systems; low tidal range and energy (Davies 1980; Woodroffe 2002) and are linked in different combinations over varying time and spatial scales. For mid -high North Atlantic coasts, which often contain a wide range of *spits* and associated sedimentary geomorphological forms, these include especially the controls of wind-wave actions, currents and linked radiative shoreline stresses; storm-surges, tides, relative sea-level (RSL) changes and other hydrodynamic contributors; 3-D coastal configuration, near- to offshore bathymetry and sedimentary accommodation space; sediments availability, sizes, supply and flux rates. These drivers in coastal systems vary in number and mass-energy significance. For each driver though there will be wide variations in its operation, integration with other controls and in feedbacks within systems over time, particularly at the 'real' operational scale of the local coast. Further, coastal systems will undergo significant process and domain changes over time (e.g., Orford et al. 1996; Jennings et al. 1998). These are likely to occur as stepwise type domain crossovers, through transitive and intransitive thresholds responses to forcing 'events', or to grouped process changes at different timescales, leading to the development of new domain states (Dury 1981; Carter and Woodroffe 1994; Delaney et al. 2012). These generic coastal process concepts are valuable in helping define the development of cohesionless sedimentary structures, such as spits. It is the local pattern in the operation of these controls, coupled with the integration of their behaviour with stochastic events and process functioning, that is critical in understanding the development of spit-barrier systems within coastal geomorphology.

The genesis and even the concept-definition of '*what is a spit*' is part of this reality of understanding what coastal systems are and how they work. Spits are a common depositional geomorphological feature of many World coastlines, comprising swash and drift aligned accumulations of cobble to sand sized sediments. In spite of their readily recognisable shape, and the almost intuitive understanding of their structural development from their shapes, as a process-response to a range of

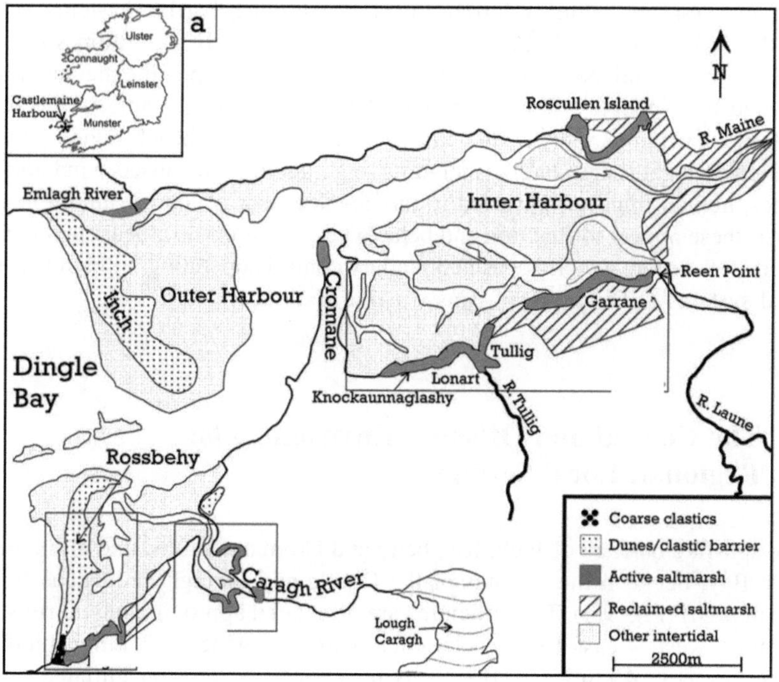

Fig. 9.1 Location map of the Rossbehy, Inch and Cromane Spits, southwest Ireland (Source: Delaney et al. 2012)

coastal hydrodynamic drivers with sediments, *definition and process* controversy may still exist. Similar issues occur elsewhere in coastal and wider geomorphology, of 'what's in a name?' (e.g., lagoons, tombolos and glacigenic features; viz., e.g., Finkl 2004; Goudie 2004). Many spit-type forms occur on Ireland's coasts (Davies and Stephens 1978; Carter 1988). These result primarily from the widespread provenance of coastal fine sized sediments (fine – coarse sands, as well as gravel – boulder sized materials), derived generally from earlier glacigenic sediments (e.g., Edwards and Warren 1985; Knight et al. 2004). The nearshore – offshore and Shelf zones are important sources for these sediments and their re-working onshore into the contemporary coastal systems (Devoy et al. 1996; Cooper 2007). The development of distinct gravel beach-barrier beach systems with associated 'spit forms' is also particularly notable in Ireland, with sediments again derived from glacigenic materials. Their significance in Ireland's beach systems and dynamics, as a former paraglacial coastal environment, has been studied in detail since the 1980s (e.g., Orford and Carter 1984; Orford et al. 1988).

This question of nomenclature, '*what is a spit*', particularly where the landform inherits characteristics influenced by former terrains (e.g., glacial, former coastal), and the effects of other process controls (e.g., aeolian, fluvial, glacial) applies to the spits of Inch, Rossbehy in Dingle Bay, southwest Ireland (Fig. 9.1). Studies of these structures show them to be complex in origin and functioning and that Inch Spit and

Rossbehy Spit are better referred to as beach-dune barrier systems (Cooper et al. 1995; Delaney et al. 2012; O'Shea and Murphy 2013). A third structure at Cromane lies behind these barriers. Though this feature is spit-like in plan view and is fronted by a cobble – boulder beach, it is clearly recognisable in the field as glacigenic in origin (Devoy 2009). In each of the two main beach-dune barriers the operation of local scale factors (including, e.g., geology, localised wind and wave regimes, human impacts) have led to the development of almost unique features. Further, these appear to function and behave very differently, at least in the present day, under the more generic coastal boundary controls operating within the framing coastal systems of Dingle Bay – Castlemaine Harbour.

9.2 The Coastal and Offshore Environments: Regional: Local Settings

These spit-like barriers of Inch, Rossbehy and Cromane are located at the head of Dingle Bay, between the uplands of the Dingle and Iveragh Peninsulas (c. 600–1,000 m high) (Fig. 9.2). The structures are orientated approximately north – south and offset west – east from each other, with Cromane c. 2 km behind Inch. Rossbehy was c. 4 km long prior to its breaching and erosion initiated in 2008,

Fig. 9.2 Southwest Ireland Peninsulas, showing topography, Ria-type embayments and link to the approximate southwest to northeast aligned regional structural. The Inch and Rossbehy Spits are shown at the *top-centre* of the photograph (Source: LANDSAT image)

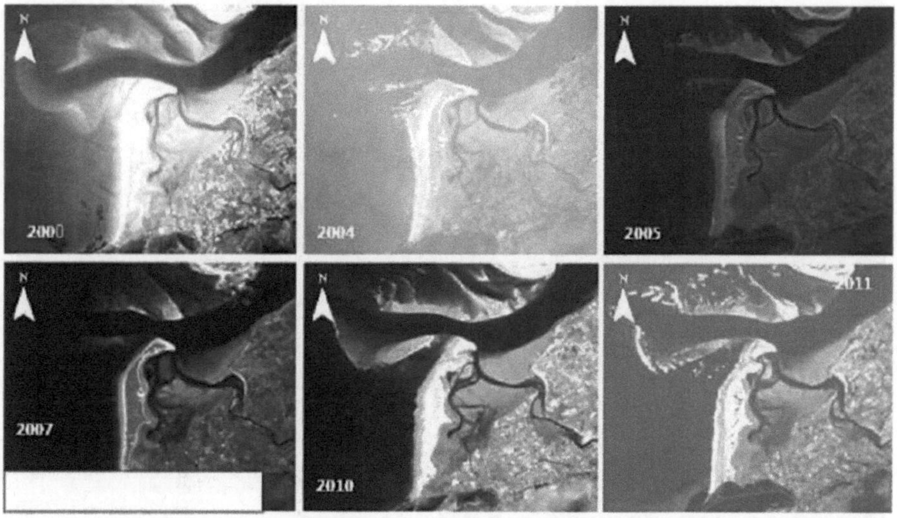

Fig. 9.3 Photograph sequence (2000–2011) of the changing Inch – Rossbehy ebb-tidal delta positions (bars and main tidal inlet channel) and the two beach- dune-barrier distal ends (Source: O'Shea and Murphy 2013)

projecting northwards from the Iveragh Peninsula (Delaney et al. 2012; O'Shea and Murphy 2013). To the north, Inch is c. 5.5 km in length and aligned NNW – SSE from Dingle toward Rossbehy. The spits are separated by an ebb-tidal delta system (Fig. 9.3), which is now undergoing radical change in shape and functioning (O'Shea et al. 2011; O'Shea and Murphy 2013; Gault et al. 2011). Together these barriers divide the high wave energy and coarse sedimentary areas of Dingle Bay (Sala 2010; Marine Institute 2014) from the lower energy and predominantly finer silt-clay sized sediments, wetlands and freshwater modified environments of Castlemaine Harbour (Delaney et al. 2012). Similar features are found on neighbouring coasts, e.g., southwards on the Iveragh Peninsula at Derrynane Bay and northwards in Tralee Bay at Derrymore (Davies and Stephens 1978).

The Dingle Bay-Castlemaine dune-barriers occur within the embayed regional coastal setting of southwest Ireland. This is characterised by crenulate, rock dominated coasts; defining c. 10–40 km long and relatively narrow c. 4–12 km wide bays and linked river valleys, separated by high peninsula-structured uplands (Fig. 9.2) (Devoy 2009). These embayments result from the approximate NE – SW trending Hercynian-Armorican age structural folding of the region. This has created deep synclinal basins, composed primarily of Devonian and Carboniferous rocks of sandstones, quartzites, shales and slates (Whittow 1974; Pracht 1996; Higgs 2009). These valley-folds were submerged during the Holocene RSL rise, with near- to off-shore water depths today of 50–100 m. The embayments are still often referred to, incorrectly, as *rias* (Goudie 2004; Devoy 2009). The elongate shape of these embayments results in the strong compartmentalisation, localised transport and storages of the coastal sediments.

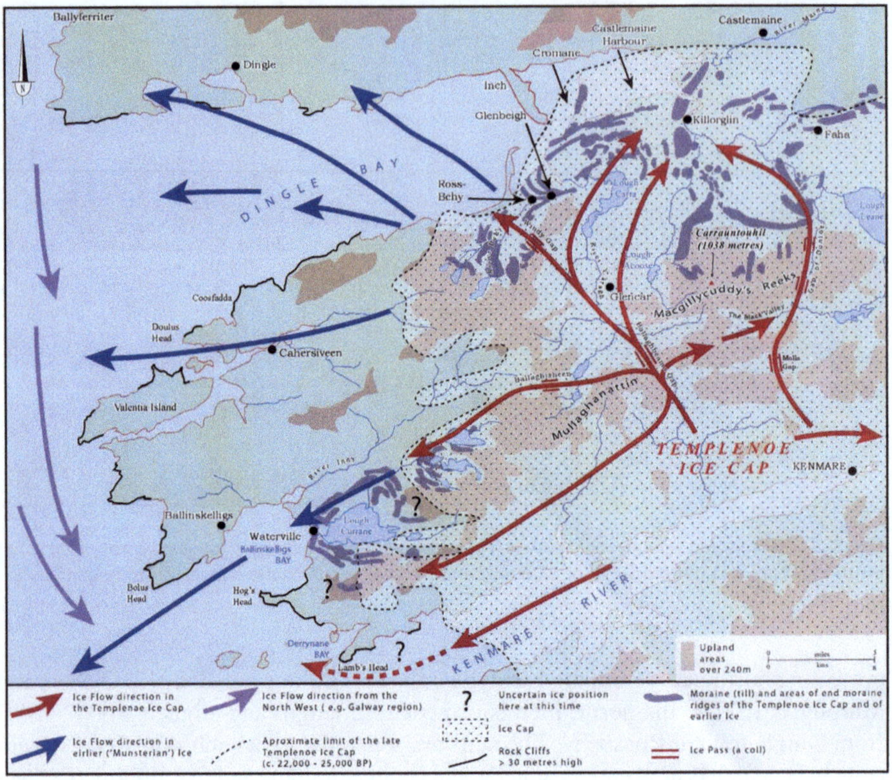

Fig. 9.4 Late Quaternary (Marine Isotope Stages 2–4) ice movements and the glaciation of Southwest Ireland, affecting the present day coasts of Dingle Bay and Castlemaine Harbour (Source: Devoy 2009)

Study of the late Quaternary pattern of glaciation in southwest Ireland shows that this area was heavily affected by ice action (Edwards and Warren 1985; Knight et al. 2004). Valley-type glaciers passed westwards through Castlemaine – Dingle Bay (Fig. 9.4), leaving an over-deepened glacial trough, with gorge-like channels incised into the pre-glacial seabed. On deglaciation, these features were infilled with glaci- and fluvi-glacial materials (Shaw et al. 1994) and later by Holocene re-worked coastal-marine sediments. Geotechnical and field surveys of Inch (Cooper et al. 1995; Vial 2008) Rossbehy and Cromane, show that they are built upon ice recessional end-moraines. These date from the final stages of ice retreat eastwards from Dingle Bay at c. 16–18 k years BP, hence, the approximate N – S alignment of the barriers (Devoy 2009). The seabed sediments within Dingle Bay and Castlemaine Harbour provide consequently a wide particle size range (sands – boulder sizes) for supply onshore. Hydrographic (e.g., sidescan sonar, multibeam) and seismic surveys of the Bay show the partitioning of these sediments, with fine – medium sized sands concentrated on the northern side and in front of the barriers

(Shaw et al. 1994; INFOMAR 2014). Extensive areas of coarser sands and sub-angular – rounded gravels -boulders occur predominantly southwards. This sedimentary separation may result from the initial glacial deposition and glaci-fluvial sorting, but has probably also been affected by subsequent Holocene RSL changes and wave action. Erosion and cliffing of the Iveragh's coastal glacigenic sediments, under the dominant westerly storm wave action continues to re-enforce this sedimentary pattern (O'Shea and Murphy 2013). The hydrometric surveys also indicate the preservation of biogenics, possibly of fresh to brackish-water, lagoonal-type deposits, in shallow 'basinal' structures within the sediments depths of >-10 to -15 m ODB (Ordnance Datum Belfast). The biogenics are likely to date from lateglacial to early postglacial times (c. 13,000–9,000 years BP). They probably record the position of former back-barrier environments developed behind earlier beach-barrier structures at times of lower RSL (Devoy 2009; Delaney et al. 2012).

9.2.1 Boundaries and Drivers

The geological structural and coupled earlier geomorphological controls, with the inheritance of extensive glacigenic sediments and the now submerged glaciated terrains, condition the primary character of the Inch-Rossbehy-Cromane Spits and barrier systems. Further, Holocene RSL rise and regional isostasy (Devoy 1991, 1995; Devoy et al. 1996; Edwards and Brooks 2008; Brooks et al. 2008), coupled with the mid-late Holocene growth of extensive fringing coastal wetlands in Castlemaine Harbour (Delaney et al. 2012) have also been important contributors in their development. Additionally, late Holocene (post c. 1000 AD/BCE) human influences, with the heavy post eighteenth century impact of people on these coasts has been particularly significant in the barrier environments (Duffy and Devoy 1999; Delaney et al. 2012). These include, farming practices of pasture development and animal grazing of the dunes, as well as local woodland clearances, adding to the effects of longer term post Neolithic regional vegetation changes. Times of temporary occupation have occurred, most recently during the Great Famine (Foley 2009; Crowley et al. 2012) and evidenced in shell middens of this period and of earlier ages preserved within the barrier sediments, most notably on Inch. Further, the abstraction of sands and gravels for building materials and agriculture, together with present day tourist impacts have all contributed directly to changes in the morpho-sedimentary development of the dune-barriers (Devoy et al. 1995; Duffy and Devoy 1999; Vial 2008; Science Direct 2013).

West of the barriers water depths are relatively shallow, averaging < c. -10 m ODB in the nearshore zone (c. 0–5 km from the barrier shorelines), but shelve westwards to depths of >-70 m ODB 35–40 km offshore (Fig. 9.5). For Castlemaine Harbour, water depths are generally >-5 m ODB and the wetland environments here receive substantial freshwater inputs from the Maine and Laune rivers (Fig. 9.1), smaller streams from the adjacent uplands and direct rainfall

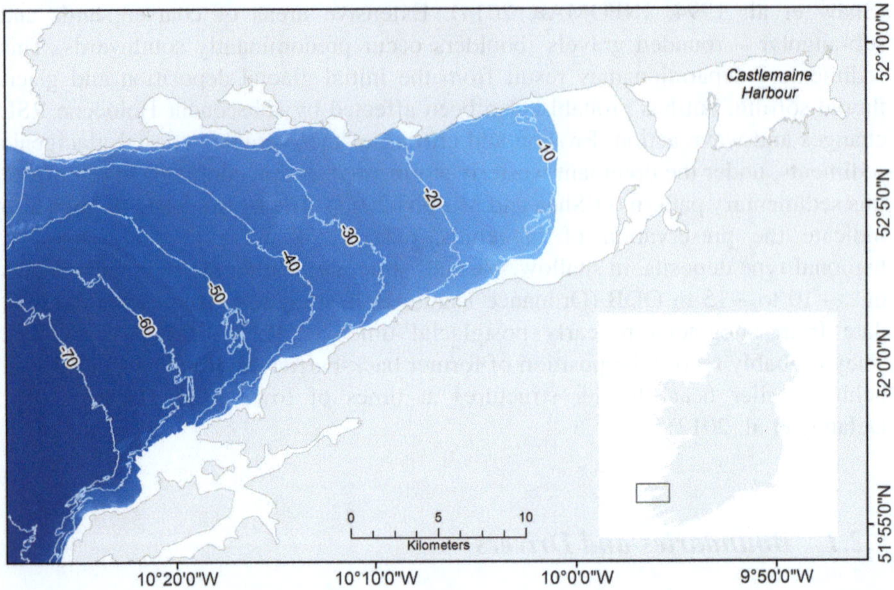

Fig. 9.5 The bathymetry of the Dingle Bay and Castlemaine Harbour area (Data Sources: Marine Institute, INFOMAR 2012)

(current regional average rainfall >1,200 mm/year) (Duffy and Devoy 1999; Delaney et al. 2012; Met Éireann 2014). Tidal ranges within Dingle Bay are of meso- to macro-tidal type (2–4 m), with a Spring Tidal range of c. 3.2 m. Until the 2008 Rossbehy breaching event, tidal exchange with Castlemaine was maintained via a c. 1.5 km wide ebb-tidal delta. Measurement of main channel tidal currents during mid- flood conditions show velocities of c. 0.9 m/s, reducing to 0.6 m/s at mid-ebb, with averaged tidal current velocities again lower at c. 0.3 m/s southwards in the swash aligned zones of the barrier (O'Shea and Murphy 2013).

Waves heights reaching the beaches facing into Dingle Bay are often >2 m Hs, with mean significant wave heights of c. 2.8 m H*s* and periods (Tz) of 7 s (Sala 2010). Both Inch and Rossbehy are subject to the impacts of onshore high wind – wave energy conditions from the dominant Atlantic SW – W directed swell waves and storm surges (Jenkins et al. 2009; Marine Institute 2014). The bathymetry and the approximate N – S orientation of the barriers lead to changes, through wave refraction, in alongshore currents, wave alignment and sediment transport on the barrier shorelines. These conditions are now being modified further by the rapid alterations in the ebb-tidal delta geometry. Broadly, alongshore current and sediment drift patterns remain operating from S to N in direction (Cooper et al. 1995; O'Shea and Murphy 2013). The wind regime affecting the barriers is conditioned primarily by the SW-W onshore winds, from Atlantic storm tracks (Lozano et al. 2004; Sweeney et al. 2002, 2008; Dunne et al. 2008; Met. Éireann 2014), though this may alter under climate warming (Dwyer 2012; IPCC 2014a, b). Well developed land-sea breezes and turbulence created by air down-draughts and eddying from the neighbouring uplands often complicate this wind pattern (Sherman et al. 1995; Duffy and Devoy 1999; Devoy 2012). These local winds create significant wave action in the areas of longest fetch within Castlemaine Harbour. Significant back spit-barrier and marsh front erosion occurs under these conditions, with accompanying sand and coarse sediments transport (sands – gravel sizes) occurring.

9.2.2 The Spit-Barriers: Significance for Science and People

The Dingle Bay – Castlemaine Harbour area has attracted persistent attention as a valuable location for coastal geomorphological and wider science research, both primary and applied in nature, and for student teaching. Contributory factors have been the prominence of the Spits, their accessibility and the adjacent extensive wetland environments. The dunes on Inch have also been seen as ecologically and floristically significant and have been classified as Special Areas for Conservation and Scientific Interest (SAC/SSIs) (National Parks and Wildlife Service 2011), though this status is now under re-examination (Science Direct 2013). Geologically, however, the area remains of high science value. For the wider public community, the visual attractiveness of the area is seen as high. This has been important in attracting not only a persistent growth since the 1980s in tourist and other recreational users to Inch and Rossbehy's beaches and dunes, but as a draw for coastal housing (both first and second homes) and small business developments (Cooper and Boyd 2010). This growth in buildings in the immediate back-barriers areas (>100 situated below <10 m ODB) has added significantly to the area's infrastructures (including those of, e.g., roads, harbours, jetties). Earlier buildings were linked more to the needs of the local primary industries, of farming, fishing, aquaculture and only later to tourism. This people-attraction factor is of particular importance now, due to the vulnerability of these environments to coastal flooding, erosion and cliffing-slope instability and collapse (Devoy 1992, 2008; Kerry County Council 2001a; Sweeney et al. 2008; Cooper and Cummins 2009; Cooper and Boyd 2010; Dodds et al. 2010; O'Connor et al. 2010). The consequences of this vulnerability for investment have been costly, in terms of demands for increased coastal protection works and earlier in the nineteenth and twentieth centuries in extensive land reclamation schemes (Duffy and Devoy 1999; Kerry County Council 2001b, 2003, 2009; Lynch 2009, *pers. comm.*; Kerry's Eye 2010; Gault et al. 2011).

From the coastal science and management viewpoints, a series of key studies have been undertaken at these sites, beginning in the 1960s (e.g., Guilcher and King 1961; Guilcher 1966; Regnaud 1989; Froger 2003). Since the 1990s the barriers and the wider area have formed the focus for a series of large scale EU funded Climate and Environment, Framework IV, VI and InterReg programmes, dealing with coastal palaeoenvironmental reconstructions, hydrodynamics, process functioning and management (e.g., projects SEA LEVELS, IMPACTS, Conscience and CoastAdapt). The area has become effectively a coastal research laboratory (Devoy et al. 1995; Delaney et al. 2012; Conscience 2010; CoastAdapt 2013; O'Shea and Murphy 2013).

9.2.3 Changes in the Barriers and Current Issues

Until recently the barriers appear to have been relatively stable in form since their first precise mapping in the 1840s (Ordnance Survey of Ireland 2014). Subsequent surveys of these systems (Cooper et al. 1995; Kerry County Council 2001a, 2003; Vial 2008; Sala 2010; Kandrot 2013, 2014) have recorded significant alterations in the shoreline positions of Inch and Rossbehy. In particular, changes of the distal ends of both barriers have occurred since c. 2000, as well as in longer term, decadal scale oscillations in sediment accumulation and erosion of the beach-barrier fronts and of linked foredune areas. In the past, these movements may have been part of a regular cycle of change in coastal progradation and retreat, linked to the process functioning of these systems. Human impacts on the dunes and wetlands during nineteenth century Famine times would also have been important in creating sediment movements (Duffy and Devoy 1999; Hanna et al. 2008; Hickey 2011; Crowley and Sheehan 2009; Crowley et al. 2012). The growth in people and tourism-related pressures on the dune-barriers since the 1990s, with the impacts of increased vehicle use, horse riding and linked human intrusions on the beach and into the fronting foredunes, may now be a further contributory factor to recent changes. There is even renewed local demand at Inch to develop parts of the northern, stabilised dune and back-dune environments as a golf course (Science Direct 2013).

At Rossbehy, large changes in barrier morphology and sediment distribution have occurred since the breaching of its distal end by a storm surge in 2008 (Fig. 9.6). The breach area has subsequently continued to expand southwards, with erosion of the main dunes and re-organisation of the hydrodynamic, tidal regime and sedimentary movements of the northern end of the barrier. This erosion and the wider system changes at Rossbehy, with probable future repercussions also for Inch and Cromane, have caused great public concern (Kerry County Council 2001b, 2009; Kerry's Eye 2010). The local residents and regular visitors to the area are most affected, some with livelihoods at stake from the disruption of beach-based tourism, aquaculture and related inshore fishing, small port activities, and for Kerry County Council (KKC) and the Office of Public Works (OPW) as coastal managers. As noted, many houses as permanent residences and holiday homes are situated adjacent to the wetlands in Castlemaine Harbour and lie close to the current high water mark (HWM). Continued morphological change, or even destruction of Rossbehy, coupled with wider coastal flooding and erosion under RSL rise and climate warming would threaten this housing, the linked coastal activities and the wetlands (Devoy 2008; IPCC 2014a, b; Cooper et al. 2009). To some degree these fears for the future of Rossbehy and Inch have already been realised in the extensive

Fig. 9.6 Rossbehy Spit, showing the dune-barrier pre the 2008 storm breaching of the distal end and post the 2008 breach event: (**a**) oblique ENE view along the western shore face; (**b**) plan aerial view, the *red line* marking the post immediate breach shoreline (KCC survey 2008); (**c**) view north along the barrier length to the breach area and showing the dune-field on the Spit; (**d**) development of the breach in 2009; (**e**) breach area at full tide 2010, showing development of drift aligned sediment transport (Sources **c–e**: open access from web and KCC); (**f**) composite aerial views (Sources: OPW and OSI) of the distal end of Rossbehy and the breach area in 2012, together with the ebb-tidal delta area between the Inch and Rossbehy Spits (Source **f**: Kandrot 2013)

damage from the intense back-to-back storms (c. 950–960 Hp) of December 2013 and January 2014 (Irish Examiner 2014).

Consequently, the functioning of both barriers continues to be monitored for morpho- and hydrodynamic sedimentary changes; to chart the progression in the contemporary patterns of the on- and off-shore environments and as a basis for the numerical modelling of inlet and barrier type coastal 'behaviours' (O'Shea and

Murphy 2013; Kandrot 2014; Ranasinghe et al. 2013). In conjunction with the earlier studies, few such concerted coastal process research programmes have been undertaken in Ireland. The results will be of significance in contributing to decisions on coastal management responses to such rapid changes, both locally and possibly nationally (Gray et al. 2014). Importantly, these studies will add to the understanding of the coastal processes operating on these compartmentalised coasts. Particular significance here is in projections of the likely future responses of these dune-barriers under climate warming, with RSL rise and storminess impacts. The current changes, coupled with the projected increase in Atlantic storm magnitude and probable increased frequency for the period 2021–2060, RSL rise (range of c. 0.05–0.95 m by 2100 for IPCC 2014a RCP scenarios) and other impacts of climate warming, will probably result in continuing *coastal squeeze* and the partial to possible complete destruction of these barriers (Devoy 2008; Science Direct 2013). Numerical modelling studies to check the accuracy of this view are now being undertaken (O'Shea 2015; Kandrot 2015). In essence, the barrier changes now underway are an illustration of the likely outcomes to be expected in many such coastal systems worldwide, marking the end of the Holocene and the beginning of the Anthropocene (Zalasiewicz et al. 2010).

9.3 The Barrier Morphologies and Environments

9.3.1 The Inch Spit

Inch comprises a sand dominated beach- and dune-barrier and forms the largest dune system in Ireland (c. 12,000 Ha). A series of shallow intertidal embayments are developed at the rear of the dunes, on the eastern side and contain sequences of fine-sized (<250 μm) sediments. Prominent vegetated and semi-vegetated sand dunes, both foredune and transverse parabolic in type, have developed here (Carter 1988; Orford et al. 1997, 1999; Science Direct 2013). These reach ground heights of generally >10 m, and often higher to >20 m ODB, with maximum heights c. 30–35 m ODB in the central areas.

Currently, wave refraction and alongshore drift mechanisms move sediments from Dingle Bay into the nearshore zone at Inch and also at Rossbehy. The resultant beach system, with backing foredunes, is dissipative in form and predominantly swash aligned, maintaining a wide (>c. 250 m at low water mark) and low angle (<c. 2°) beach face with average sediment particle sizes of 235 μm (Vial 2008; O'Shea and Murphy 2013; Kandrot 2013). In turn, these controls on the barrier operate through tidal cycle variations and seasonal wave energy-conditioned movements of sediments (timescales of months–years). The significance of this is in the operation of storm events, the impacts of which are superimposed upon these daily controls on sediment movements and appear to be particularly important in developing the longer term morphology and supply of sands into the dunes

(Sherman et al. 1994; Cooper et al. 1995, 2007; Devoy et al. 1995; Lozano et al. 2004; Conscience 2010).

The dunes are the main visible morphological elements of Inch (Science Direct 2013) and are composed of fine – medium sized sands (c. 90–300 μm). The sands are quartz dominated, but also contain a variable proportion (c. 3–10 % volume) of broken shell fragments ($CaCO_3$), from offshore, intertidal and within-dune shell communities. The dune sands, whilst containing broken marine shells, are not seen as part of the machair sand systems developed further north in western Ireland, The Western Isles (Hebrides) and western Scotland. The barrier lacks the extensive back-dune cover sand sheets of the machair proper (Ritchie 1979; Gilbertson et al. 1999; Hansom and Angus 2001). The sands are generally well-rounded, having undergone probably numbers of sedimentary recycling stages over at least the last c. 200,000 years. The sediments have moved from source to sink to re-activation through river, glacial and marine actions and contain minimal shape evidence of more recent movements by wind action.

Sands move into the dune-barrier under westerly and south-westerly onshore wind and wave action, primarily during phases of storminess (Orford et al. 1997, 1999; Kandrot 2013). The dunes are accreting sands though in western beach-face areas and in the southern portion of the barrier (between c. 2 and 5.5 km, N – S down-barrier length). These areas support broken, or little long-term vegetation cover. Marram (*Ammophila arenaria*) and Lyme (*Elymus arenaria*) grasses are dominant, with large areas of exposed and moving sands. This southern area of Inch forms the 'active' zone (Fig. 9.7) of dune slope and morphological changes (Table 9.1). Drift alignment of the beaches at end of the barrier occurs, associated with the adjacent ebb-tidal delta. To the north, from its junction with Dingle and in the areas immediately southwards (c. 0.0 2.5 km) the dune-field is essentially vegetated and stabilised. Away from the shore-face, the dunes currently receive little, if any, new sands and have inherited much of their form from late Holocene phases of barrier development (Cooper et al. 1995; Wintle et al. 1998; Orford et al. 1997, 1999). Any changes here today in dune morphology result primarily from the internal recirculation of sands. Human controls on dune slopes and erosion mechanisms through farming practices are dominant, mainly from animal grazing and associated vehicle movements, together with the activities of burrowing animals.

On the western, Dingle Bay side of the barrier, the dunes are organised from the HWM into a series of N – S shore-parallel foredune ridges, variably two to four are recognised (Orford et al. 1997). The incipient foredunes, where sands are actively accumulating occur along the length of the barrier, but are concentrated particularly in the southern section, in the last c. 3 km. Behind these areas occur older and stabilised dune ridges, with more established vegetation cover and separated by intervening low elevation swales. The foredunes are backed throughout the length of Inch by the main dune-fields (Fig. 9.7). These are comprised of transverse (approximately E – W aligned) transgressive parabolic dunes, with flanking low ground of the dune corridors (Cooper et al. 1995; Orford et al. 1997, 1999). The basal surfaces of these corridors, and of other lower elevation inter-dune areas often

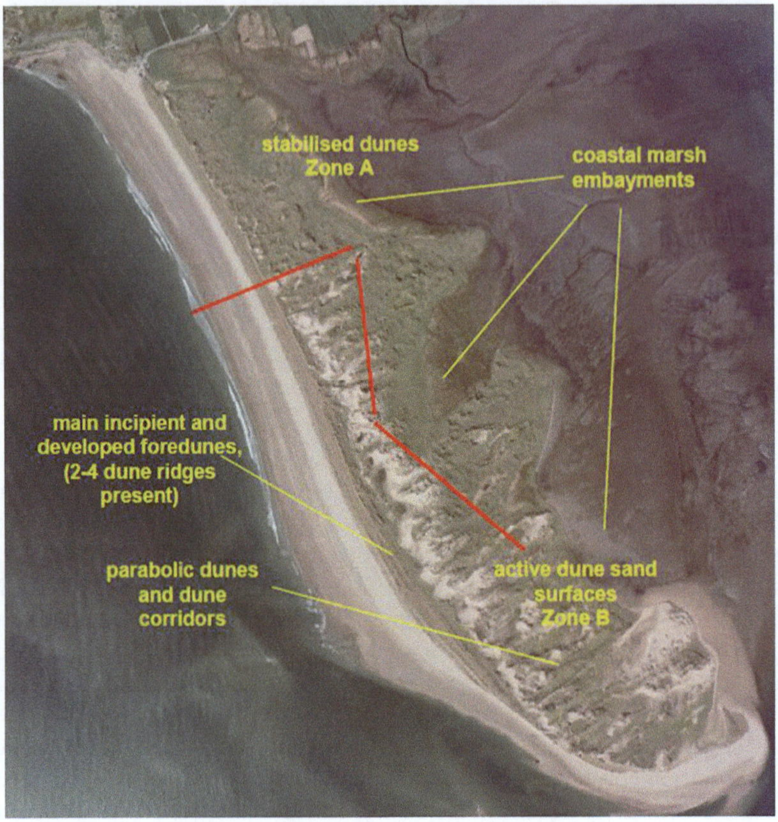

Fig. 9.7 Inch Spit dune-barrier morphometric type areas and environmental zonations (Science Direct 2013) (Photograph Source: Office of Public Works and Ordnance Survey of Ireland 2012)

now support extensive dune slack environments (Science Direct 2013). These form important dune wetlands, dependent on the seasonal height of the local water-table, as well as sites for other sand-adjusted open habitat plant communities.

Behind the main dune-barrier, on the Castlemaine Harbour side of Inch, ground heights are generally low, at c. 3–5 m ODB. In the stabilised dunes (Fig. 9.7) this low ground forms a bench-type morphology in the trailing edge of degraded parabolic dunes. The 'bench' is of variable width, commonly c. 100–400 m wide, but is less well developed in the 'active' dunes at the southern end of Inch. Marsh and poorly drained areas form part of this low ground surface and are replaced seawards by intertidal saltmarsh and mudflats, developed particularly within three main embayments. The embayments were formerly and, in some locations, continue as intertidal areas subject to marine inundation. These areas have been embanked and drained in the past, with poorly maintained and remnant protecting barriers, sluices and drainage ditches still visible (Science Direct 2013). The built structures are part of the drainage and wetland reclamation works in Castlemaine Harbour, begun in the nineteenth century (Duffy and Devoy 1999; Delaney et al. 2012).

Table 9.1 The main geomorphological and morpho-environmental area types and characteristics comprising Inch Spit. From Devoy RJN in, Science Direct (2013)

Area Types 1	Areas of low, degraded and stabilised dune field. Maximum variation in ground elevations (highest to lowest surfaces) are c. 5 m, with areas generally <10 m ODB in height (see OS map, 1:50,000). Slope angles are between c. 5–8°, developed as low, undulating sand surfaces, with some discrete sand hills present. Slope mass movement *terracetting* and small scale slumps are widespread. Large dug pits and other hollows, also with some 'levelled' areas in the dunes/sands, have been created by sand mining and other human impacts (e.g., nineteenth century sand 'borrowing'). The dunes/sand surfaces are well-vegetated, mostly with complete vegetation cover. No evidence of recent dune blow-outs due to wind erosion mechanisms
2	Active shore parallel incipient and foredune systems, developed along the western shore edge of Inch above levels of MHWMST (Mean High Water Mark of Spring Tides) comprising sand dune ridges and separating swale areas. Between 2 and 4 foredune sequences have been noted along the length of Inch. These are variably developed in width and depth, but are all aligned c. N – S. Maximum ground elevation variations (*swale* bottoms to ridge crests), c. 5–6 m, with ground heights generally <10 m ODB
	The base of the swales show exposures of the earlier cobble barrier underlying the dunes and which formed under RSL rise, with wave/storm action, on the basal glacigenic surface of Inch. These are well developed in the central shore section of the dune-barrier (southern end of the proposed golf course, in Zone A). Shell middens are present on these exposed cobbles. Many are probably nineteenth century 'Famine middens', dating from c. 1840 to 1860s, but possibly some are from earlier historical times, or are even Neolithic and Bronze Age in origin. Shell middens have also been noted from other parts of Inch, from the parabolic dune corridors and from the residual sand hills on the eastern side of the dunes
3	Poorly drained and full coastal wetland environments (supra-tidal to high marsh – saltings). This area is generally below c. 3 m ODB, with low angle slopes (<3°). The higher elevation water-logged zones here comprise sandy, minerogenic-rich biogenic sediments and overlie the sand dunes, often representing former intertidal areas on the eastern side of Inch. Present day active marine wetlands (saltmarsh and mudflats) form on the seawards side of these higher level 'freshwater/brackish water' marshes. The former intertidal/high tide areas were subject to embanking and draining works in the nineteenth and early twentieth centuries as part of land reclamation schemes. The drainage and protection structures (e.g., low earth banks/walls, ditches and sluices) have now been 'let-go' and areas have reverted to zones of poor drainage and marsh, characterised by *Juncus* spp., Carex spp. and Poaceae dominated vegetation
4	A low, undulating 'sand-bench', formed at the rear the main dune areas (parabolic dunes). Low residual sand hills/dunes may also occur here. Ground heights are characteristically c. +2–5 m ODB, with variations in surface elevations of c. 1–2 m and variable low slope angles of <c. 2–5°. This represents a transitional environment of low, irregular sand surfaces, separating the areas of distinct dune forms (Type 5) and the sedimentary fines environments of the former and the present inter-tidal zones (Type 3). The area forms a 'corridor' of low ground on the eastern side of Inch, with vegetation characterised by rough pasture (grasses and sedges dominant as vegetation cover) and calcicole meadowland type plants. The core of a proposed golf course is planned in and around these areas

(continued)

Table 9.1 (continued)

5	Stabilised and well vegetated dunes, developed to heights of >20 m ODB, though most areas lie between c. 10 and 20 m, with variations in ground surface elevations commonly of c. 5–10 m (high points to low points). Steep slope angles characterise this area, with slopes of 5–>15°. Slope terracetting is common, generally as large-scale features and affecting areas of most dune slopes. These indicate the continued mass movement of the dune forms, due to the original dune morphology and steep primary slopes. The terracette features measure at c. >0.3 × 0.3 m height and width and are controlled by factors including, e.g., climate variables, inherited/original dune slope angles, sediment size, vegetation cover, soil – organic sediment cover, pore water pressure, burrowing animal activity. Slope movements (under gravity) will operate and accelerate if animal and human activity cause reductions, or other changes, in vegetation cover ('climate warming' controls could be significant here in future by 2050). Dune blowouts and erosion scars developed under wind action occur, especially on the seaward facing slopes. Vegetation is dominated by Marram (*Ammophila arenaria*) and Lyme (*Elymus arenaria*) grasses. Mosses (e.g., *Sphagnum* spp.), other grasses, sedges and a diverse herb flora form in the sheltered microclimate at the base of the Marram and Lyme grasses. This ground flora develops an important vegetation mattress on the dunes and reinforces the initial stabilisation of the sand dunes, which has resulted from the development of the primary vegetation cover. Vegetated dunes of this type extend across Inch from W – E, though are dominant now on the western and central sections of the dune barrier in zone B. Shell midden materials have been recorded as emerging through animal burrowing from the eastern side of Inch in these areas
6	Dune *slacks* in the main dune field areas (i.e. Types 5 and 7) are common; also links to the swale environments of Type 2 areas. These represent relatively extensive areas (order of 10^{2-3} m^2) of low, 'flat' ground (slope angles <1–3°) formed between the main sand dune ridges and hills. Drainage from the dune slopes may develop seasonal standing water bodies here and the slacks support extensive sand-adjusted and potentially wetland open habitat plant communities. These are formed as an integral part of dune field morphodynamic mechanisms (e.g., possibly as extensive blowouts, more commonly as part of parabolic dune formation) under wind action. Where these inter-dune hollows intersect with the watertable then areas of longer term standing water may form (not observed on Inch). The occurrence of the wet sand sediments reinforces environmental and linked slope 'stability' in these areas
7	Active parabolic dune fields (southern end of Inch, last c. 2.5–3 km), with dune heights commonly reaching >20–30 m ODB and steep slope angles (commonly >15–20°). This terrain differs from that of Type 5 by having larger areas of incomplete vegetation cover, extensive blowouts and sand is clearly dynamic under current onshore dominant wind action. 'Linear' parabolic dune corridors, formed in between side-marginal, residual dune ridges are aligned c.W – E across Inch in the direction of dominant wind blow. At the western shore-face foredunes (Type 2) front these areas (representing contemporary deposition and sediment storage sources for sand inputs and for dune re-organisation into the main dune areas). At the southernmost end of this type area contemporary coastal wave action has led to the reworking of the dune sands and the development of wave refracted, spit sedimentary structures. The area is likely to be the most vulnerable to immediate erosion with the break-up of the Rossbehy barrier

(continued)

Table 9.1 (continued)

8	Contemporary intertidal areas of saltmarsh (vegetation dominants include variably, *Spartina* spp., *Salicornia* spp., *Festucia* spp. (grasses), Carex spp.(sedges), *Armeria maritima, Plantago* spp., *Aster tripolium*: i.e., common species of the saltmarshes in Ireland and Britain). The saltmarsh vegetation is semi-salinity adjusted by the seaward dipping, cross-shore gradient from high to low tide levels. Slope angles are very low (<1–2°) though and the vegetation is often graded more by the occurrence of isolated areas of higher ground. Beyond the saltmarsh extensive sand- and mudflats are formed, developing offshore into Castlemaine Harbour

9.3.2 The Rossbehy Barrier and the Ebb-Tidal Delta

The Rossbehy beach and dunes (Fig. 9.1) form a separate barrier system, but an important element of the boundary control to the linking ebb-tidal delta and possibly to the long term maintenance of Inch. Full tidal and wave action travelling into Castlemaine Harbour and the eastern shore of Inch, is controlled by the geometry of the ebb-tidal channel (Fig. 9.3), with Rossbehy forming a mixed wind-wave and tide regime dominated barrier (Fitzgerald et al. 2000; O'Shea and Murphy 2013). The breaching of the adjacent distal end of Rossbehy occurred in 2008 at the narrowest part of this northern end of the dune-barrier (width then c. 0.35 km). A near-breach storm event also occurred here in 1842 (Delaney et al. 2012). Extensive erosion of the Rossbehy dune face southwards is occurring currently, with rates of c. 25 m/year since 2010 recorded at the northern end of the barrier close to the breach zone, as well as losses in areas of back-barrier sedimentation (O'Shea et al. 2011; O'Shea and Murphy 2013; Kandrot 2014, *pers. comm.*)

The Rossbehy dune-barrier is less complex than Inch, but is fronted similarly by a wave-dominated, dissipative and swash aligned beach, with relative tidal range (RTR) of 2.5 and a Dean's parameter (Fall velocity, Ω) of 6 (Sala 2010). Fine-medium sized sands are dominant on the beach, though extensive spreads of gravels and cobble sized sediments occur. The dunes behind are, as at Inch, characterised by fine – medium sized sands (<c. 350 μm) with massive bedding. The barrier probably results from the development of a series of foredunes, which have undergone some re-organisation but comprise now essentially a single dune-field, with little evidence of secondary parabolic dune development. The dunes commonly reach heights of 5–16 m ODB and have developed to a maximum width of c. 0.5 km midway along the pre-2008 barrier length. Many of the dune forms are heavily degraded through the direct impacts of people entering the dunes and from animal grazing. At Rossbehy village, the southern end of the barrier is composed of a cobble-boulder beach, developed from mid-tidal levels to above HWM, where there is a prominent berm structure. This cobble-boulder beach has been re-structured and re-enforced with additional boulder-sized material by Kerry County Council (KCC) to form part of coastal protection measures. These were destroyed in the 2013/2014 storms, with consequent public demand for their reconstruction; together with much of the coastal defences for the west and south

of Ireland (Examiner 2014). The beach forms the narrowest point of separation of the barrier from mud – sand flats and remnant wetlands behind. These were formerly embanked and reclaimed in the 1840s, but subsequently allowed to revert to intertidal areas (Duffy and Devoy 1999; Delaney et al. 2012). Sand dune development begins c. 0.5 km from the start of the boulder beach.

9.3.3 *Cromane*

Behind Rossbehy and Inch occurs a similar spit-like structure within Castlemaine Harbour (Fig. 9.1), aligned approximately N – S and c. 3.4 km in length. It differs from the two main fronting barriers in having no sand dune cover. A relatively steep, reflective beach of coarse sand and gravels, with an upper beach-berm area of cobble sized materials fronts the structure. The beach has developed eastwards, topped by the present roadway, as a thin veneer of these coarse clastics overlying basal glacigenic sediments. These form the core of the spit and represent an ice-recessional end-moraine feature (Devoy 2009). The beach-barrier and moraine ridge are at c. 3–4 m ODB in height and are backed by lower biogenic and silt to clay-rich organic dominated marshes cut by inter-tidal creeks. Spring Tides regularly flood these wetlands and associated saltings, which occur at uniform levels of c. 2 m ODB and are composed of c. 1–4 m of homogeneous monocot peats, with occasional thin bands of sands and clay-silts (Cott et al. 2012; Delaney et al. 2012). Human impacts on these areas is clear, with plough furrowed surfaces and dug, linear drainage ditches joined into the natural inter-tidal creeks (Duffy and Devoy 1999). The dumping of overburden on the northern, more fragmented end of the marshlands began in the 1990s, as a prelude to intended housing and road development, though construction has not occurred to date. The wetlands link eastwards into Castlemaine Harbour, forming thicker sequences of similar biogenic sediments above the underlying glacigenics (Delaney et al. 2012). The beach at Cromane continues south and westwards towards the Caragh River (Fig. 9.1), grading into finer beach sands and a narrow zone of low sand dunes (c. 5–6 m in height), containing thin interbedded organics and nineteenth century Famine shell middens.

9.4 Holocene Spit-Barrier Development in Dingle Bay-Cromane Harbour

9.4.1 *Large- to Meso-scale Models of Holocene Barrier Development*

Conceptual and behavioural models at macro- to meso-scales for the development of coastal sedimentary barriers, under the conditions of rapid (>10 mm/year) to decelerating RSL rise (to < c. 2–3 mm/year) are well established (e.g., Belknap and Kraft 1977; Kraft and John 1979; Kraft and Chrzastowski 1985; Thom 1984; Cowell and Thom 1994; Roy et al. 1994; Cooper 2007; Cooper and Pilkey 2007a; Cooper et al. 2007a, b; Backstrom et al. 2007). Barrier systems at these time and spatial scales are controlled primarily by, the rate of RSL rise, tidal range, barrier sediment volume (as expressed in barrier width, height and sediment supply), sediment size, wave climate, storm magnitude and frequency, the degree of wave overtopping and overwash and of barrier-foot erosion (Carter 1988; Roy et al. 1994; Jennings et al. 1998; Orford and Carter 1984; Orford et al. 1992). Under these conditions barrier structures in the Holocene have moved progressively on-land over time, dependent particularly on the RSL factors operating locally, the slope-gradient of the offshore zone and its degree of 'roughness', the 3D seabed geometry and the available sedimentary accommodation spaces. The process of movement has been referred to broadly as barrier *roll over*. The changes of associated individual coastal sedimentary features over time, such as spits, will also be controlled by a similar combination of controls.

Validation of the operation of these controls has been undertaken on many medium to high energy coastlines (e.g., of the North and South Atlantic, Northern European Shelf Seas, eastern Australia, New Zealand and southern Africa), together with numerical modelling of barrier process functioning (e.g., de Vriend 1991; de Vriend et al. 1993a, b; Cowell and Thom 1994; Roy et al. 1994; Cowell et al. 2003a, b; Hanson et al. 2003; Aagaard et al. 2004; Stive 2006; Cooper 2007; Geleynse et al. 2010; Tung et al. 2009; Roelvink et al. 2009; Rossington et al. 2011). Dependent on other key characteristics of the coastal setting in which the barriers occur, e.g., estuarine, deltaic, arctic, paraglacial, tectonic-earthquake conditioned (Carter and Woodroffe 1994; Woodroffe 2002), then other types of broad-scale process model may also apply, e.g., estuary-tidal basin (Roy et al. 1980; Heap et al. 2004), erosion front (Orford et al. 1991; Carter and Orford 1993; Greenwood and Orford 2007), tectonic-tsunami (Nichol et al. 2007). In spite of the understanding of coastal behaviours at these scales, and of the conditions under which eventual barrier breaching and breakdown occurs, the specific hydrodynamic mechanisms operating at the meso- to micro-scales (decades to days) and their feedbacks to the macro- levels is less clear (Carter and Woodroffe 1994; Cooper et al. 2007). At these levels many other different coastal process factors become more important, or dominant (e.g., as linked to beach radiative stresses, tides, wave climate, storminess, sediment supply and coastal geometry). Barriers may not *roll over* as such, but undergo degrees of lateral spatial translation and morphological change, particularly where breaks in the barriers and consequent tidal and wave forcings occur. A useful summary of the high resolution hydrodynamic modelling needed and detail of the range of possible processes and outcomes for barriers is given in Cronin (2010), O'Shea and Murphy (2013) and Cooper and Jackson (this volume).

9.4.2 *Castlemaine Harbour: Palaeoenvironmental Evidence of the Spit-Barriers*

The coastal wetlands and offshore sedimentary records of Castlemaine Harbour (Fig. 9.1), evidence the genesis of these two spit-like barriers (Devoy et al. 1995; Delaney et al. 2012).[1] These data show the likelihood that the two barriers probably formed a single sedimentary structure until c. 3,000 years BP. But they were not necessarily in their present separate and moraine-stranded position till post this time. Coring, geotechnical and related palaeoenvironmental studies within the harbour show the sedimentary sequences here to overlie an irregular glacigenic surface, though at the coastal margins found commonly at depths of ≥ c. −4 m ODB.

The stratigraphic evidence shows a complex and changing west to east sedimentary sequencing, of predominantly inorganic-rich biogenics interleaved with clay-silts and thin sands. Grass and sedge dominated peat, with detrital and *in situ* wood (i.e., rooted pine stumps and other woody parts of the leaves and fruits of pine, oak, alder and hazel) developed first on the underlying glacigenic surfaces. The basal organic sediments, showing essentially plant communities growing under rising sea levels and consequent high groundwater conditions, are dated at western sites within Castlemaine to c. 6,000–4,500 years BP and at heights from −2.7 to −1.2 m ODB. To the east and higher on-land, the timing of the onset of this sedimentary and environmental pattern becomes relatively later, though there are significant intra site differences. These basal biogenic sediments become transitional to a monocot peat, of sedges, *Phragmites* and other grasses. This unit is replaced, sometimes with a sharp boundary contact evident, by an organic-rich clay-silt. In western areas particularly, stratigraphies show the inclusion of significant percentages of coarse silts – sands in the sediments, some sites recording distinct sandy units present. This widely recognisable sandy stratigraphic unit and boundary change is dated to c. 2,900–3,000 years BP between sites and often shows channelization and erosion of the underlying organics. After c. 3,000 years BP there is a marked alteration in the sediments. The earlier dominant peats are replaced by heterogeneous clay-silts and biogenics, formed in intertidal to high marsh settings. Full discussion of the site stratigraphies and palaeoenvironmental reconstruction is given in Delaney et al. (2012).

The initial fresh – brackish water environments, represented by the basal biogenics, are replaced in most areas at c. 3,000 years BP by a brackish-marine environmental signal in the progressively minerogenic rich sediments. Evidence of an earlier strong brackish water – marine signal is common at many sites, however, from the central – western end of Castlemaine Harbour. In the outer harbour area of the Caragh River, close to Rossbehy (Fig. 9.1), a strong marine influence is detectable in these basal mid-Holocene sediments, dating from at least

[1] "Radiocarbon dates are given as calibrated ^{14}C ages (Delaney et al. 2012)".

4,700 years BP. These marine influenced palaeoenvironmental signals may have resulted from frequent marine overwash events of the seaward protecting sand-barriers in Dingle Bay (i.e., proto Inch and Rossbehy), or of temporary breakdowns in these barriers under storm conditions and RSL rise. Similar situations in Ireland have been recorded in the late Holocene to the present day at other coastal barrier sites, e.g., Tacumshin (Orford and Carter 1984), Ballycotton Bay (Carter et al. 1989) and Greystones and Wexford Harbour, eastern Ireland (Sinnott 1999).

9.4.2.1 Rossbehy Early Barrier Evidence and Coastal Morphodynamics

The dating to c. 6,100 years BP (6,275–5,950 BP, UB-3813) at −1.22 m ODB of organic clay-silts beneath the modern beach at Rossbehy indicates that blocking beach-barrier structures were seaward then of their present separated positions, providing sheltered and brackish-fresh water dominated back-barrier environments (Fig. 9.8a). This view is supported by the presence of younger, thin interbedded peat and dune sands found beneath the beach at the northern end of the present boulder (coarse clastic) barrier. The upper and lower contacts of this peat at +0.94–0.97 m and +0.0.89–0.92 m ODB gave ages of 508–470 BP (UB-3811) and 728–685 (UB-3812), respectively. The sediments are comparable to the stratigraphies also recorded behind the present dune-barrier and probably represent a former dune slack. They show the marine erosion of the back-dune edge in the last c. 400 years, with the coarse clastic- and dune-barrier moving progressively on-land: an interpretation further supported by historical evidence for barrier *roll over*.

Behind Rossbehy biogenic deposition has continued landward of the salt marsh margins almost to the present day. In the more seaward areas, the presence of channelised sands and silts in cores shows that after c. 2,873 BP (UB 3815) there was an abrupt change to high energy, marine conditions with tidal scouring of the marsh – intertidal areas (Fig. 9.8b, c). The ‘main palaeo-barrier breaching’ phase probably occurred at this time, possibly in the area of the pre-2008 tidal channel, allowing marine inundation and the rapid deposition of intertidal sediments. In coring transects behind the dune-barrier a basal sequence of freshwater organics (peat) was found underlying the late Holocene intertidal sediments in the sample core (Fig. 9.8a). This peat was not identified elsewhere at Rossbehy and may represent an isolated, allochthonous balk of sediment. The peat may have been rafted into position following storm-linked erosion of the neighbouring marsh edge and then compressed by the subsequent and possibly rapid deposition of over 2 m of silts and sands. The upper eroded surface of the peat unit, and ^{14}C dating to c. 2,900 years BP, shows that these sediments could also have been associated with the main barrier breaching. Alternatively, it is possible that this organic unit might be the remnant of an in situ stratigraphy; indicative of freshwater dominated sediments formed within a local, glacigenic basinal topography that provided protection from RSL rise (Delaney et al. 2012).

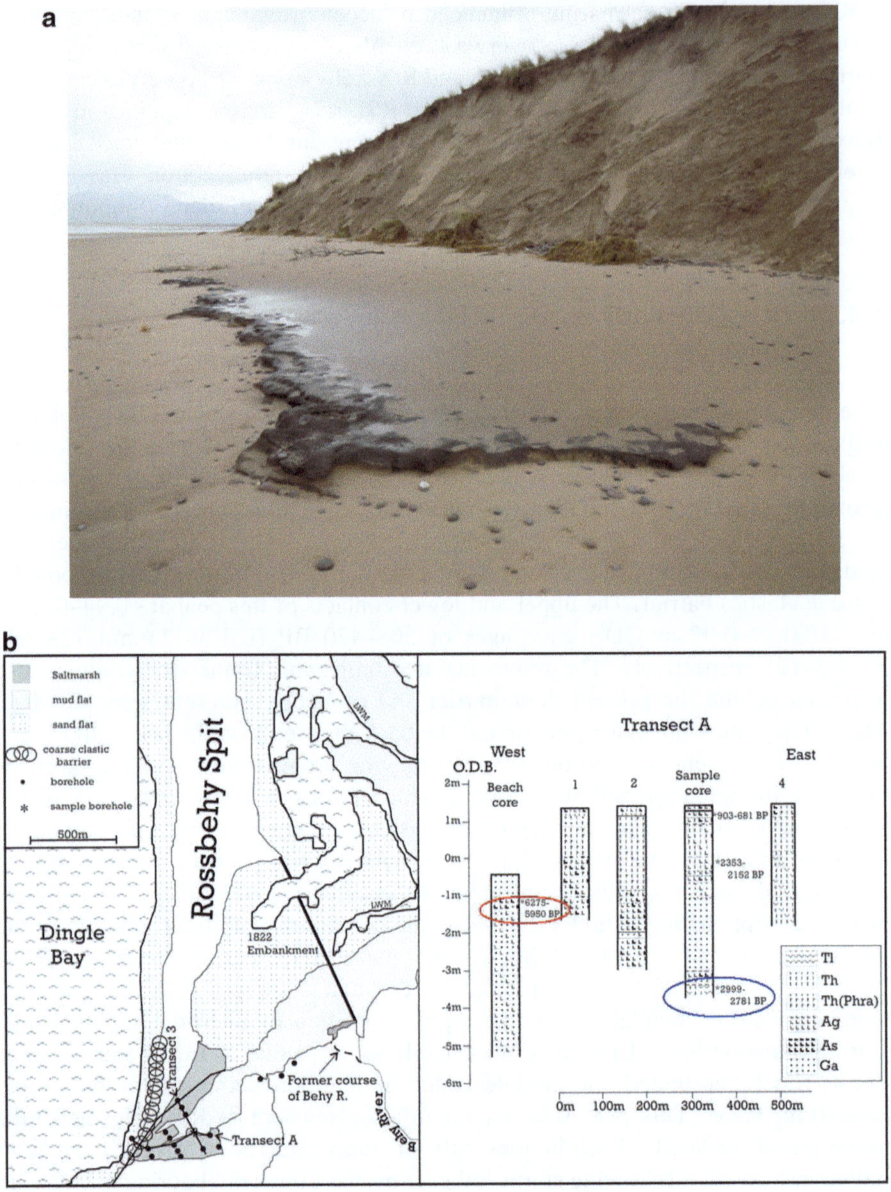

Fig. 9.8 (**a**) Clay-silt-rich biogenics (peat) on foreshore at start of the Rossbehy dune-barrier, ^{14}C dated to c. 6,100 years BP; (**b**) Rossbehy Spit, the main across-barrier core Transect (Delaney et al. 2012), showing beach to back-barrier stratigraphy. The foreshore organics shown in (**a**) are *circled* in core 1 and the possible storm rafted peat 'moorlog' *circled* in core 4; (**c**) Palaeoenvironmental changes at Rossbehy from Transect A (**b**), showing summary relative percentage frequencies of diatom habitat – life-form assemblages (Delaney et al. 2012)

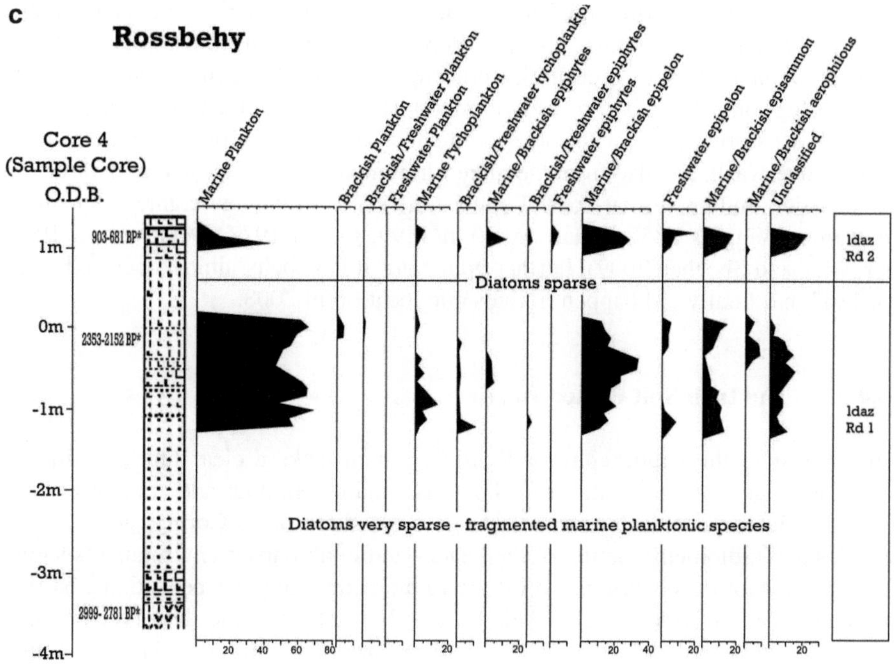

Fig. 9.8 (continued)

The main barrier breakdown at c. 3,000 years BP was most likely associated with a major storm, or cluster of storm events (Jennings et al. 1998; Orford et al. 1999; Stone and Orford 2004; Clarke and Rendell 2006), acting against a slowing rate of RSL rise, possibly as low as <0.75 mm/year in this region (Carter et al. 1989; Delaney et al. 2012). This coastal change in the outer harbour was on a scale large enough to have impacted rapidly the entire Castlemaine Harbour area. The timing of the barrier breaching corresponds approximately with a 'cycle' of increased wetness, storminess and on-shore coastal sediment movements recorded on European Atlantic coasts and in northern Europe (Lamb 1995; Mackay et al. 2003; Devoy et al. 1996; Clarke and Rendell 2006; Gilbertson et al. 1999; MacClenahan 1997; Dawson et al. 2004; Ritchie and Angus 2012). A case for wider regional to hemispherical scale teleconnection drivers to such events, such as a reduction in solar activity linking to NAO and ENSO type oscillations, sea surface temperature (SST) and Arctic sea ice changes has been made, but remains speculative (e.g., Pestiaux et al. 1988; Finkl 1995; van Geel et al. 1996; Campbell et al. 1998; Bond et al. 2001; Dawson et al. 2003, 2004; Allen 2005; Berner et al. 2008; Sabatier et al. 2012; Gomez et al. 2012).

Subsequently, the transition in intertidal areas at Rossbehy into mixed biogenic and fine-sized minerogenic sedimentation, evidences a later re-formation of the dune-barrier post c. 2,800 years BP. Sheltered conditions were re-established, allowing salt marsh formation. The complex interleaving of biogenic, clay-silt and sandy sediments in this final phase of deposition, following the coastal systems' *tipping point* of barrier breaching, indicates probable frequent changes in dune morphology and back-barrier environments (Fig. 9.8b). These resulted possibly from direct storm action, with some sedimentary evidence for further breaching and barrier overwash, but also from the increasing impacts of people and agriculture on this barrier, and on similar coastal environments, from the late eighteenth century onwards (Wiggins 1852; Allanson-Winn 1899; Quinn 1977; Devoy et al. 1995; Crowley and Sheehan 2009). Further breaching at Rossbehy almost occurred again in 1842 and finally did happen in the same location in 2008.

9.4.2.2 The Inch Spit Palaeo-Barrier

In contrast to the stratigraphy of Rossbehy, Inch lacks a clear linkage with the wetland palaeoenvironmental record of Castlemaine Harbour and appears to have operated in the late Holocene as a large scale dune field. Coring, geotechnical studies and radiometric dating of the beach- and dune-barrier environments shows no evidence for the occurrence of extensive biogenic sediments beneath the barrier, or interleaved in the massively bedded dunes. Evidence of dune slack type organic-rich sediments (peats) is also absent. Only minor, discontinuous thin and poorly humified vegetation units and proto-soil horizons occur within the barrier (Table 9.1) (Cooper et al. 1995; Wintle et al. 1998; Science Direct 2013).

The dune-barrier, as at Rossbehy, has developed over the position of a glacial recessional end-moraine. This landform, and the wider occurrence of glacigenics (fine sized – coarse clastics) in the region, have been an important immediate source of sediments into the beach – intertidal environments in Dingle Bay (Devoy et al. 1995; Shaw et al. 1994; Vial 2008; Sala 2010). The end moraine at Inch has formed a stable base for the subsequent development of the dunes and provides a transient location and 'stage' in the progress of the dune-barrier sands from the sea toward the land. Barrier movement here was again probably in the form of coastal sediment *roll over* (e.g., Devoy 2009; Delaney et al. 2012). The immediate surface of the glacigenic sediments appears to have been initially wave-eroded, under mid- to late Holocene rising sea-levels. This resulted in the development of marine re-worked and wave refracted gravel – cobble sized sedimentary ridges, with the inclusion of broken marine shell material identified in some locations. These ridge surfaces are now overlain by the dunes, but are often exposed in the swales of the older, stabilised foredunes on the western side of the dune-barrier. They have also been identified beneath the parabolic dunes at the southern end of Inch. Radiocarbon dating of included oyster shells gives a minimum age of c. 1,100 years BP, a *terminus ante quem* for the building of the ridges in this area of the barrier (Wintle et al. 1998). These marine re-worked sediments may exist under much of Inch, but this has not been proven by ground-work to date (Devoy et al. 1995; Orford et al. 1999). Depths of sands overlying the glacigenic and cobble ridge surfaces appear to be >5 m in the western beach areas, though sands thickness varies seasonally. Eastwards, into the main dunes, the height of this contact is unknown. Sand depths, below the basal low-ground elevations within the dunes are likely to be irregular, but relatively shallow and are estimated to be between 3 and 10 m thick.

Some sand dune cover may have begun to develop on Inch prior to the construction of the main coastal barrier system, both here and at Rossbehy. Though no clear dating for this phase exists, it is inferred from the palaeoenvironmental data from Castlemaine that it would have occurred before c. 3,000 BP and the timing of the main barrier breakdown. Dating of the dune sediments along the N – S length of the barrier has been undertaken using Optically Stimulated Luminescence (OSL) techniques (Wintle et al. 1998). These show dates from the main dune ridge surface sediments, the parabolic dune corridors and a short beach core section, from heights of c. −3 m – c. 18 m ODB, of no older than c. 600 years BP. The oldest dates (c. 500–600 years old) were found from the currently 'stabilised' northern end and from the deepest sediments sampled. The data indicate the occurrence of substantial sand movements within the dunes during this recent timeframe, but with the dates showing a conformable vertical surface height sequence of dune ages in the range c. 368–151 years old. A second 'cluster' of dates from sands within three parabolic dune corridors gave an age of c. 150 years old. These may evidence a later distinct phase of sand mobilisation, possibly from a major storm event (e.g., the Big Wind 1839), or also from the impact of people living on the dunes at this time of the Great Famine and in subsequent famine periods. Radiocarbon dating of shell middens exposed at the back edge of a parabolic dune gave a calibrated age of 340–500 years BP (UB 3970), possibly relating to the Famine period, or times of older human use of the dune-barrier. Sand movements were probably a combination of both factors. The *Big Wind* of 1839 is certainly recorded as having a major impact on Inch, with blowing sand in the storm driving cattle on the dunes to distraction and forcing them to take to the water and swim across to Rossbehy (Shields and Fitzgerald 1989; Carr 1993; Devoy et al. 1995).

This OSL and linked ^{14}C dating survey, indicates the likely occurrence of a number of phases of dune erosion and sand re-deposition on Inch since the building of the cobble beach ridges at c. 1,100 years BP, or earlier. The dating, though limited in extent and giving no direct age for the basal dune sands, does suggest that repeated sand movements have operated as the key driving mechanism for dune morphological changes and regeneration over the last c. 600 years and probably longer (Wintle et al. 1998; Orford et al. 1999), with the present dune surfaces as the most recent expression of this process. Documentary records do also show though the occurrence of a possible storm breaching event in the north of Inch during the eighteenth century, approximately in the area of its junction with Dingle and the intertidal marsh embayment at the rear of the Spit (Fig. 9.7). The breach was 'healed' by subsequent sand movements. No other evidence for direct breaching of this barrier has been found so far (Smith 1756; Devoy et al. 1995; Delaney et al. 2012, *pers. comm.*).

9.5 Contemporary Spit-Barrier Functioning

Past studies of the dunes-barriers and the other coastal systems in Castlemaine Harbour (Vial 2008; Sala 2010; Duffy and Devoy 1999; Delaney et al. 2012) show these environments are impacted not only by Atlantic wave and tidal action, but also to the influence of strong local wind turbulence and the generation of local wind waves over short fetch distances. Together, these controls have caused significant erosion of the barrier shorelines and of the marshes in the area. Under future climate warming, changes in these controls are likely to happen (Sweeney et al., 2002, 2008; Dunne et al. 2008; Devoy 2008; Dwyer 2012; IPCC 2007, 2014a), but their effects on the dune-barriers and on wider shoreline erosion is unknown. Projections of future bathymetric, RSL, storminess, wave refraction, sediments supply and linked controls from climate warming, and the changes now occurring at Rossbehy, have yet to be completed (O'Shea and Murphy 2013; Kandrot 2014), or even commissioned (O'Dwyer and Gault 2014). However, any increases in wind speeds above critical storm size threshold speeds of c. >17 m/s are likely to create barrier sand blows, dependent on the degree of vegetation cover (Carter et al. 1992; Lozano et al. 2004; Sherman et al. 1994; Houser et al. 2008; Houser and Hamilton 2009; Hickey 2011).

Pressures from recreation and tourism activities are currently minimal in the main dune-barrier areas of Inch, though less so at Rossbehy. People impacts, however, from seasonal and daily recreation on the western beach-front and foredunes of Inch are heavy, especially in the summer months. Much of this activity does not penetrate far into the main dunes (maximum distance of <200–300 m), except through the impacts from beach- dune-buggy vehicles and some walking and horse riding. Access by vehicles and horse-riding into the dunes has grown extensively in recent years (post c. 2,000), particularly in the central sections of Inch, with significant erosional and negative consequences on the vegetation and dune surfaces. Means to limit general public access into the dunes, particularly the prohibition of vehicle and horse riding access is now essential.

How the supply of offshore cobble – sands sized sediments and shell material to these dune sands will alter under the impacts of climate warming is also unknown. It is likely that the warming of coastal waters on this North Atlantic margin will increase local shell productivity, which will have knock-on effects in the sediments supplied from the beach into the dunes (IPCC 2007, 2014a). But, much will depend on whether any of this material is moved actively into the dunes in the future. Its immediate importance will be in affecting the nature of plant communities and the linked ecology on the dunes.

9.5.1 *The Inch Barrier*

The dune sediments, nourished from the western shore-face, appear to be continuing to move W – E across the southern 'active' areas of Inch (Figs. 9.3 and 9.9), with drift-aligned development of spit and linked beach sediments at the distal end of the barrier (O'Shea and Murphy 2013; Kandrot 2014). Archival record (maps, photographs and historical information), with contemporary monitoring and process studies of the dune-barrier system (including the use of satellite, LIDAR and TLS imagery), together evidence three major phases of progradation and recession of the last 3 km of the western shore face areas (as upper beach and foredune ridges) since 1841 (Orford et al. 1997, 1999; O'Shea et al. 2011; O'Shea and Murphy 2013).

a

Fig. 9.9 (**a**) Inch Spit and Rossbehy Spit shoreline movements from OSI map, aerial photograph and field survey data since 1842 (Source: O'Shea and Murphy 2013). More detailed survey data for Inch pre-2000 are given in Orford et al. 1997; (**b**) Conceptual-behavioural model of Inch functioning (Source: Orford et al. 1997)

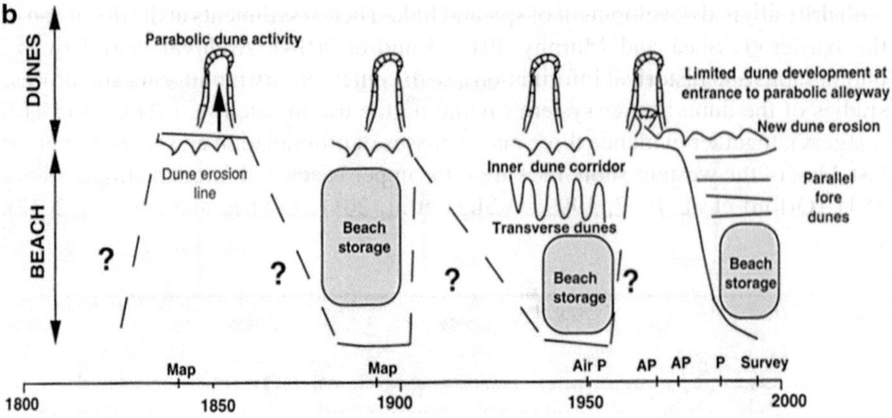

Fig. 9.9 (continued)

These beach and foredune sediment accumulation – retreat 'cycles' vary in size and significance N – S along the dune-barrier and appear to be out of phase with movements recorded at Rossbehy (Fig. 9.9a, b). Studies on Inch show horizontal distance changes in shoreline position of maximally c. 300 m, with sediment volume movements of c. 18,000 m^3/year during periods of progradation (Cooper et al. 1995). Times of shoreline retreat toward the inner parabolic dune edge within the barrier are identified as beginning at c. 1840, c. 1900 and c. 1960. Beach and incipient foredune recovery occurs subsequently at decadal timescales. Since the early 1900s, sediment accumulation and dune growth have been dominant, though shorter term oscillations along the barrier do occur and comparison of the 2000 and 2006 shoreline data shows distinct distal end erosion (Vial 2008; O'Shea and Murphy 2013; Kandrot 2014). From these accumulation areas, sediments move subsequently into the main dune field under regular wind and storm-event actions, are 'processed' through the parabolic dunes and probably re-enter the marine environment on the Castlemaine Harbour side of Inch. A proportion of these sediments then return to Dingle Bay via the outlet ebb-tidal delta, which serves as a short term (decadal scale) *sink* and *source* area . From here they are again circulated northwards alongshore by wave, current and tidal actions (Orford et al. 1997, 1999; Vial 2008).

Examination of past wind and tide data suggest that large storm events (with wind speeds >30 m/s) may have triggered these mesoscale erosional phases, operating on a quasi-60 year cycle (Orford et al. 1997, 1999). However, analysis of storm surge magnitudes on Atlantic coasts since c. 1915 show no increased significance for potential maximal storminess (*storm surge potential*) and wave heights coupled with these events (Orford et al. 1992; Stone and Orford 2004). It is likely that these large storms are only partly the cause of the beach-barrier movements over the last 150–200 years. Orford et al. (1997, 1999) suggest that the reworking and build up at decadal scales of sediments in the ebb-tidal delta and a linked steepening of the Inch shore face under longer term wave action may be more critically involved. Development of an increased shore face gradient and consequent nearshore water depths would allow greater storm impacts and in which the larger events would become more effective in causing system changes. This largely conceptual processes model, represents the beach and dunes at Inch as effectively a semi-closed and self-nourishing barrier, in which mesoscale storm events initiate the transitive crossing of the systems' threshold and resilience factors, leading at present to repeating cycles of barrier face erosion and recovery. Elements of this sedimentary processes model (Fig. 9.9c) still require validation, through sediment monitoring and process-morphodynamic studies. The dramatic changes occurring in the shape, extent and sediment volume storages of the ebb-tidal delta system may well now alter radically the former apparent 'cyclical' operation of Inch. Under future climate warming projections, the impacts of RSL rise and North Atlantic storm magnitudes (IPCC 2014a, b), coupled with the repercussions of the 2008 barrier breaching at Rossbehy, may lead to much greater erosion of Inch and breakdown of the main parabolic dunes.

9.5.2 The Rossbehy Barrier and the Ebb-Tidal Delta

This spit-barrier, represented by a single main dune ridge, is far less stable than Inch, though it too has been effectively fixed in its location in the late Holocene on the underlying glacial end-moraine (Devoy 2009). Historical and recent observations of the barrier, and of the adjacent ebb-tidal delta sedimentary system, show that this barrier has also undergone a series of distinct shoreline position and morphological changes since 1841 (Fig. 9.9b) (O'Shea et al. 2011; O'Shea and Murphy 2013). Movements, however, appear to be on a non-cyclical, infrequent and stochastic basis. Mapping, air photographic data with present field studies show that there is little indication of direct sediment transfers between Rossbehy and Inch, though the linking role of the ebb-tidal banks is probably critical in the sedimentary functioning of Inch (Orford et al. 1997, 1999; O'Shea and Murphy 2013). Study of these data show that the 1,841 shape of Rossbehy was similar to its 2000-pre 2008 form, with a clear distal, thin end-neck, but that it was positioned seaward of its later location. Subsequent surveys (in 1894 and 1977), show barrier build-up and movement eastwards and on-land in the intervening 83 year period. Little change in overall barrier shape seems to have occurred though, in spite of the attempts at inning the back-barrier marshes (Delaney et al. 2012). There is indication of some erosion of back-barrier areas, but apart from some occasional photographs, no accurate data on barrier movements in this period exist. Since 1977, erosion of the front edge of the dune-barrier has occurred and has increased in the period 2000–2006, together with significant sediment losses in the ebb-tidal delta. The breaching of the distal end of Rossbehy in December 2008 (Kerry County Council 2009) in the thinned barrier area of 1841 (Fig. 9.6), was not caused by an exceptional storm surge event (Met. Éireann 2014). It resulted more from the combination of, the occurrence of full westerly (c. 240–270°) onshore wind-wave conditions, barrier orientation and the timing of tidal height – wave interaction with ebb-delta bar shape and exposure (O'Shea et al. 2011; O'Shea and Murphy 2013).

Contemporary morphodynamic process functioning of the dune-barrier has been recognised as being organised into two main components, linked by a 'hinge' zone of variable width and position over time (Sala 2010; O'Shea and Murphy 2013) (see Fig. 9.4). A northern end current drift aligned system is linked and driven by the operation of the ebb-tidal delta and main tidal channel into Castlemaine Harbour. Post c. 2000 this is expanding in influence along the remaining length of the barrier (Fig. 9.3). This system changes southwards through the 'hinge-line', into the formerly dominant swash aligned zone. Sedimentary pattern changes in this zone, as dune-shoreline erosion and progradation, have been largely balanced alongshore by regular seasonally driven onshore to offshore sedimentary movements. Map, photographic and field data show that the growth now of the drift aligned system results probably from the changes in sediment flux and higher current speeds (>0.9 m/s) within the ebb-tidal delta and main channel, most recently since c. 2000. Growth of the ebb bars pre by c. 2000 led to sediment starvation of the nearshore zone along the barrier front, with consequent deepening water and increased effective wave erosion of the dunes and loss of this sediment offshore, possibly to the ebb-bars. This macro-scale cannibalisation of the barrier probably facilitated the effectiveness of the 2008 storm event, and added further to subsequent sediment loss from the barrier with southwards propagation of the breach area since 2009.

A notable effect of the breaching has been a change in sediment size distribution and patterning on this now drift aligned end of Rossbehy. Extensive gravel-cobble sized sediment bars, ridges and berm have developed in the expanding breach area. Transport of these coarse clastics appears to be alongshore from the south of the barrier, though these materials may also be partly sourced from re-working of the glacigenics beneath the barrier and exposed in the breach (Kandrot et al. 2014). Post 2009 erosion of the remaining length of the dune-barrier core adjacent to the breach and southwards has been fast, at rates of c. 30–50 m/m width of dune/year and continues to accelerate. Since June 2012 precise Terrestrial Laser Scanning (TLS) measurements show a further 90 m of shoreline and dune loss, with >40 m of this recession occurring in the storm events of December 2013 to January 2014. The breach zone has more than doubled in width since 2010, from c. 650 to 1400 m (Kandrot 2014) (Fig. 9.10). The area has joined with the former main tidal inlet channel, where tidal current speeds remain high in spite of the increased bed roughness created by the areas of coarse gravel – cobble lags and bars.

Analysis of ongoing barrier sediment movements indicates that they are probably linked closely with the observed sequential morphodynamic changes in the ebb-tidal bars and tidal inlet channel (O'Shea and Murphy 2013). The complex of large sediment ebb bars and associated "S" curved channel system seen in 2000 (Fig. 9.3) is coupled to extensive dissipative beach development and dune accretion: this ebb-tidal system shape re-enforcing wider-scale sediment deposition. Subsequent straightening of the tidal channel and loss of ebb bars between 2004 and 2007 led

a

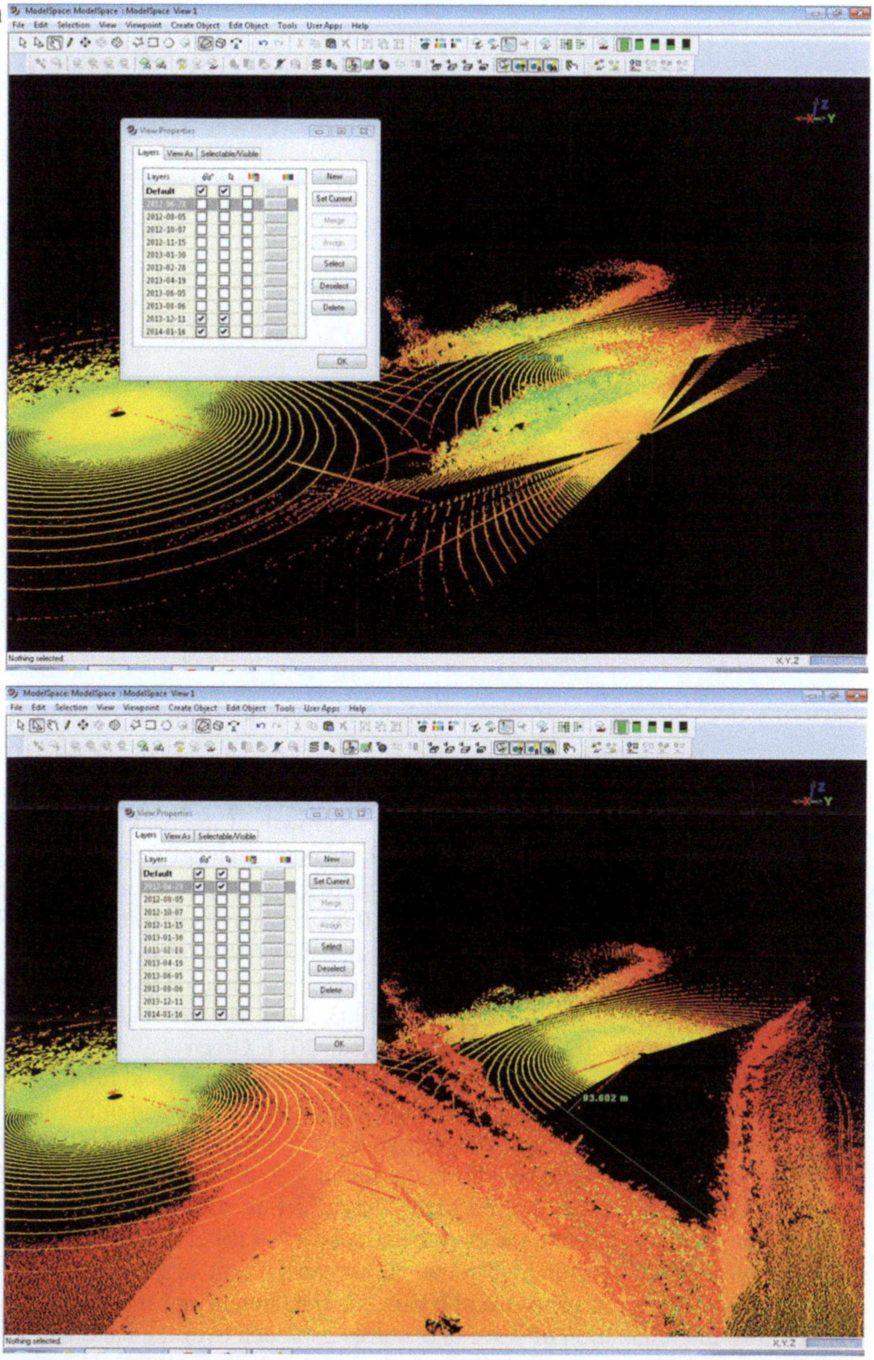

Fig. 9.10 (**a**) Screenshot graphics of TLS measurements of the rate of erosion and breach widening at Rossbehy 2012–2014 (Kandrot 2014); (**b**) Shoreline and barrier-end recession of the Rossbehy Spit from archive records and TLS measurements (Source: Kandrot 2014)

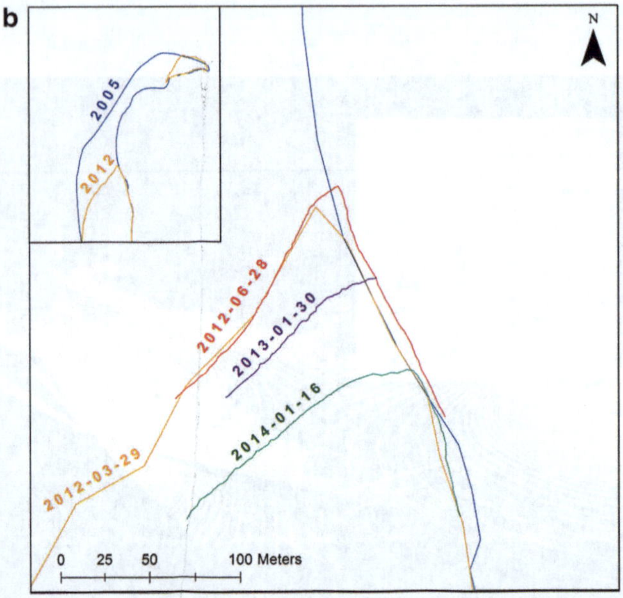

Fig. 9.10 (continued)

to the reverse effect, with increased tidal current speeds and widespread erosion of the different sedimentary components within the system complex. Aerial photographs of the delta post 2010 (Fig. 9.3) show re-growth of the back-barrier channels, outer ebb bars and a return to a curvilinear main channel structure, though erosion of the former distal-end island of Rossbehy and outer channel area has continued. Sediment deposition now looks as though it may be returning to a self-balancing, and by analogy with similar spit- and dune-barrier systems the Rossbehy barrier may now stabilise and re-grow, though this view remains speculative (e.g., Michael and Howa 1997; Vila-Concejo et al. 2003, 2004; Houser et al. 2008; Giese et al. 2009; Cooper et al. 2007; Cooper and Jackson 2011; Green et al. 2013; O'Shea 2015; Kandrot 2015). In spite of this, erosion rates in the system and tidal channel inlet speeds continue to remain high. Present monitoring of wave and tidal currents, coupled with wave refraction and other large scale morphodynamic modelling of the Dingle and Castlemaine coasts shows the situation to be complex. The ebb-tidal delta system with the barriers doesn't fit any single setting elsewhere. The question of what is the primary driver to the growth and 'decay' of the ebb-tidal delta and its sedimentary connection with Inch remains, for the present largely unanswered. Clearly, the extent and height of the ebb bars, the tidal height and seasonal regime, coupled with the orientation of wave and storms approach over the bars, will have continued significant effects on the functioning of Rossbehy and the wider coastal system. What will happen to the barrier under climate warming remains unknown.

Consideration of increased RSL (worst case IPCC projection of >0.9 m), greater frequency of large magnitude storms (as experienced in December 2013 – January and February 2014) and a likely continued reduction in new onshore and land-derived sediment inputs, should be built into the numerical modelling of this coast. As stated at the beginning of the chapter, these Spits (dune-barriers) are unique, as are all coasts at real operational scales. They don't fit any single, other contemporary spit- dune-barrier model (Fitzgerald et al. 2000; Ranasinghe et al. 2013).

9.6 The Future of the Spit-Barriers and Conclusions

Coastal morphodynamics studies since the 1980s show that single, dominant controls (e.g., RSL, sediment supply, storm surges) are progressively less likely to have operated in isolation by the late Holocene in driving coastal systems' functioning. Decreasing RSL rise, coupled with increased variations in sediment supply sources and rates, result in the progressive operation of a complex of coastal process control factors and system feedbacks. These work at the site level to create unique coastal geomorphological units, functioning within a wider regional scale framework of controls (e.g., Carter and Woodroffe 1994; Devoy 1982, 1987; O'Shea and Murphy 2013; Jennings et al. 1998; Klein et al. 2006; Swift et al. 2006; Cooper et al. 2007; da Silva 2009; Houser 2009). Development of the Inch and Rossbehy dune-barriers within the compartmentalised sedimentary coast of Dingle Bay, provide a clear demonstration of this process of coastal 'evolution'. Onshore sediment movements in Dingle Bay during the Holocene resulted in the fixing of these structures on former end-moraines. A third end-moraine behind them at Cromane is covered by a thin veneer of modern beach sediments, with thicker sedimentary organic sequences behind. This barrier forms more a part the low energy sandy mudflat and wetland systems of Castlemaine Harbour, which developed in the fresh-brackish water environments maintained by the two fronting and morphodynamically active dune-barriers (Delaney et al. 2012).

In the mid- Holocene and earlier, when RSL was regionally below c. −10 m ODB, these barriers probably formed a single structure across the bay, moving progressively on-land under sea-level rise (SLR) (Devoy 2009). By c. 3,000 years BP large scale breaching of this main barrier had occurred, with the Inch and Rossbehy structures forming in their present position, or soon after. This coastal change in the outer harbour is shown through palaeoenvironmental reconstructions to have been rapid and on a scale large enough to have impacted the entire harbour area. However, whether a particular storm event was the cause, or a longer period of increased storminess, coupled with reducing sediment supply and development of the organic-rich wetland environments of the inner harbour and falling rates of SLR, is uncertain (Delaney et al. 2012). The timing of the barrier breaching corresponds approximately with a 'cycle' of increased wetness, storminess and on-shore coastal sediment movements recorded on European Atlantic coasts and in northern Europe (Dawson et al. 2003, 2004; Clarke and Rendell 2006; Gilbertson et al. 1999). A case for wider regional to hemispherical scale teleconnection drivers to such events, such as a reduction in solar activity linking to NAO and ENSO type oscillations, Sea Surface Temperature (SST) and Arctic sea ice changes has been made, but remains speculative (Pestiaux et al. 1988; Finkl 1995; van Geel et al. 1996; Campbell et al. 1998; Bond et al. 2001; Dawson et al. 2003, 2004; Allen 2005; Gomez et al. 2012).

In the future at Rossbehy, the barrier breaching that occurred in 2008 and has continued in the now rapid erosion of the dune-barrier (c. 1,400 m by 2014), may result in its complete disintegration if left to itself and without human intervention in providing protection measures. Such intervention is likely to be a double edged sword though, in terms of wider scale coastal systems' repercussions in Dingle Bay and Castlemaine Harbour (Cooper et al. 2009; Cooper and Pilkey 2007b). The erosion at Rossbehy is also likely to alter the recorded quasi-periodicity in storm driven control of sediment movements in the active barrier and shoreline areas of Inch. It presages the possible development of similar morphodynamic changes throughout Inch, though this barrier is much larger than Rossbehy and with probably greater in-built resilience to change. Forcing factors will result particularly from the now radical alteration in the geometry and behaviour of the ebb-tidal delta between the two barriers, changed seabed bathymetry, in tidal forcing and elevations, wave set up and consequent storm surge penetration into Castlemaine Harbour. These changes, coupled with local wind-wave climate factors, are likely to expose progressively the eastern side of Inch to more extreme wave action. But this is again, speculative without more detailed morphodynamic modelling process-based study (O'Shea 2015; Kandrot 2015).

Additionally, and independent of the present barrier changes, then with the onset of climate warming as currently developing (IPCC 2007, 2014a); these dune-barriers will be forced to adjust morphodynamically to the altered local – regional climate-linked boundary controls (e.g., SLR, increased frequency of large magnitude waves – storm surges, wind regime, precipitation and sediment supply). This is estimated as very likely within the next 20–50 years. Modelling studies of other inlet-interrupted coasts emphasise the significance of climate warming in future (Ranasinghe et al. 2013), with SLR seen as accounting for 20–50 % of modelled coastal changes, but with SLR-induced removal of sediment accommodation spaces and precipitation with river discharges as also being critical to the re-organisation of coastal systems. For Inch, such changes may result in the re-activation of the now stable northern part of the dune-barrier areas, with probable increased erosion of the southern active areas in the next decade. At present though, little change of this southern shoreline is in evidence (Kandrot 2014). It is also possible that breaching of the Inch barrier will now occur in the next decades, following future climate warming induced changes in the coastal climate and physical system boundary controls, but at present this is again speculative. In the short term, much will depend on the immediate movements and re-organisation of sands in the Rossbehy-Inch ebb-tidal delta (O'Shea et al. 2011; O'Shea and Murphy 2013). A further important factor in these future barrier changes may be the continuing degree of destructive human intervention in these systems, e.g., uncontrolled tourism-recreational uses, agriculture and sand mining, leading to breaks in the vegetation cover and/or direct erosion of the dunes (Science Direct 2013).

In the long term (i.e., twenty-first century), based upon the Holocene pattern of dune-barrier *roll over* under SLR, and consequent upon the impacts of future climate warming and the outcome of the morphodynamic changes occurring at Rossbehy, then the present dune-barrier sediments are likely to move increasingly landwards under *coastal squeeze*. Linked to this will be significant losses of the seaward edges of the fringing wetlands in Castlemaine or, worst case scenario, their complete removal given their low height in relation to Spring Tides and susceptibility to erosion (Devoy and Duffy 1999; Delaney et al. 2012). The barrier sediments may be part dispersed around Castlemaine Harbour, possibly into smaller new 'true' coastal spit structures. Some sediment losses seawards into Dingle Bay is also likely, into storage in offshore bars. Alternatively, systems' inertia resulting from the inheritance factors of the existing barrier dimensions and sediment volumes, the glacigenic core ground heights, and of morphodynamic feedbacks in sedimentary adjustments to rising sea-levels, will help keep elements of these barriers where they are, but changed radically in morphology and size.

References

Aaraard T, Davidson-Arnott R, Greenwood B, Nielsen J (2004) Sediment supply from shoreface to dunes: linking sediment transport measurements and long-term morphological evolution. Geomorphology 60:205–224

Allanson-Winn RG (1899) Foreshore protection, with special reference to the case system of groyning. Society of Engineers, London

Allen JRL (2005) Teleconnections and their archaeological implications, Severn estuary levels and the wider region: the 'fourth' and other mid-Holocene peats. Archaeol Severn Estuary 16:17–65

Backstrom J, Cooper JAG, Jackson DWT (2007) Shoreface dynamics on the high energy coast of Northern Ireland. J Coastal Res Spec Issue 50:594–598

Belknap DF, Kraft JC (1977) Holocene relative sea-level changes and coastal stratigraphic units on the northwest flank of the Baltimore Canyon trough geosyncline. J Sed Pet 47:610–629

Berner KS, Koc N, Divine D, Godtliebsen F, Moros M (2008) A decadal-scale Holocene sea surface temperature record from the sub-polar North Atlantic constructed using diatoms and statistics and its relation to other climate parameters. Paleoceanography 23:1–15

Bond G, Kromer B, Beer J, Muscheler R, Evans MN, Showers W, Hoffmann S, Lotti-Bond R, Hajdas I, Bonani G (2001) Persistent solar influence on North Atlantic climate during the Holocene. Science 294:2130–2136

Brooks AJ, Bradley SL, Edwards RJ, Milne GA, Horton B, Shennan I (2008) Postglacial relative sea-level observations from Ireland and their role in glacial rebound modelling. J Quat Sci 23:175–192

Campbell ID, Campbell C, Apps MJ, Rutter NW, Bush AGB (1998) Late Holocene 1500 year periodicities and their implications. Geology 26:471–473

Carr P (1993) The night of the big wind. White Row Press, Belfast

Carter RWG (1988) Coastal environments. Academic, London

Carter RWG, Orford JD (1993) The morphodynamics of coarse clastic beaches and barriers: a short- and long-term perspective. J Coast Res Spec Issue 15:158–179

Carter RWG, Woodroffe CD (1994) Coastal evolution: late quaternary shoreline morphodynamics. Cambridge University Press, Cambridge

Carter RWG, Devoy RJN, Shaw J (1989) Late Holocene sea level in Ireland. J Quat Sci 4:7–24
Carter RWG, Curtis TGF, Sheehy-Skeffington MJ (eds) (1992) Coastal dunes: geomorphology, ecology and management for conservation. In: Proceedings of the 3rd European Dune Congress, Galway, Ireland. AA Balkema, Rotterdam
Clarke ML, Rendell HM (2006) The effects of storminess, sand supply and the North Atlantic Oscillation on sand invasion and coastal dune accretion in western Portugal'. Holocene 16:341–355
Clifford NJ, French JR, Hardisty J (1993) Turbulence: perspectives on flow and sediment transport. Wiley, New York
CoastAdapt (2013) CoastAdapt: the sea as our neighbour. European Union, Northern Periphery Programme (2007–2013). http://coastadapt.org
Conscience (2010) Concepts and science for coastal erosion management. http://www.conscience-eu.net/inch_beach/index.htm
Cooper JAG (2007) Temperate coasts. In: Perry C, and Taylor K (eds) Environmental sedimentology. Blackwell, Oxford, pp 263–301
Cooper JAG, Boyd S (2010) Climate change and coastal tourism in Ireland. In: Disappearing destinations: climate change and the future challenges for coastal tourism. CAB International, pp 125–143
Cooper JAG, Cummins V (eds) (2009) Coastal management in Northwest Europe. Elsevier, Amsterdam, 869 pp
Cooper JAG, Jackson DWT (2011) Geomorphology of a high energy barrier island on the rocky west coast of Ireland. J Coast Res Spec Issue 64:6–9
Cooper JAG, Pilkey OH (2007a) Quantification and measurement of longshore sediment transport: an unattainable goal. In: Balson P, Collins M (eds) Coastal and shelf sediment transport. Geological Society of London, London, pp 37–43
Cooper JAG, Pilkey OH (2007b) Rejoinder to Cowell PJ and Thom BG, 2006. Reply to Pilkey OH and Cooper JAG, 2006. Management of uncertainty in predicting climate change impacts on beaches. J Coast Res 22:232–245; J Coast Res 22:1577–1579. In: J Coast Res 23:277–280
Cooper JAG, Pilkey OH (2008) Discussion of Broker et al., 2007. Morphological modelling: a tool for optimisation of coastal structures. J Coastal Res 23:1148–1158, In: J Coastal Res 24: 814–816
Cooper A, Orford JD, McKenna J, Jennings SC, Scott B, Malvarez G (1995) Mesoscale behaviour of Atlantic coastal systems under secular climate and sea-level rise. Final Report, CEC Environment and Climate Programme No. EV5V-CT-0266
Cooper JAG, Hooke JM, Bray MJ (2001) Predicting coastal evolution using a sediment budget approach: a case study from southern England. Ocean Coast Manag 44:711–728
Cooper JAG, Jackson DWT, Navas F, McKenna J, Malvarez G (2004) Identifying storm impacts on an embayed, high energy coastline: examples from western Ireland. Mar Geol 210:261–280
Cooper JAG, McKenna J, Jackson DWT, O'Connor M (2007) Mesoscale coastal behavior related to morphological self-adjustment. Geology 35:187–190
Cooper JAG, Anfuso G, Del Rio L (2009) Bad beach management: European perspectives. Geol Soc Am, Special Paper 460:167–179
Cott GM, Jansen MAK, Chapman DV (2012) Saltmarshes on peat substrate: where blanket bogs encounter the marine environment. J Coast Res 28:700–706
Cowell PJ, Thom BG (1994) Morphodynamics of coastal evolution. In: Carter RWG, Woodroffe CD (eds) Coastal evolution: late quaternary shoreline morphodynamics. Cambridge University Press, Cambridge, pp 33–86
Cowell PJ, Stive MJF, Niedoroda AW, de Vriend HJ, Swift DJP, Kaminsky GM, Capobianco M (2003a) The coastal-tract (part 1): a conceptual approach to aggregated modelling of low-order coastal change. J Coast Res 19:812–827
Cowell PJ, Stive MJF, Niedoroda AW, Swift DJP, de Vriend HJ, Buijsman MC, Nicholls RJ, Roy PS, Kaminsky GM, Cleveringa J, Reed CW, de Boer PL (2003b) The coastal-tract (part 2): applications of aggregated modelling of lower-order coastal change. J Coast Res 19:828–848

Cowell PJ, Thom BG, Jones RA, Everts CH, Everts DS (2006) Management of uncertainty in predicting climate change impacts on beaches. J Coast Res 22:232–245
Cronin K (2010) The value of different modelling approaches to investigate estuarine morphodynamics. PhD thesis, University College Cork, Ireland
Crowley J, Sheehan J (eds) (2009) The Iveragh peninsula: a cultural atlas of the Ring of Kerry. Cork University Press, Cork
Crowley J, Smyth WJ, Murphy M (eds) (2012) Atlas of the great Irish famine 1845–1852. Cork University Press, Cork
da Silva CP (ed) (2009) Proceedings of the 10th international coastal symposium 2009, Lisbon. J Coastal Res Spec Issue 56, 2 vols
Davidson-Arnott R (2010) Introduction to coastal processes geomorphology. Cambridge University Press, Cambridge
Davies JL (1980) Geographical variation in coastal development. Longman, London
Davies GLH, Stephens N (1978) The geomorphology of the British Isles: Ireland. Methuen, London
Dawson AG, Elliott L, Mayewski P, Lockett P, Noone S, Hickey K, Holt T, Wadhams P, Foster IDL (2003) Late Holocene North Atlantic climate "seesaws" and Greenland ice sheet (GISP2) palaeoclimates. Holocene 13:381–392
Dawson AG, Elliott L, Noone S, Hickey K, Holt T, Wadhams P, Foster IDL (2004) Historical storminess and climate 'see-saws' in the North Atlantic region. Mar Geol 210:247–259
de Vriend HJ (1991) Mathematical modelling and large-scale coastal behaviour. J Hydraul Res 29:727–753
de Vriend HJ, Capobianco M, Chesher T, de Swart HE, Latteux B, Stive MJF (1993a) Approaches to long-term modelling of coastal morphology: a review. Coast Eng 21:225–269
de Vriend HJ, Zyserman J, Nicholson J, Roelvink JA, Pechon P, Southgate HN (1993b) Medium-term 2DH coastal area modelling. Coast Eng 21:193–224
Delaney CA, Devoy RJN, Jennings SC (2012) Mid to late Holocene relative sea-level and sedimentary changes in southwest Ireland. In: Duffy PJ, Nolan W (eds) At the anvil. Geography Publications, Dublin, pp 697–746
Devoy RJN (1982) Analysis of the geological evidence for Holocene sea-level movements in southeast England. Proc Geol Assoc 93:65–90
Devoy RJN (1987) Sea surface studies: a global view. Chapman Hall – Croom Helm, London
Devoy RJN (1991) The study of inferred patterns of Holocene sea-level change from Atlantic and other European coastal margins as a means of testing models of earth crustal behaviour. In: Sabadini R, Lambeck K, Boschi E (eds) Glacial isostasy, sea-level and mantle Rheology, NATO ASI Ser C: Math and Phys Sci 334. Kluwer Academic Publishers, London, pp 213–236
Devoy RJN (1992) Questions of coastal protection and the human response to sea-level rise in Ireland and Britain. Irish Geogr 25:1–22
Devoy RJN (1995) Deglaciation, earth crustal behaviour and sea-level changes in the determination of insularity: a perspective from Ireland. In: Preece RC (ed) Island Britain: a quaternary perspective. Geological Society of London, Sp Publ 96:181–208
Devoy RJN (2008) Coastal vulnerability and the implications of sea-level rise for Ireland. J Coast Res 24:325–341
Devoy RJN (2009) Iveragh's coasts and mountains. In: Crowley J, Sheehan J (eds) The Iveragh peninsula: a cultural atlas of the Ring of Kerry. Cork University Press, Cork, pp 33–44
Devoy RJN (2012) Student wind monitoring records at Inch, 2005–2012. Unpublished, Department of Geography, University College Cork
Devoy RJN, Delaney CA, McCall S, Blythe CD (1995) The interaction of natural and artificial controls in influencing the impacts of relative sea-level rise on Atlantic European coasts. In: Devoy RJN (ed) Project IMPACTS: the impacts of climate change and relative sea-level rise on the environmental resources of European coasts: final report, pp 1–21. Commission of the European Communities, Research Contract Publication, Contract No. EV5V-CT93-0258, Brussels, 2 Volumes (vol 1, p 220, vol 2, p 50)
Devoy RJN, Delaney C, Carter RWG, Jennings SC (1996) Coastal stratigraphies as indicators of environmental changes upon European Atlantic coasts in the late Holocene. J Coast Res 12:564–588

Dodds W, Cooper JAG, McKenna J (2010) Flood and coastal erosion risk management policy evolution in Northern Ireland: "incremental or leapfrogging". Ocean Coast Manag 53:779–786
Duffy MJ, Devoy RJN (1999) Contemporary process controls on the evolution of sedimentary coasts under low to high energy regimes: western Ireland. Geologie en Mijnb 77:33–49
Dunne S, Hanafin J, Lynch P, McGrath R, Nishimura E, Nolan P, Ratnam JV, Semmler T, Sweeney, C, Wang S (2008) Ireland in a warmer world, scientific predictions of the Irish climate in the 21st century. Community Climate Change Consortium for Ireland, Met Eireann. www.c4i.ie/docs/IrelandinaWarmerWorld.pdf
Dury GH (1981) An introduction to environmental systems. Heinemann, London
Dwyer N (ed) (2012) The status of Ireland's climate, 2012. Environmental Protection Agency, Dublin
Edwards RJ, Brooks AJ (2008) The island of Ireland: drowning the myth of an Irish landbridge. In: Devenport JJ, Sleeman DP, Woodman PC (eds) Mind the gap: postglacial colonisation of Ireland. Special Suppl Ir Nat J, pp 19–34
Edwards KJ, Warren WP (1985) The quaternary history of Ireland. Academic, London
Finkl CW (ed) (1995) Holocene cycles: climate, sea level, and coastal sedimentation. J Coastal Res Spec Issue 17:141–152
Finkl CW (2004) Coastal classification: systematic approaches to consider in the development of a comprehensive scheme. J Coast Res 20:166–213
Fitzgerald DM, Kraus NC, Hands EB (2000) Natural mechanisms of sediment bypassing at tidal inlets. U.S. Army Corps of Engineers, Vicksburg
Foley K (2009) The great famine in Iveragh. In: Crowley J, Sheehan J (eds) The Iveragh peninsula: a cultural atlas of the Ring of Kerry. Cork University Press, Cork, pp 217–223
French JR, Burningham H (2011) Coastal geomorphology. Progr Phys Geogr 35:535–545
Froger J (2003) Le littoral Irlandais et la gestion des zones côtières. Mémoire de Maîtrise, Université de Nantes
Gault J, O'Hagan AM, Cummins V, Murphy J, Vial T (2011) Erosion management in Inch beach, southwest Ireland. Ocean Coast Manag 54:930–942
Geleynse N, Storms JEA, Stive MJF, Jagers HRA, Walstra DJR (2010) Modelling of a mixed-load fluvio-deltaic system. Geophys Res Letters 37, L05402
Giese GS, Mague ST, Rogers SS (2009) A geomorphological analysis of Nauset beach/Pleasant Bay/Chatham Harbour for the purpose of estimating future configurations and conditions. Provincetown Centre for Coastal Studies, p 30
Gilbertson DD, Schwenninger JL, Kemp RA, Rhodes EJ (1999) Sand-drift and soil formation along an exposed North Atlantic coastline: 14, 000 years of diverse geomorphological, climatic and human impacts. J Arch Sci 26:439–469
Gomez B, Carter L, Orpin AR, Cobb KM, Page MJ, Trustrum NA, Palmer AS (2012) ENSO/SAM interactions during the late Holocene. Holocene 22:23–30
Goudie A (2004) Encyclopaedia of geomorphology. Routledge, London
Gray SRJ, Gagnon AS, Gray SA, O'Dwyer B, O'Mahony C, Muir D, Devoy RJN, Falaleeva M, Gault J (2014) Are coastal managers detecting the problem? Assessing stakeholder perception of climate vulnerability using fuzzy cognitive mapping. Ocean Coast Manag 94:74–89
Green A, Cooper A, LeVieux A (2013) Unusual barrier/inlet behaviour associated with active coastal progradation and river-dominated estuaries. J Coast Res SI 66:10–16
Greenwood RO, Orford JD (2007) Factors controlling the retreat of drumlin cliffs in a low energy marine environment: Strangford Lough, Northern Ireland. J Coast Res 23:285–297
Guilcher A (1966) Les grandes falaises et mégafalaises des côtes sud-ouest de l'Irlande. Ann Geogr 75:26–38
Guilcher A, King CAM (1961) Spits, tombolos and tidal marshes in Connemara and west Kerry, Ireland. Proc R Ir Acad 61B:283–338
Hails J, Carr A (eds) (1976) Nearshore sediment dynamics and sedimentation. Wiley, London, p 316
Hanna E, Cappelen J, Allan R, Jonsson T, Le Blanq F, Lillington T, Hickey KR (2008) New insights into North European and North Atlantic surface pressure variability, storminess and related climate change since 1830. J Clim 21:6739–6766
Hansom JD, Angus S (2001) Tir á Machair (Land of the Machair): sediment supply and climate change scenarios for the future of the Outer Hebrides machair. In: Gordon JE, Lees KF (eds)

Earth science and natural heritage: interactions and integrated management, Natural Heritage of Scotland Series No. 9. The Stationery Office, Edinburgh, pp 68–81

Hanson H, Aarninkhof S, Capobianco M, Jimenez JA, Larson M, Nicholls RJ, Plant NG, Southgate HN, Steetzel HJ, Stive MJF, de Vriend HJ (2003) Modelling of coastal evolution on yearly to decadal time scales. J Coast Res 19:790–811

Hart MG (1986) Geomorphology: pure and applied. Allen & Unwin, London

Heap AD, Bryce S, Ryan DA (2004) Facies evolution of Holocene estuaries and deltas: a large sample statistical study from Australia. Sed Geol 168:1–17

Hickey KR (2011) The hourly gale record from Valentia Observatory, SW Ireland 1874–2008 and some observations on extreme wave heights in the NE Atlantic. Clim Change 106:483–506

Higgs K (2009) The geology of the Iveragh peninsula. In: Crowley J, Sheehan J (eds) The Iveragh peninsula: a cultural atlas of the Ring of Kerry. Cork University Press, Cork, pp 16–20

Houser C (2009) Synchronisation of transport and supply in beach-dune interaction. Progr Phys Geogr 33:733–746

Houser C, Hilton S (2009) Sensitivity of post-hurricane beach and dune recovery to event frequency. Earth Surf Proc Land 34:613–628

Houser C, Hapke C, Hamilton S (2008) Controls on coastal dune morphology, shoreline erosion and barrier island response to extreme storms. Geomorphology 100:223–240

INFOMAR (2014) Integrated mapping for the sustainable development of Ireland's marine resources. http://www.infomar.ie

IPCC (2007) Climate change 2007, the physical science basis and climate change 2007: impacts, adaptation and vulnerability. Contributions of Working Groups I and II to the IPCC Fourth Assessment Report. Cambridge University Press, Cambridge. http://www.ipcc.ch/publications_and_data/ar4/wg2/en/contents.html

IPCC (2014a) Climate change 2014, the physical science basis: summary for policymakers. Contribution of Working Group I to the IPCC Fifth Assessment Report. Intergovernmental Panel on Climate Change. Cambridge University Press

IPCC (2014b) Climate change 2014, impacts, adaptation, and vulnerability: summary for policymakers. Contribution of Working Group II to the IPCC Fifth Assessment Report. Intergovernmental Panel on Climate Change. Cambridge University Press

Irish Examiner (2014) Storm chaos: disaster zones. Irish Examiner No. 59, 579, Wednesday, Feb 5 (and other news article coverage of the December 2013 – February 2014 storm impacts on Ireland's coasts). Irish Examiner, Cork. www.irishexaminer.com

Jenkins GJ, Murphy JM, Sexton DMH, Lowe JA, Jones P, Kilsby CG (2009) UK Climate projections: briefing report (and 2010 update). Met Office Hadley Centre, Exeter

Jennings S, Orford JD, Canti M, Devoy RJN, Straker V (1998) The role of relative sea-level rise and changing sediment supply on Holocene gravel barrier development: the example of Porlock, Somerset, UK. Holocene 8:165–181

Kandrot S (2013) Quantifying post storm dune recovery using terrestrial laser scanning in Dingle Bay, County Kerry. (*Pers comm* and part of PhD submission in 2014 Department of Geography, University College Cork). Chimera 28

Kandrot S (2015) High resolution monitoring techniques and the modelling of storm impacts upon coastal beach-barrier systems in Southwest Ireland. Unpublished PhD thesis, University College Cork, Ireland

Kandrot S, Devoy RJN, Cawkwell F, Gault J (2014) Monitoring and modelling micro- to meso-scale dune-barrier behaviour using terrestrial laser scanning. Abstract and poster presentation, Irish Geomorphology Group First Annual Papers and Workshop Meeting, Department of Geography, Trinity College Dublin

Kerry County Council (2001a) Rossbehy strand coastal and cliff stabilisation study report. Kerry County Council, Tralee

Kerry County Council (2001b) Development plan review. KCC Review issues document 10/12/01-5/02/02 Tralee

Kerry County Council (2003) Kerry county development plan 2003–2009. Tralee. http://www.kerrycoco.ie/planning/devplan03.asp

Kerry County Council (2009) December ordinary meeting minutes. Tralee, p 58

Kerrys Eye (2010) High tides bring flood threat. The Kerry's Eye, p 8
Klein AHF, Beaumord AC, Finkl CW, Diehl FL, Calliari LJ (eds) (2006) Proceedings of the 8th international coastal symposium 2004, Itajaí – SC- Brazil. J Coast Res Spec Issue 39, 2 vols
Knight J, Coxon P, McCabe AM, McCarron SG (2004) Pleistocene glaciations in Ireland. In: Ehlers P, Gibbard PL (eds) Quaternary glaciations: extent and chronology. Elsevier, Amsterdam, pp 183–191
Kraft JC, Chrzastowski MJ (1985) Coastal stratigraphic sequences. In: Davis RA (ed) Coastal sedimentary environments, 2nd edn. Springer, New York, pp 625–623
Kraft JC, John CJ (1979) Lateral and vertical facies relations of transgressive barriers. Am Assoc Pet Geol 63:2145–2163
Lamb HH (1995) Climate, history and the modern world, 2nd edn. Routledge, London
Lozano I, Devoy RJN, May W, Andersen U (2004) Storminess and vulnerability along the Atlantic coastlines of Europe: analysis of storm records and of greenhouse gases induced climate scenario. Mar Geol 210:205–225
Lynch K (2009) *pers. comm.* Re. cost of coastal protection work at Inch, Kerry County Council, Ireland
MacClenahan P (1997) Geographical variations in the Holocene chronology of Western European coastal dunes in relation to climate, sea-level and human impact. Unpublished PhD thesis, University of Ulster, Coleraine
Mackay AW, Battarbee RW, Birks HJB, Oldfield F (eds) (2003) Global change in the Holocene. Edward Arnold, London
Marine Institute (2014) Wave buoy data for Ireland. http://www.marine.ie/home/Search?qt=Wave%20buoy%20data
Masselink G, Hughes MG, Knight J (2011) Coastal processes and geomorphology. Hodder Education, London
Met Éireann (2014) The Irish meteorological service online information and data sources. http://www.met.ie/default.asp
Michael D, Howa HL (1997) Morphodynamic behaviour of a tidal inlet system in a mixed-energy environment. Phys Chem Earth 22:339–343
National Parks and Wildlife Service (2011) Castlemaine Harbour SAC (site code 343): conservation objectives supporting document, coastal habitats. National Parks and Wildlife Service, Dublin
Nichol SL, Goff JR, Devoy RJN, Chague-Goff C, Hayward B, James I (2007) Lagoon subsidence and tsunami on the west coast of New Zealand. Sed Geol 200:248–262
O'Connor MC, McKenna J, Cooper JAG (2010) Coastal issues and conflicts in Northwest Europe: a comparative analysis. Ocean Coast Manag 53:727–737
O'Dwyer B, Gault J (2014) Climate information platform for Ireland Phase 2 (ICIP2). Coastal and Marine Research Centre, Beaufort Research, University College Cork (Environmental Protection Agency Ireland Project, online details at http://cmrc.ucc.ie/)
O'Shea M (2015) Monitoring and modelling the morphodynamic evolution of a breached barrier beach system. Unpublished PhD thesis, University College Cork, Ireland
O'Shea M, Murphy J (2013) Predicting and monitoring the evolution of a coastal barrier dune system post breaching. J Coast Res 29:38–55
O'Shea M, Murphy M, Sala P (2011) Monitoring the morphodynamic behaviour of a breached barrier beach system and its impacts on an estuarine system. OCEANS' 11 I.E. Conference, Santander
Ordnance Survey of Ireland (2014) The historical mapping of Ireland: first edition six-inch maps 1829–1842. Ordnance Survey of Ireland, Phoenix Park, Dublin. http://www.osi.ie/Products/Professional-Mapping/Historical-Mapping.aspx
Orford JD, Carter RWG (1984) Mechanisms to account for the longshore spacing of overwash on a coarse clastic dominated barrier beach in southeast Ireland. Mar Geol 56:207–226
Orford JD, Carter RWG, Forbes DL, Taylor RB (1988) Overwash occurrence consequent on morphodynamic changes following lagoon outlet closure on a coarse clastic barrier. Earth Surf Proc Landforms 13:27–35

Orford JD, Carter RWG, Jennings SC (1991) Coarse clastic barrier environments: evolution and implications for Quaternary sea level interpretation. Quat Int 9:87–104

Orford JD, Hinton AC, Carter RWG, Jennings SC (1992) A tidal link between sea-level rise and coastal response of a gravel-dominated barrier: story head, Nova Scotia. In: Woodworth P, Pugh DT, de Ronde J, Warrick RG, Hannah J (eds) Sea-level changes: determination and effects. Geophys Monogr Ser 69, Am Geophys Union, Washington, DC, pp 71–79

Orford JD, Carter RWG, Jennings SC (1996) Control domains and morphological phases in gravel-dominated coastal barriers of Nova Scotia. J Coast Res 12:589–604

Orford JD, Cooper JAG, Smith B (1997) LOICZ: the human factor as an influence on the Irish coast. In: Sweeney J (ed) Global change and the environment. Royal Irish Academy, Dublin, pp 88–107

Orford JD, Cooper JAG, McKenna J (1999) Mesoscale temporal changes to foredunes at Inch Spit, southwest Ireland. Z Geomorph NF 43:439–461

Pestiaux P, Duplessy JC, van der Mersch I, Berger A (1988) Paleoclimatic variability at frequencies ranging from 1 cycle per 10,000 years to 1 cycle per 1,000 years: evidence for non linear behaviour of the climate system. Clim Change 12:9–37

Pracht M (1996) Geology of Dingle Bay Geological Survey of Ireland, Dublin p 58

Quinn AM (1977) Sand dunes: formation, erosion and management. An Foras Forbartha, Dublin

Ranasinghe R, Duong TM, Uhlenbrook S, Roelvink D, Stive M (2013) Climate-change impacts assessment for inlet-interrupted coastlines. Nat Clim Change 3:83–87

Regnaud H (1989) Lan dune bordier Inch. In: Les littoraux. Caen, Groupe Francais de geomorphologie, second forum, pp 21–24

Ritchie W (1979) Machair development and chronology of the Uists and adjacent islands. In: Boyd JM (ed) The natural environment of the Outer Hebrides, vol 77B, Proc Roy Soc Edinburgh., pp 107–122

Ritchie W, Angus S (2012) Studies in machair evolution. Aberdeen University Press, Aberdeen

Roelvink D, Reniers A, van Dongeren AP, van Thiel de Vries J, McCall R (2009) Modelling storm impacts on beaches, dunes and barrier islands. J Coast Eng 55:1041–1051

Rossington SK, Nicholls RJ, Stive MJF et al (2011) Estuary schematisation in behaviour-oriented modelling. Mar Geol 281:27–34

Roy PS, Thom BG, Wright LD (1980) Holocene sequences on an embayed high-energy coast: an evolutionary model. Sed Geol 26:1–26

Roy PS, Cowell PJ, Ferland MA, Thom BG (1994) Wave-dominated coasts. In: Carter RWG, Woodroffe CD (eds) Coastal evolution: late quaternary shoreline morphodynamics. Cambridge University Press, Cambridge, pp 121–186

Sabatier P, Dezileau L, Colin C, Briqueu L, Bouchette F, Martinez P, Siani G, Raynal O, von Grafenstein U (2012) 7000 years of paleostorm activity in the Northwest Mediterranean Sea in response to Holocene climate events. Quat Res 77:1–11

Sala P (2010) Morphodynamic evolution of a tidal inlet mid-bay barrier system. Master's thesis, University College Cork, Ireland

Science Direct (2013) Assessment of habitats in Inch Spit, Co. Kerry in relation to a proposed golf course. Report ER13-03, Cork, p 76 (App V)

Shaw J, Forbes DL, Beaver DE, Wile BD (1994) Marine geological surveys in Dingle Bay, Co. Kerry, southwest Ireland: cruise report 93–303. Geological Survey of Canada, Dartmouth, Open File 2980

Sherman DJ, Jackson DWT, Namikas SL, Wang J (1995) Aeolian sand transport to the coastal dune complex at Inch, Co. Kerry, Ireland. Geomorphology Research Report No. 3, Geography Department, University of Southern California, p 43

Shields L, Fitzgerald D (1989) The 'night of the Big Wind' in Ireland, 6–7 January 1839. Ir Geogr 22:31–43

Short AD (1999) Handbook of beach and shoreface morphodynamics. Wiley, New York

Sinnott AM (1999) Sea-level and related coastal change in south and southeast Ireland. PhD thesis, University College Cork, Ireland

Smith C (1756) The ancient and present state of the county of Kerry. Dublin

Stive MJF (2006) Morphodynamics of coastal inlets and tidal lagoons. J Coast Res Spec Issue 39:28–34

Stone GW, Orford JD (2004) Storms and their significance in coastal morpho-sedimentary dynamics. Mar Geol 210:1–362

Sweeney J, Donnelly A, McElwain L, Jones M (2002) Climate change indicators for Ireland. ERTDI Report 2, Environmental Protection Agency, Johnstown Castle Estate, Co. Wexford, Ireland

Sweeney J, Albanito F, Brereton A, Caffarra A, Charlton R, Donnelly A, Fealy R, Fitzgerald J, Holden N, Jones M, Murphy C (2008) Climate change: refining impacts for Ireland. Environmental Protection Agency, Johnstown Castle Estate, Co, Wexford

Swift LJ, Devoy RJN, Wheeler AJ, Sutton GD, Gault J (2006) Sedimentary dynamics and coastal changes on the south coast of Ireland. J Coast Res Spec Issue 39:234–239

Thom BG (1984) Transgressive and regressive stratigraphies of coastal sand barriers in eastern Australia. Mar Geol 7:161–168

Thorn CE (1988) Introduction to theoretical geomorphology. Unwin Hyman, London

Tung TT, Walstra DJR, van de Graaff J, Stive MJF (2009) Morphological modelling of tidal inlet migration and closure. In: Proceedings of the 10th international coastal symposium (ICS 2009), 13–18 Apr 2009, Lisbon, 2 vols and in J Coastal Res Spec Issue 56:1080–1084

van Geel B, Burrman J, Waterbolk HT (1996) Archaeological and palaeoecological indications of an abrupt climate change in the Netherlands and evidence for climatological teleconnections around 2650 BP. J Quat Sci 11:451–460

van Rijn LC (1998) Principles of coastal morphology. Aqua Publications, London

Vial T (2008) Monitoring the morphological responses of an embayed high energy beach to storms and Atlantic waves. Master's thesis, University College Cork, Ireland

Vila-Concejo A, Ferreira A, Matias A, Dias JMA (2003) The first two years of an inlet: sedimentary dynamics. Cont Shelf Res 23:1425–1445

Vila-Concejo A, Ferreira A, Ciavola P, Matias A, Dias JMA (2004) Tracer studies on the updrift margin of a complex inlet system. Mar Geol 208:43–72

Whittow JB (1974) Geology and scenery in Ireland. Pelican Books, London

Wiggins J (1852) The practice of embanking lands from the sea, treated as a means of profitable employment of capital; with examples and particulars of actual embankments and also practical remarks on the repair of old sea walls. J. Weale, London

Wintle AG, Clarke ML, Musson FM, Orford JD, Devoy RJN (1998) Luminescence dating of recent dunes on Inch Spit, Dingle Bay, southwest Ireland. Holocene 8:331–339

Woodroffe CA (2002) Coasts: form, processes and evolution. Cambridge University Press, Cambridge

Zalasiewicz J, Williams M, Steffen W, Crutzen P (2010) The new world of the Anthropocene. Environ Sci Technol 44:2228–2231

Chapter 10
Polish Spits and Barriers

Kazimierz Furmańczyk and Stanisław Musielak

Abstract The landscape of the areas occupied at present by the Polish coast was shaped by the Scandinavian ice sheet. The coast is dominated by morainic uplands and the lowland relief occurs in fragments of glacier lake sedimentary remains, on the bottom of ancient river valleys. About 5 Ka BP, the sea level was by 2.5 m lower that it is at present, the shoreline being shifted about 1–2 km seaward. As a result of a prolonged impact of the sea on the "soft" shore, the profile of the latter is generally even, with a dominance of barrier forms, specific for fluctuations of the sea level in the Holocene. At that time, underwater long shore bars were produced in shallow inshore areas as a result of interaction between the wave activity and the seafloor. As the sea level rose, the bars were gradually shifted towards the shore. Peninsulas can be found in the Gulf of Gdańsk only, but they have evolved from barriers as well.

The Polish coastline borders the southern part of the Baltic Sea, an area for millennia covered by an ice sheet. The southern Baltic coast was shaped by phenomena and processes associated with ice sheet retreat and the accompanying changes in sea level (Kostrzewski and Musielak 2008).

The Polish coastal zone is built of till clays as well as fluvioglacial and marine sands, the latter forming, via aeolian transport, coastal dunes. It is only at the foot of Pleistocene cliffs that a coarser material, gravel and even boulders, has been deposited.

Generally, the materials building the Polish coast are "soft", very susceptible to erosion. As a result of a prolonged impact of the sea on the "soft" shore, the profile of the latter is generally even, with a dominance of barrier forms, specific for fluctuations of the sea level in the Holocene. Peninsulas can be found in the Gulf of Gdańsk only, but they have evolved from barriers as well.

The landscape of the areas occupied at present by the Polish coast was shaped by the Scandinavian ice sheet. The coast is dominated by morainic uplands characterised by undulating relief, in places planar or hilly. The lowland relief occurs in fragments of glacier lake sedimentary remains, on the bottom of ancient

K. Furmańczyk (✉) • S. Musielak
University of Szczecin, Institute of Marine and Coastal Sciences, Mickiewicza 18, 70-383 Szczecin, Poland
e-mail: kaz@univ.szczecin.pl

G. Randazzo et al. (eds.), *Sand and Gravel Spits*, Coastal Research Library 12,
DOI 10.1007/978-3-319-13716-2_10

river valleys serving as conduits for the ice sheet meltwater, and on coastal accumulation plains (Marsz 1984).

While melting, the ice sheet was releasing huge amounts of water. During the last 10,000 years, the basin-like depression remaining after the ice sheet retreated filled up with water to eventually become the Baltic Sea. Initially, the sea level was about 60 m lower than it is at present, the coastline being located 20–100 km north of the present seashore. During the first Baltic Holocene transgression persisting for about 2,000 years, the sea level was rising at a rate exceeding 20 mm/year (Rosa 1984). About 8.5–5.0 Ka BP, the completely ice-free reservoir connected with the Atlantic, which resulted in a rapid rise (by almost 25 m) of the sea level. About 5 Ka BP, the sea level was by 2.5 m lower that it is at present, the shoreline being shifted about 1–2 km seaward (Uścinowicz 2003).

The sea level, although constantly on the rise, fluctuated. At that time, underwater longshore bars were produced in shallow inshore areas as a result of interaction between the wave activity and the seafloor. As the sea level rose, the bars were gradually shifted towards the shore (similarly to what has been observed on the Atlantic US coast by Leatherman 1988). During periodic lowering of the sea level, the bars emerged above the water. The next sea level rise would erode the bars which would move landwards. About 4.5 ka BP, a somewhat longer period of stabilisation, and even lowering of the sea level, occurred. At that time, the emergent bar surfaces were affected by aeolian processes resulting in the formation of chains of sandy barriers parallel to the shore, in the form of dune bars. Another sea level rise moved the barriers towards the shore. At the same time, in areas of today's spits, the sea level rise produced a rise in the groundwater level and wetting of coastal depressions, conditions leading to the development of peatbogs. When the subsequent sea and groundwater level rises were faster than peat accretion, coastal lakes emerged. The lakes were filled with barrier dune sands which moved following the sea level rise. On occasions, the landward-shifting barriers cut off parts of marine embayments. When the sea level rise was faster than the rate of aeolian aggradation bar formation, the barriers were broken and inlets, connecting the coastal lake with the sea, or flood deltas were formed (Rosa 1984).

On its both ends, a barrier usually adjoins morainic plateaus or terminal moraines the sandy-clayey slopes of which are eroded by the sea. The material produced is transported along the shore and feeds the neighbouring barriers. It is carried by wind-driven waves and currents. The size of a dune-bearing barrier and dune thickness height depend on the amount of sand supplied by seafloor erosion in the coastal zone or by erosion of the neighbouring cliffs.

The processes involved in evolution of barrier shore and peninsulas were taking place on the Polish coast throughout the Holocene, and are continuing. Generally, those processes underpin the development of all sandy barriers on the Polish coast (Fig. 10.1), both large (the Łebska, Dziwnów, and Wisła spits) and small (spits of the lakes Bukowo and Jamno). Dominant in their formation was sediment movement transverse to the shore, generated by sea level changes, the longshore bedload transport being important during the barrier stage of evolution (Musielak 1988).

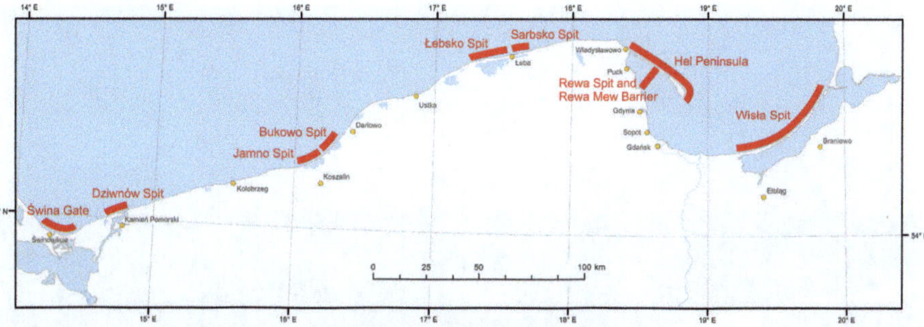

Fig. 10.1 Peninsula, spits and selected barriers of the Polish coast

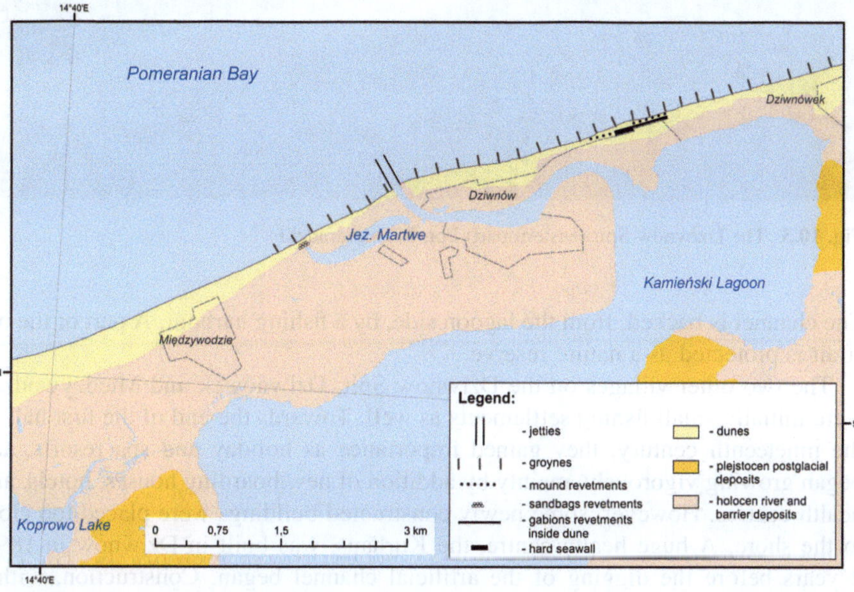

Fig. 10.2 The Dziwnów Spit "Mierzeja Dziwnowska"

The Dziwnów Spit is located in the western part of the Polish coast (Figs. 10.2 and 10.3) and separates the Kamieński Lagoon from the Pomeranian Bay. The Dziwna Strait connecting the Lagoon with the Bay is one of the conduits transporting water from the Szczecin Lagoon, via the Kamieński Lagoon, into the Baltic Sea. It was already in early Middle Ages that the Vikings used the Dziwna Strait to reach their settlement of Jomsborg (today's Wolin). The Strait was first mentioned by the name of Divenov In historical sources in 1243.

The Strait was frequently closed by intensive longshore sediment transport, which diminished the inlet's utility for commercial transport and leaving it mainly to be used by fishermen from settlements in today's Dziwnów. Eventually, an artificial channel, existing until today was dug in 1892–1900 (Furmańczyk 2013).

Fig. 10.3 The Dziwnów Spit (eastern part) Fot. P. Domaradzki

The channel is backed, from the lagoon side, by a fishing harbour. A part of the old strait is protected as a nature reserve.

The two other villages on the Dziwnów Spit, Dziwnówek and Międzywodzie, were initially small fishing settlements as well. Towards the end of the first half of the nineteenth century, they gained importance as holiday and spa resorts, and began growing vigorously, mainly by addition of new boarding houses, hotels, and health centres. However, some newly constructed buildings were placed too close to the shore. A huge health centre, the Kurhaus, was built in Dziwnów in 1890, 2 years before the digging of the artificial channel began. Construction, within 1892–1900, of the channel and of breakwaters protecting its mouth resulted in coastal processes along the entire spit being modified. At the turn of the year 1913, a heavy storm raising the water level significantly by: 1.59 m in Kołobrzeg and 1.98 m in Świnoujście (neighbouring harbors) eroded the belt of low dunes near the Kurhaus. The water surged up the dunes and destroyed the prominent building. The water-borne destruction affected many more parts of the shore and buildings then, and left some buildings in serious condition. To protect the most endangered stretch of the shore and objects it supported, groynes and a revetment shielding the threatened buildings were built in 1918–1924 east of the channel, along an about 10 km section of the shore.

The channel divides the Dziwnów Spit into two parts, eastern and western. The eastern part is narrow (300–500 m wide). It is covered by low (not higher than 4 m) dunes and protected mainly by groynes, a seawall, and gabions; it is also artificially nourished. The western part of the Spit is wide, its high dunes rising up to 12 m; it is only in the area of the former Dziwna Strait that the dunes are lower than 4 m and protected by geotextile bags revetment. The shore stretch from the channel to the former strait is strengthened with groynes, the remaining part bearing no protective constructions. Dudzińska-Nowak (2006) showed some parts of the shore to be prone to heavy erosion visible as a change in the dune base location in excess of 0.7 m/year. The present shores of the Dziwnów Spit are naturally nourished by transverse bedload transport and longshore transport of sediment from cliffs located east and west of the Spit. The beach at the foot of the dunes is about 30–50 m wide, the shore being flanked by two to three underwater longshore bars. The villages on the Spit support a population of about 4,100, more than 70,000 tourists visiting the area in summer.

Bordering the sea on one side and the lagoon on the other, the Dziwnów Spit is highly attractive for tourists (Fig. 10.3). Its location is ideal for sailing, angling, kayaking, and beach recreation. Those amenities are supplemented by beautiful natural landscapes of the Kamieński Lagoon and the nearby Wolin National Park with its 90 m high sandy-clayey cliffs.

The spit situated between the islands of Usedom and Wolin, the so-called Świna Gate, is one of the largest spits on the Polish coast. Although its origins are much more complex, it evolved basically as described above. The spit consists of two parts separated by the Świna Strait. The Świna Gate evolved mainly on account of the longshore sediment transport and changes in the shoreline, as described in detail by Keilhack (1912), Rosa (1984), and Osadczuk (2004).

The origin of the accumulative coastal formations located in the western part of the Gulf of Gdańsk, the **Hel Peninsula** and the **Rewski Spit**, is more complex (Fig. 10.4). The Hel Peninsula has been usually presented as a simple scythe-shaped formation built by the sediment eroded from a morainic hill to the west and transported along the shore (Zenkowicz 1955; Słomianko 1968). In reality, however, the origins of the Hel Peninsula are more complex (Pazdro 1948; Musielak 1989; Tomczak 2005) and, similarly to the origins of the Rewski Spit (Bohdziewicz 1959; Musielak 1967), have not been adequately elucidated yet. On account of their spatial and genetic connectivity, the two spits are here dealt with together.

When, more than 5,000 years ago, the sea level was low, the present inner Puck Bay (the so-called Puck Lagoon) was a terrestrial part of the ground moraine, an island being located to the east. The gradually rising sea level, and its fluctuations, generated two barriers on the margins of the Lagoon: the Hel Peninsula barrier and that of the Rewa Mew (Fig. 10.5). The Hel Peninsula barrier was fed with material eroded from cliffs located west of it. The accumulated sand formed dunes which, despite the successive sea level rise stages, remained emergent.

The material which eventually formed the Rewa Mew barrier (known locally also as Ryf Mew) (Bohdziewicz 1959; Musielak 1967) was supplied primarily by the transverse sediment transport, with a small contribution of sediment eroded from the Oksywie cliffs to the south. The small dunes that had formed on the Rewa Mew barrier were gradually broken and inundated during successive stages of sea level rise. At present, the barrier is very narrow, its fragments periodically emerging from the water (Fig. 10.6). Its southern part, intercepting material from the

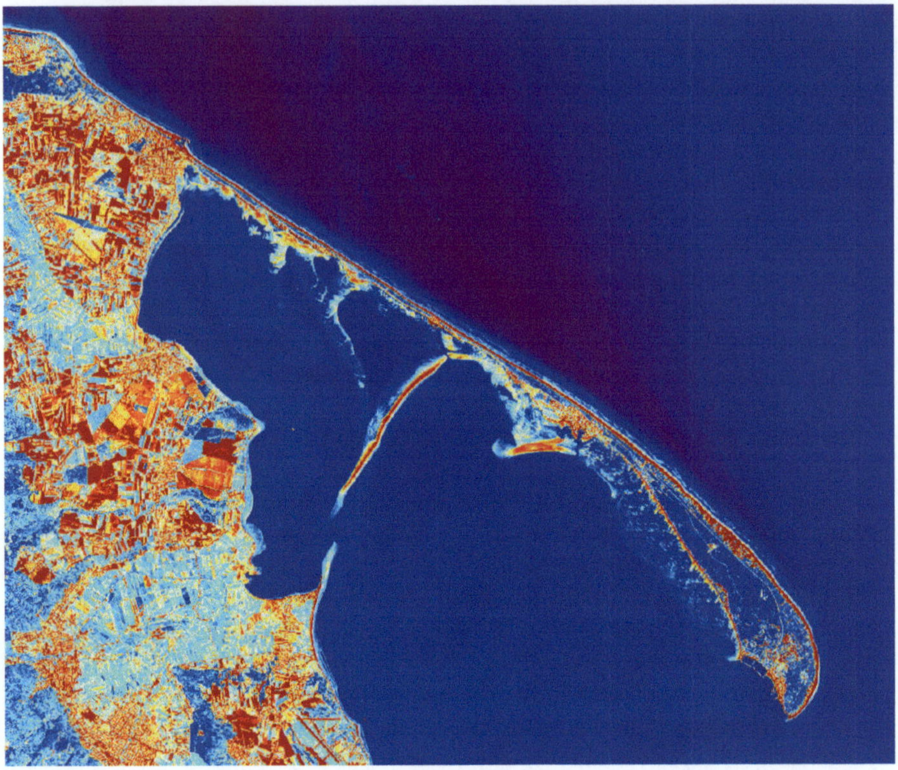

Fig. 10.4 The Hel Peninsula, Rewa Spit, and Rewa Mew Barrier (Landsat ETM+)

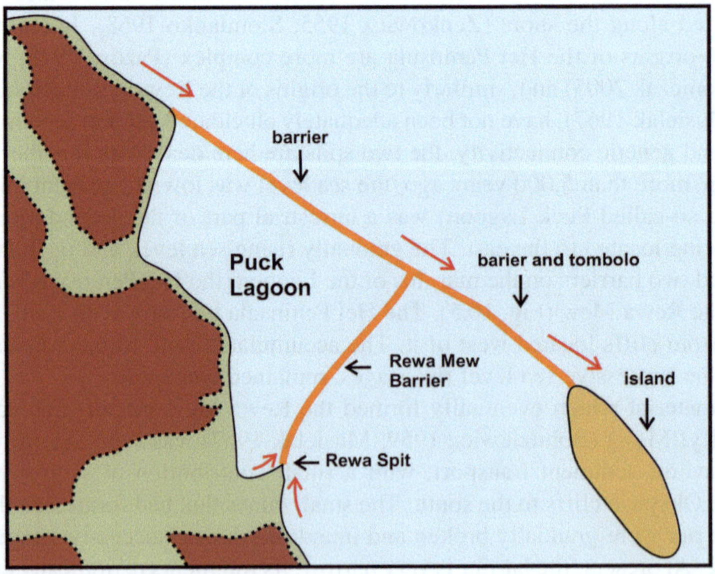

Fig. 10.5 Stages in the formation of the Hel Peninsula, Rewa Spit, and Rewa Mew Barrier in the western part of the Gulf of Gdańsk (scheme)

Fig. 10.6 The Rewa Mew Barrier Fot. P. Domaradzki

longshore bedload transport, features a small, emergent sandy peninsula called the Rewski Spit, locally known as "Szpyrk" (Musielak 1980).

At a low water level, the Rewski Spit (Fig. 10.7) can be up to 1 km long and up to 100 m wide at its base. It grades into an extensive (about 500 m long) shoal. On the cross-section, the Rewski Spit shows two longshore bars up to 0.5 m wide at the base, separated by a flattened area (Musielak 1967). The Rewski Spit shores are natural, unprotected. The Spit is very susceptible to effects of sea level change and wave action. It is fed predominantly by the longshore bedload transport from the Babie Doły cliff to the south, the sediment transported from the west contributing somewhat to the nourishment as well. The base of the Spit features a small village of Rewa with a population of 900. Initially, it was inhabited by farmers and fishermen; for almost 20 years, it has been functioning as a suburb of Gdynia as well as a holiday resort. The natural, sandy Spit is to some extent used as a beach, but it lacks any tourist infrastructure. The beach infrastructure is to be found at the Spit's base, north and east of Rewa.

The island located east of the Puck Lagoon was undergoing various transformations resulting eventually in its connection ('tombolo') with the mainland to form the Hel Peninsula (Musielak 1989). The Peninsula, with its very low dunes at the base, in the barrier part, was shaped by a combination of sea level rise and the longshore transport of sand eroded from the Kępa Swarzewska cliffs to the west. In the past, the basal fragment of the Peninsula was frequently broken by storm surges forming flood deltas in the barrier's hinterland.

Fig. 10.7 The Rewa Spit Fot. P. Domaradzki

The Dahlberg map of 1696 (Fig. 10.8) depicts the Hel Peninsula, in its western, barrier part, as a chain of islands, which emerged due to overflow (Dahlberg 1696) although, the earlier Getkant map of 1637 (Fig. 10.9) shows a solid peninsula there (Getkant 1637).

In the seventeenth century, the Hel Peninsula featured, in today's villages of Chałupy and Kuźnice, two forts: Władysławowo and Kazimierzowo, the Peninsula itself being in those places broken a few times by storm surges. Occasionally, the Peninsula served as a haven for pirates operating in the area.

Later on, the Peninsula became inhabited by fishermen operating from the villages of Hel, Jastarnia, Kuźnice, and Chałupy. The Peninsula dwellers used various means to maintain and raise the dunes (Wunsche 1904). The Peninsula was also supplied with sand from the longshore bedload transport, the sand being eroded from cliffs to the west. The natural bedload transport was broken in 1939, after a port was built at the Peninsula base (Fig. 10.10). Consequently, the base became increasingly prone to shore erosion. Longer and longer stretches of the shore were being strengthened with groynes (Basiński 1961), artificial beach nourishment being used as of 1989. During the last 10 years, the dune ridge in the most endangered places was reinforced with gabions. Along its seaward shore, the Peninsula measures 35 km in length, about 13 km featuring groynes; artificial nourishment is applied to a 25 km long stretch.

At present, the Peninsula is subject to substantial coastal dynamics-related processes. The dune base of the unprotected seaward shore is known in places to retreat on the distance of about 40 m during a year (1991/1992). However, during

Fig. 10.8 The Dahlberg map 1696

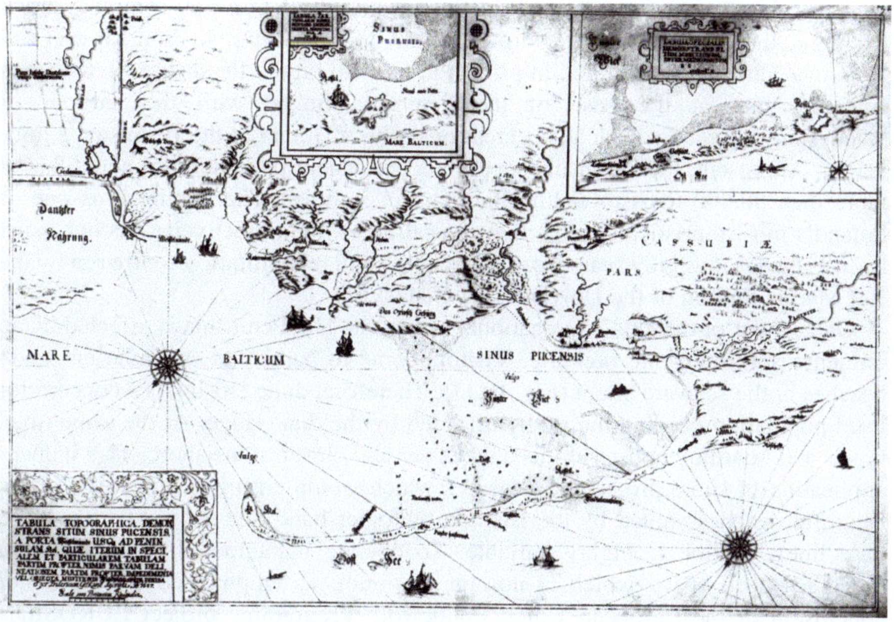

Fig. 10.9 The Getkant map 1637

Fig. 10.10 The Hel Peninsula Fot. P. Domaradzki

34 years (1957–1991), the total erosion there did not exceed 64 m (Furmańczyk 1995). The changes on the sheltered Puck Bay shore are small and result mainly from ice scour in winter.

When the port was constructed at the base of the Peninsula, a railway track leading to the navy harbour at its tip was built, as were fishing harbours in Hel and Jastarnia. The railway resulted in growth in tourism on the Peninsula. New resorts emerged: that at the base of the Peninsula linked with the harbour of Władysławowo, and a typically recreational one of Jurata at the Peninsula's mid-length. At present, the Peninsula supports a permanent population of about 8,000, more than 80,000 tourists visiting the area in summer. The Peninsula is one of Poland's most touristically attractive coastal areas; the attractiveness is enhanced by, i.e., the famous grey seal research, breeding and rehabilitation centre run by the Hel Marine Station of the University of Gdańsk.

On account of the area's recreational assets, the Hel Peninsula is affected by an immense anthropogenic pressure. Visitors come to bathe and sunbathe on sandy beaches of the seaward shore (Fig. 10.11). Therefore, dune erosion is a concern for local population needing the safety afforded by the dune ridge. At the same time, hotels and boarding houses are located closer and closer to the shore. This induces application of shore protection measures which would ensure maintenance of the beautiful beaches valued by tourists. On the other hand, the shore line (the dune base line) location is highly variable. To operate unhampered, natural coastal processes need space which is now at premium, as emphasised by the major conclusion of the European Union Framework Programme project EUROSION (Doody et al. 2004).

Fig. 10.11 Sunbathers on a Hel Peninsula beach 2005 Fot. K. Furmańczyk

The single-track railway line and the narrow, single-line road running along the Peninsula are efficient constraints for the number of visitors to the Peninsula. While this transport infrastructure is sufficient off season, it substantially hampers the traffic during the summer season: the time it takes to leave the Peninsula on cloudy days or when vacation periods turn is very long indeed.

Another characteristics of the Peninsula is its very shallow (about 1–1.5 m depth) inshore zone on the Puck Bay side. This area is very attractive to windsurfers for whom camping sites have been built. The sites have changed the natural environment of the inshore zone on the Bay side, which supports important fish spawning grounds and nurseries, fairly substantially (Fig. 10.12).

Maintenance of the shores at the base of the Peninsula presents the most difficult problem. The Peninsula there is in places less than 200 m wide. The longshore bedload transport which supplies sediment to this part of the coast has been disrupted by the harbour breakwaters; the sediment with which underwater longshore bars would have been built is deficient, and there is no space for natural development of dune ridges. The shore is being strengthened with groynes, artificial dunes with gabions inside, and an earth bar in the certain part. The entire endangered section is periodically artificially nourished. Given adverse climatic effects and sea level rise, the situation there could become critical.

Fig. 10.12 Camping sites on the Hel Peninsula Fot. P. Domaradzki

The existing fishing harbours and havens are being successively modernised and enlarged by adding marinas; this enhancement of tourist-related aspects of those ports gradually changes their nature.

The shore protection measures already in place have not eliminated the threat of erosion, but have significantly reduced it (Furmańczyk and Łęcka 2005).

Of key importance at the present stage of development of all sandy barrier forms at the Polish coast are changes in sea level (Wiśniewski and Wolski 2009) and its accelerated rise (Rotnicki et al. 1995) as well as ways of coastal management, the shore protection measures applied in particular.

As demonstrated by research carried out at an about 80 km long section of the shore in the western part of the Polish coast (Dudzińska-Nowak 2006a), the shore protection measures applied there, while reducing the erosion at the dune base, have increased the total length of eroded sections of the shore. It follows then that erosion has diminished, but affects a larger area. In addition, the magnitude of sediment accumulation has significantly decreased on accumulative shore stretches, which can be associated with climatic change effects (mean sea level rise) and extension of the protected shore length.

References

Basiński T (1961) Kształtowanie się obszarów erozyjnych na zakończeniu grup ostróg na Helu w latach 1949–1956, Materiały do monografii polskiego brzegu morskiego, vol 01, PWN, Gdańsk-Poznań

Bohdziewicz L (1959) Budowa geologiczna I procesy dynamiczne w strefie brzegowej w Orłowie I Rewie. In: Rocznik Pol. Tow. Geol., vol XXIX, Warszawa, pp 347–357

Dahlberg (1696) Tabula Geographica. In: Puffendorf S (ed) De Rebus a Carolo Gustavo, vol 1, pp 148–600

Doody P, Ferreira M, Lombardo S, Lucius I, Misdorp R, Niesing H, Salman A, Smallegange M (2004) Living with coastal erosion in Europe. Sediment and space for sustainability. Results from the EUROSION study. European Commission, p 40

Dudzińska-Nowak J (2006) Coastline long-term changes of the selected area of the Pomeranian Bay. In: Tubielewicz A (ed) Coastal dynamic, geomorphology and protection. Eurocoast-Littoral, Gdańsk, pp 163–170

Dudzińska-Nowak J (2006a) Zmiany morfologii jako wskaźniki tendencji rozwojowych brzegu. Ph.D. thesis, University of Szczecin, Poland, 225 pp

Furmańczyk K (1995) Coast changes of the Hel Spit over the last 40 years. In: Rotnicki K (ed) Polish coast, past, present and future, Journal of Coastal Research, Special Issue 22. Quaternary Research Institute, Poznań, pp 193–196

Furmańczyk K (2013) Poland. In: Pranzini E, Williams A (eds) Coastal erosion and protecion in Europe. Routledge, pp 81–95

Furmańczyk K, Łęcka A (2005) Ochrona brzegu na odcinku Władysławowo Jurata. In: FurmańczykK (ed) ZZOP w Polsce, Problemy erozji brzegu. INOM US Szczecin, pp 165–170

Getkant (1637) Tabula Topographica, vol 1, pp 116–300

Keilhack K (1912) Die Verlandung der Swinepforte. Jahrbuch der Koniglisch Preussischen Geologischen Landesanstalt. Bd. XXXII, T.2. Berlin, pp 209–244

Kostrzewski A, Musielak S (2008) Współczesna ewolucja rzeźby wybrzeża Południowego Bałtyku. In: Starkel L (ed) Współczesne przemiany rzeźby Polski. IGIGP Uniwersytetu Jagielońskiego, Kraków, pp 327–348

Leatherman SP (1988) Barrier island handbook, Coastal Publications series. The University of Maryland, Maryland

Marsz A (1984) Główne cechy geomorfologiczne. In: Augustowski B (ed) Pobrzeże Pomorskie. GTN, Ossolineum, Gdańsk, pp 42–66

Musielak S (1967) Niektóre procesy brzegowe w okolicach Rewy. In: Zeszyty Geograficzne WSP, R. IX, Gdańsk, pp 233–243

Musielak S (1980) Współczesne procesy brzegowe w rejonie Zatoki Gdańskiej, Peribalticum, Ossolineum, Gdańsk

Musielak S (1988) Morfolitodinamika Bieriegowoj zony biespriliwnogo moria, Habilitation thesis, MGU, Moskwa, 301 pp

Musielak S (1989) Uwagi dotyczące genezy Półwyspu Helskiego w świetle nowszych badań. In: Studia i Materiały Oceanologiczne, vol 56, Ossolineum, pp 311–321

Osadczuk K (2004) Geneza i rozwój wałów piaszczystych Bramy Świny w świetle badań morfometrycznych i sedymentologicznych, vol 552, Rozprawy i Studia. Wydawnictwo Naukowe US, Szczecin

Pazdro Z (1948) Półwysep Hel i jego geneza. In: Technika Morza i Wybrzeża, No. 1, Gdańsk, pp 07–13

Rosa B (1984) Rozwój brzegu i jego odcinki akumulacyjne. In: Augustowski B (ed) Pobrzeże Pomorskie. GTN, Ossolineum, Gdańsk, pp 67–112

Rotnicki K, Borówka RK, Devine N (1995) Accelerated sea levelrise as a threat to the Polish Coastal Zone – quantification of risk. J Coast Res Spec Issue 22, 111–135

Słomianko P (1968) O niektórych osobliwościach struktury potoków rumowiska przy brzegach piaszczystych, Archiwum Hydrotechniki, vol 15, No. 3, PWN, Warszawa, pp 331–341

Tomczak A (2005) Wybrane zagadnienia z przeszłości geologicznej i przyszłości Półwyspu Helskiego. In: Cyberski J (ed) Stan i zagrożenia Półwyspu Helskiego. GTN, Wydział Nauk o Ziemi, Gdańsk, pp 13–58

Uścinowicz S (2003) The Southern Baltic relative sea level changes, glacio-isostatic rebound and shoreline displacement, Państwowy Instytut Geologiczny, Prace Specjalne No. 10, Warszawa

Wiśniewski B, Wolski T (2009) Katalog wezbrań i obniżeń sztormowych poziomów morza oraz ekstremalne poziomy wód na polskim wybrzeżu. Wydawnictwo Naukowe Akademii Morskiej, Szczecin, 156 pp
Wunsche H (1904) Studien auf der Halbinsel Hela. Inaugurial Dissertation. Universitat Leipzig. Druck Heinrich, Dresden, 79 pp
Zenkowicz WP (1955) Niektóre zagadnienia brzegów polskiego Bałtyku, Technika i Gospodarka Morska, No. 9, Gdańsk, pp 222–226

Chapter 11
Tidal Flat-Barrier Spit Interactions in a Fetch-Limited, Macro-tidal Embayment, Lubec, Maine, USA

Joseph T. Kelley, Daniel F. Belknap, and J. Andrew Walsh

Abstract This report describes two sand and gravel spits and associated tidal flat environments in a fetch-limited, macrotidal setting in Lubec, Maine, USA. The spits have been remarkably dynamic since the late eighteenth century despite the low wave energy. The beaches were originally sourced from erosion of glacial and post-glacial bluffs, but the contemporary spits are apparently growing from clasts reworked from former barrier sites on the tidal flat. Attached algae coupled with strong tidal currents permits landward-directed floating and dragging of cobble-sized clasts that could not otherwise move, underscoring the potential importance of algal-assisted transport. This paper underscores the unexplored potential of algal transport across macrotidal flats as a mechanism to permit barriers to transgress in a punctuated manner from one location to another.

11.1 Introduction

> Sometimes these plants (seaweeds) attach themselves by their root-like bases...which are not in fact roots, for they serve only for support-to shells which lie prone or are fixed upon the bottom. More commonly they adhere to a pebble left on the sea-floor by the melting glacial sheet, or drifted out in the "pan-ice" which in winter forms along the sea margins. All these sea-weeds have floats which hold them upright in the water, and as they increase in size, they pull on their bases with constantly augmenting force. As the waves roll over them, they increase the tugging action, until finally, in some time of storm, the plant lifts the stone from its bed and floats it in the water, buoyed up by the vesicles of air contained in its fronds. The plant and upturned stone are together borne in by the heave of the sea onto the shore. Coming into the breakers, the weed is quickly beaten to pieces, and the pebble enters the mill where so many of its fellows have met their fate. The close observer after a storm may find any number of these bowlders along a pebbly shore which still show traces of the sea-weeds which bore them to the coast...On a quarter of a mile of the Marblehead (MA) beach I have estimated that as much as 10 tons of these seaweed-borne pebbles

J.T. Kelley (✉) • D.F. Belknap • J.A. Walsh
School of Earth and Climate Sciences, University of Maine, Orono, ME 04469-5790, USA
e-mail: jtkelley@maine.edu; Belknap@maine.edu

G. Randazzo et al. (eds.), *Sand and Gravel Spits*, Coastal Research Library 12,
DOI 10.1007/978-3-319-13716-2_11

came ashore in a single storm. Many of the beaches, which are so adequately provided with pebbles from the neighboring shores where the waves are attacking the firm land that they could not be maintained from that source alone, are sufficiently fed by the means of supply afforded by the action of marine plants. (N.S. Shaler 1895, p. 55)

Gravel barriers are much less studied than their sandy counterparts, yet along rocky shores and in formerly glaciated regions (paraglacial coasts), gravel barriers are common and often still front the eroding deposits which sourced them, or they are attached as spits to those sources (FitzGerald and Van Heteren 1999; Boyd et al. 1987). Although gravel barriers in differing wave and tidal regimes have been reviewed, the range of conditions under which gravel barriers occur has not been fully explored (Antony and Orford 2002). Gravel barriers in meso-macrotidal (2 to >6 m) regions have appeared in some case studies and reviews (Short 1991; Antony and Orford 2002; Masselink and Short 1993; Orford et al. 2012), but gravel barriers in fetch-limited embayments, where tidal amplitude is large relative to wave height (Type 3 barriers in Short's (1991) nomenclature), are very rarely discussed. In such locales, a tide-dominated flat environment commonly occupies the lower foreshore, and a wave-dominated beach rests in the mid-high tide level (Antony and Orford 2002). There seems to be general agreement that in such settings "increasing tide range retards the rate at which sediment transport and morphological changes take place" (Masselink and Short 1993; p. 788). This is because there is limited time for waves to break on the beach.

In wave-dominated, paraglacial settings, where tides seem less important, sand and gravel barriers are often very dynamic and shift position relatively rapidly as old sediment sources are depleted and new ones exposed (Boyd et al. 1987). Details on mechanisms of sediment transfer from one barrier location to another are scarce, but presumably abetted by extreme storms with high wave energy (Orford et al. 1996, 2003). The dynamics of sediment transport and morphological change in fetch-restricted, paraglacial embayments with large tidal ranges is presumably much less, but relatively unexamined. Although most reports on meso-macro tidal barriers describe sandy mid-low-tide environments, the lower beach-tidal flats of macrotidal, paraglacial gravel barriers are seldom described.

In this study we describe two sand and gravel barriers and an associated tidal flat in a highly fetch-restricted, macrotidal setting. The barriers have been historically very dynamic despite the restricted wave energy. We examine the lower foreshore-tidal flat in detail and consider its interaction with the beach to find mechanisms capable of affecting rapid shoreline change.

11.2 Geological Setting

Lubec, Maine is near the easternmost point in the United States (Fig. 11.1), along the border with Canada. The embayment is sheltered from swells from the Atlantic Ocean (Gulf of Maine) by West Quoddy Head and from local waves by Campobello Island and other, local headlands. The most common summer winds are from the

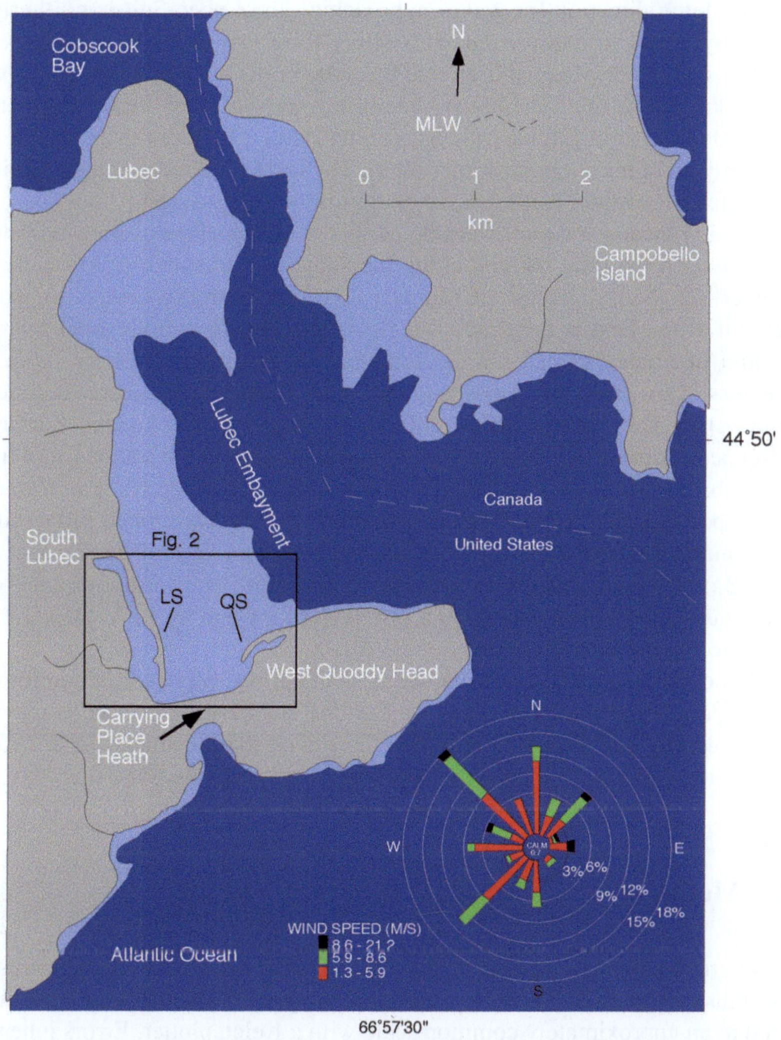

Fig. 11.1 Location map for Lubec, Maine (Modified from Walsh 1988). The wind rose is from http://www.wrcc.dri.edu/cgi-bin/wea_windrose.pl?laKEPO. Some small intertidal islands in Canada were left out for the sake of simplicity

south-southwest, while the winter-fall (and annual average wind) wind is mostly from the northwest. The strongest winds are from the east during winter storms that occur several times per year (http://www.wrcc.dri.edu/cgi-bin/wea_windrose.pl?laKEPO; Hill et al. 2004).

The Spring tidal range is just less than 7.0 m and tides are semi-diurnal. Wave height in the Lubec Embayment is fetch limited because storm waves generally come from the east (Fig. 11.1). When large storms do occur in the winter, much of the embayment is typically choked with sea ice, further impeding wave attack.

Maximum fetch direction for waves approaching the Lubec Spit is northerly, for which the typically strongest winds of 6–9 m/s occur less than 3 % of the time. The fetch across the embayment at high tide from the north is about 4 km, but it is only about 1.7 km at low tide. The highest likely waves impacting the spit are between 0.2 and 0.3 m (Coastal Engineering Research Center 1984), but these would occur infrequently and for a short duration (<2 h). The large tidal range interacts with the limited fetch to severely restrict wave energy in this embayment.

Paleozoic igneous and metamorphic rocks define the shape of the environment, and crop out on the tidal flat and along the shoreline (Bastin and Williams 1914; Osberg et al. 1985). Glacial till directly overlies bedrock. Owing to isostatic depression of the land at the time of deglaciation, marine submergence led to the deposition of muddy glacial-marine sediment over till. Radiocarbon dates from marine fossils just above till deposits in the Lubec Embayment average around 15.5 (calibrated) ka B.P. (Dorion et al. 2001). A coarse-grained, layered sand-gravel deposit uncomformably overlies the glacial-marine mud in the southeastern upland of the embayment (Figs. 11.2 and 11.3). This regressive deposit was formed by waves as post-glacial sea level fell across the area. A wave-eroded hill of till and bedrock mark the highstand shoreline in this area just east of the present coast (Fig. 11.3a). The wave-cut, sand and gravel platform is uniformly eroding on its seaward side today. Sea level fell to a lowstand of −60 m by 12.5 (cal.) B.P., and has risen to the present day (Kelley et al. 2010).

To the southwest of the raised marine sand and gravel deposits, an ombrotrophic bog (Carrying Place Heath) borders most of the southern side of the embayment. The peat deposits of this bog unit overlie glacial-marine mud and are rapidly eroding, with a 3 m scarp exposed today along the border with the tidal flat.

11.3 Methods

Paleogeographic reconstructions of the Lubec Embayment were made from a time series of historic maps and vertical aerial photographs (Walsh 1988). All maps were reduced to an approximately common scale with a Kelch plotter. Errors inherent in antique maps range from survey errors and use of an uncertain datum to a promotional bias in emphasizing valuable landforms on old maps, and render these maps useful for depicting only gross landform changes. The entire spit and tidal flat were also surveyed with a conventional theodolite with observations gathered from every significant slope change and at all contacts between landforms (approximately every 50 m across the area; Walsh 1988). The most recent maps of the area are available from Google Earth since 1996.

Thirty-seven bottom samples were collected from throughout the area and subjected to grain size analyses by settling tube and pipette (Folk 1974). More than 50 Dutch and vibracores were collected, mostly from the salt marsh.

To evaluate the role of algal dragging of clasts, 16 stations were established across the tidal flat. At each station six algal-colonized clasts (*Fucus vesiculosis*)

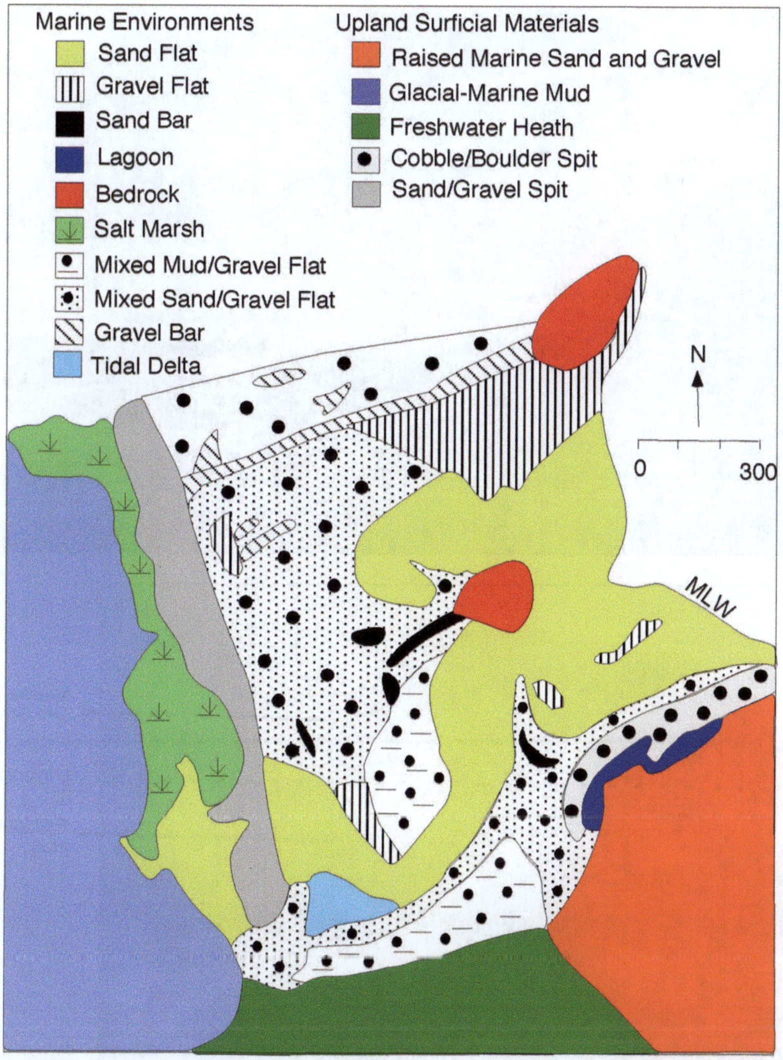

Fig. 11.2 Quaternary landforms of the Lubec Embayment (Modified from Walsh 1988)

were painted with fluorescent orange paint, labeled and sealed with a durable marine varnish. The clasts were selected from the tidal flat and ranged between −5 and −7 phi in size (32–128 mm); great care was spent to avoid harming the algae during drying and painting. Clasts were monitored for eight consecutive low tides after the June 22, 1986 deployment, and then at 1–3 week intervals for the next 2 months, followed by a final check on October 25, 1986. Clast motion was measured with a tape measure from the clast position to a stake located at the station from which the clast originated.

Fig. 11.3 Late Quaternary deposits: (**a**) Aerial photo of late Pleistocene raised shoreline and wave-cut platform, Lubec, Maine. *Dashed line* shows wave-eroded shoreline position. A wave-cut platform lies seaward of the paleo-shoreline. Note the abundance of algae attached to pebbles and cobbles on the modern high-tide shoreline; (**b**) photo of raised sand and gravel nearshore marine deposit. Section is located on Fig. 11.3a with *arrow*

11.4 Results

11.4.1 Time Series Changes in the Lubec Embayment

One of the earliest (ca. 1785) high-resolution maps of the Lubec Embayment (1:48,000) shows a peninsula that projects northward, the opposite direction of the modern Lubec Spit (Fig. 11.4a). It is curved in outline, and more than 2 km long. Other maps from this period also depict a beach extending northward (Walsh 1988).

By 1805, this spit was broken up, and only an island with a trailing intertidal bar remained near the former spit's connection to the mainland (Fig. 11.4b). North of the island, a new spit is shown in the shape of the present spit, though more seaward. This new spit appears to have attached to the island by 1830 (Fig. 11.4c), and

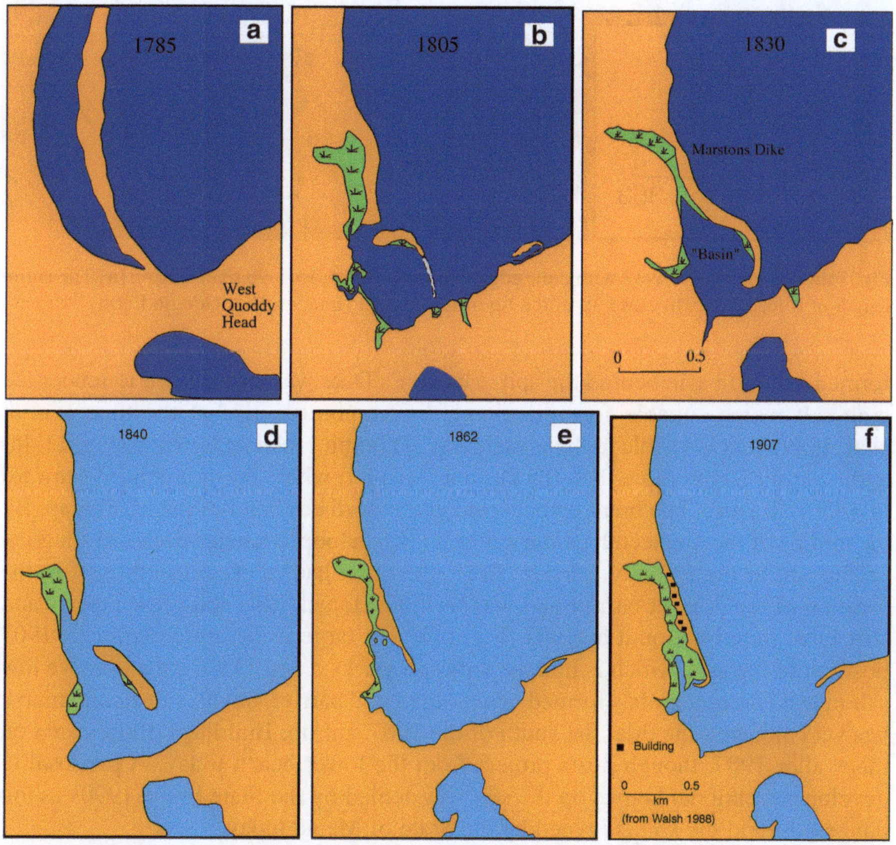

Fig. 11.4 Historic maps of Lubec Embayment: (**a**) 1785 (Putnam et al. 1785); (**b**) 1805; Fifth Division, 1805); (**c**) 1830 (Colby 1881; date of publication, not survey); (**d**) 1840 (Walling 1861; date of publication, not survey); (**e**) 1862; (U.S. Coast Survey 1862); (**f**) 1907 (U.S. Geological Survey 1908 date of publication, not survey) (Modified from Walsh 1988)

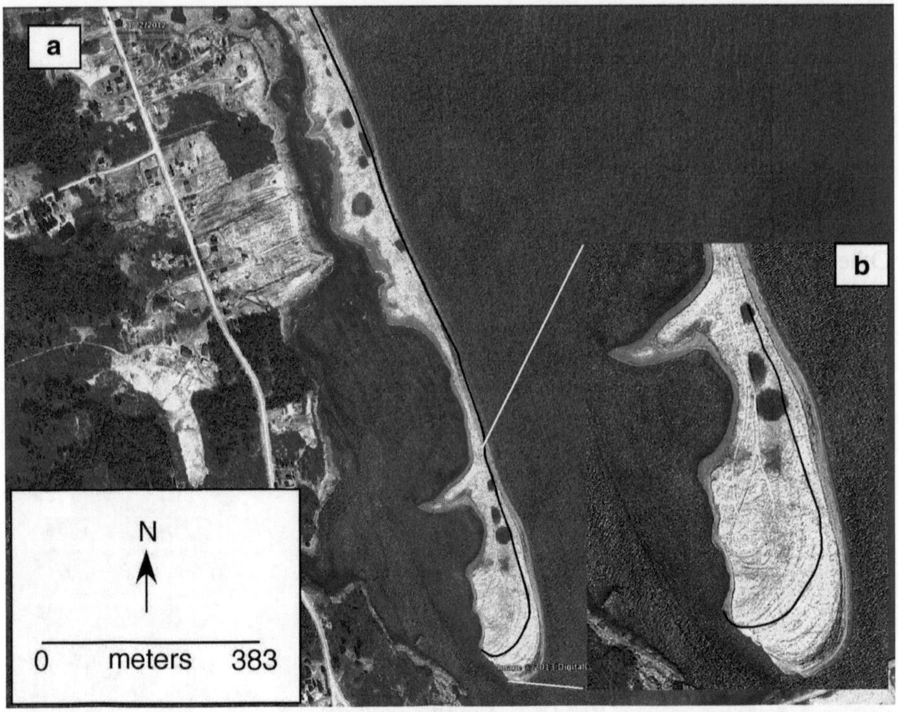

Fig. 11.5 Lubec spit 3-22-12 with dune edge digitized from 5-15-1-96 (*black line*). (**a**) The entire length of Lubec Spit; (**b**) close up of the tip of Lubec Spit (Images from Google Earth)

formed a broad basin behind the spit. Marston's Dike was constructed to render the high salt marsh suitable for agriculture, a common practice at the time (Smith et al. 1989), and is visible to the present day. This spit broke up by 1840 (Fig. 11.4d) and a barrier island more than 0.5 km long was left where the spit existed seaward of 1830s "Basin". The breakup occurred at the northern bend of the 1830 map. By the middle of the nineteenth century (Fig. 11.4e), a looped barrier enclosed a lagoon on the site of the Quoddy Barrier. The Lubec Spit had re-established itself in the location of the present barrier and was well developed. This spit grew rapidly and structures were built on it by 1907 (Fig. 11.4f). A large recurvature marks the 1907 position of the spit, which is notable today (Figs. 11.5 and 11.6). An extensive low salt marsh community has formed behind the 1907 barrier, but no substantial marsh has yet colonized the tidal flat south of the 1907 spit tip. Buildings disappeared on maps after 1907, though posts project from the lower beach today. A proposal to develop recreational homes on the spit was denied by the State in the 1990s as too dangerous, and the spit is owned by the State of Maine today.

Lubec Spit has continued to grow at a rate up to 3 m/year between 1996 and 2012, and averaging 0.4 m/year of growth at its terminus since 1907. This growth has required at least 1100 m^3/year of sand and gravel. At the same time, the spit has widened near its tip, while narrowing slightly along most of its length (Fig. 11.5). Quoddy Spit is more difficult to assess. Some historical maps fail to depict this spit,

Fig. 11.6 Air photo of the modern Lubec Spit. An *arrow* marks the 1907 spit terminus. The freshwater bog is at the *bottom* of the photo. Scale varies across the image, but it is about 120 m from the seaward edge of dune vegetation to the landward edge of vegetation on the 1907 recurve tip

while others only show the most recent spit and not the lagoon-enclosing barrier. Since 1996, there has been no obvious growth, but its unvegetated tip is impossible to discern from the adjacent tidal flat.

11.4.2 Sedimentary Environments and Geomorphology of the Lubec Embayment

Gravel is very abundant in the Lubec Embayment. It occurs as the dominant component of the gravel flat, gravel bar, and Quoddy Spit environments (Figs. 11.2 and 11.7). In the northern part of the study area, a topographically high gravel flat crops out over a large area (Fig. 11.8a). Here it is a very poorly sorted, bimodal deposit (modes near −2 phi and <4.5 phi; 4 mm- >24 mm) (Table 11.1). Boulders are rare, with many well rounded pebbles and almost no mud. Many of the larger clasts are colonized by algae and are typically embedded in the deposit. No bedforms exist on the surface of the gravel flat, and pits that occur across the gravel flat (Fig. 11.8a) were dug by people harvesting the soft-shell clam, *Mya arenaria*.

Gravel, grading from boulders and cobbles (proximal end; Fig. 11.8b) to cobbles and pebbles (distal end, Fig. 11.8c), form the Quoddy Spit (Figs. 11.2 and 11.7). The spit crest varies from 7.4 m (proximal) to 6 m (distal) in height with a steep

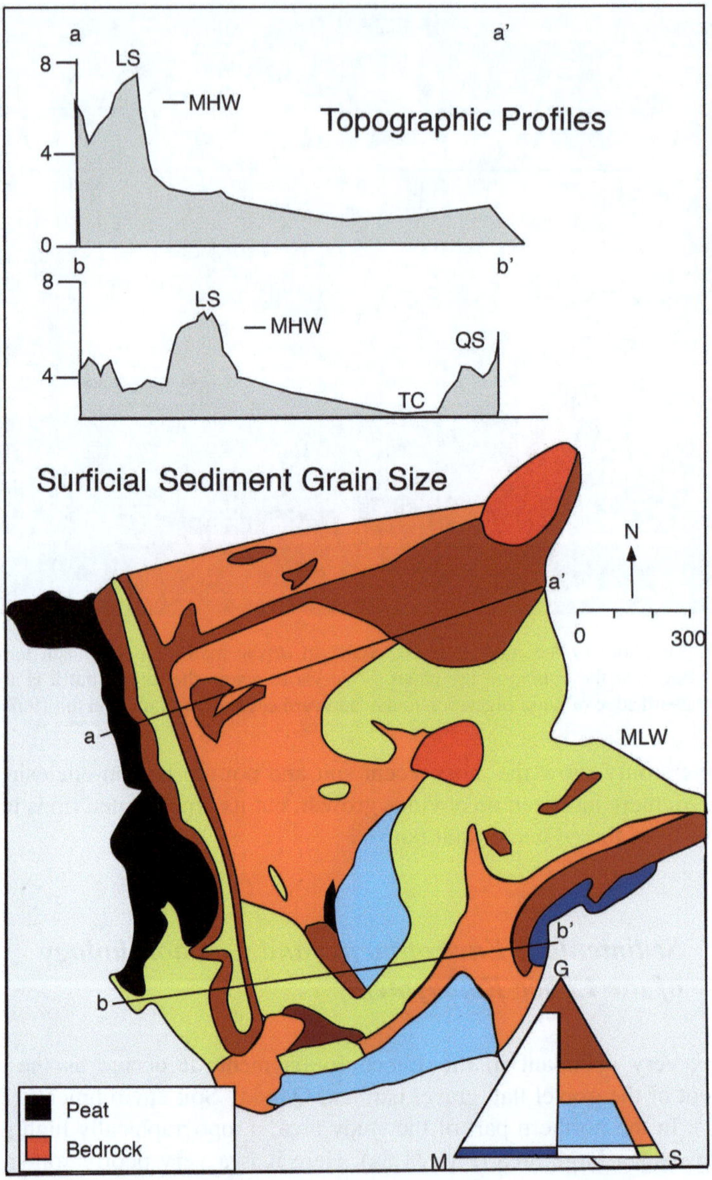

Fig. 11.7 Surficial material of the Lubec Embayment (Modified from Walsh 1988). Topographic cross sections are located on map as a-a' and b-b'. LS means Lubec Spit, QS means Quoddy Spit, TC means tidal channel, MHW is the approximate location of mean high water. The triangular diagram depicts the relative amounts of Sand, Gravel and Mud. Topographic profiles end at approximate low water mark. The map scale precludes showing the location of the numerous bedforms and scattered bedrock outcrops

Fig. 11.8 Photographs of environments within the Lubec Embayment: (**a**) Gravel flat from the northwest part of the embayment. Water-filled depressions were created by people harvesting shellfish (*Mya arenaria*); the shovel is 50 cm high (Modified from Walsh 1988); (**b**) Quoddy Spit, proximal end. The sediment consists of boulders and cobbles; the freshwater peat block in the foreground is approximately 0.3 m long. Note the steep boundary with the lagoon and the more gentle slope (*left*) to the tidal flat with its abundance of algae-covered clasts; (**c**) The distal end of the Quoddy Spit, with a steep slope into the lagoon. Here the sediment ranges from pebbles to cobbles; (**d**) Intertidal tombolo-like bar connecting the Lubec Spit with a bedrock outcrop 1 km seaward. It is used as a low-tide road by shellfish harvesters today, but is not a human construction. Note the patchiness of the grain size and abundance of algae; (**e**) A gravel-sand flat with abundant oscillation ripples and covered by algae-covered cobbles; (**f**) Air photo showing tip of Lubec Spit and surrounding environments. Note the bend that the tidal channel makes around the ebb-tidal delta (*arrow*) and the raised (*lighter color*) channel margin linear bars. The freshwater bog, crossed by a road, is to the *right*

slope (>80 % grade (>40°)) into the lagoon and a gentle one (10 % grade (5.7°)) onto the tidal flat (Figs. 11.8b, c). No sedimentary structures occur on the spit, and algal-covered clasts line the lower beach face (Figs. 11.3a and 11.8b). Algae with attached clasts and freshwater peat blocks are also common on the spit.

Table 11.1 Sediment grain size

Sample	Environment	Mean (phi)	Sorting (phi)	%Gravel	%Sand	%Mud
1	Sandflat	2.2	0.4	0.0	99.5	0.5
2	Sandflat	2.1	0.5	0.35	99.5	0.1
3	Sandflat	2.0	0.5	1.7	98.3	0.1
4	Sandflat	1.7	0.5	0.5	99.5	0.0
5	Mixed Flat	2.0	1.0	13.3	86.3	0.4
6	Mixed Flat	0.06	2.8	31.8	68.0	0.3
7	Mixed Flat	1.37	1.4	13.5	86.1	0.4
8	Mixed Flat	−0.38	2.6	39.4	60.4	0.2
9	Mixed Flat	2.12	1.4	6.9	86.9	6.1
10	Mixed Flat	−0.1	2.2	36.1	63.2	0.7
11	Sand/Grav Spit	−0.07	2.0	38.3	61.7	0.0
12	Sand/Grav Spit	−0.47	2.5	38.3	61.5	0.0
13	Sand/Grav Spit	−1.53	2.4	62.3	37.7	0.0
14	Sand/Grav Spit	−2.7	1.8	87.8	12.2	0.0
15	Sand/Grav Spit	−2.1	0.7	91.7	8.3	0.0
16	Sand/Grav Spit	−2.7	1.3	90.8	9.1	0.1
17	Sand/Grav Spit	−1.4	2.5	69.5	30.5	0.0
18	Sand/Grav Spit	−3.4	1.9	84.4	15.6	0.0
19	Sand/Grav Spit	−2.4	2.6	68.2	31.5	0.3
20	Sand/Grav Spit	−2.6	2.2	79.8	20.2	0.0
21	Gravel Flat	−2.1	2.6	68.1	31.8	0.1
22	Gravel Flat	−2.6	2.3	78.5	21.5	0.0
23	Gravel Flat	−1.7	2.7	59.5	40.3	0.2
24	Gravel Flat	−1.9	2.5	65.3	34.5	0.2
25	Gravel Flat	2.0	2.6	61.5	38.5	0.0
26	Gravel Bar	−3.5	0.7	99.2	0.8	0.0
27	Gravel Bar	−0.9	2.4	50.5	49.3	0.2
28	Gravel Bar	−2.3	2.3	73.0	26.9	0.1
29	Mixed Mud/Grav	−2.6	2.2	67.8	32.1	0.1
30	Mixed Mud/Grav	3.7	5.7	5.8	67.4	26.8
31	Mixed Mud/Grav	2.6	1.5	3.1	83.7	13.2
32	Dune	1.7	0.7	0.02	99.84	0.14
33	Dune	2.0	0.4	0.02	99.9	0.08
34	Salt Marsh	5.2	3.9	3.0	10.0	87.0
35	Salt Marsh	5.6	3.7	0.0	11.3	88.7
36	Lagoon	7.0	3.3	3.0	12.0	84.4
37	Lagoon	5.9	3.5	0.8	8.1	91.2

Sand and gravel bars (Fig. 11.2) are topographically raised, curvilinear landforms up to 1.5 m above the surrounding tidal flat. They appear to be relatively static landforms and not migratory on decadal time frames. Pebble-sized gravel is a common sediment size, with subordinate coarse sand and rare boulders, making the features poorly sorted. In many areas sand is segregated from gravel and is most abundant over large areas. The 1 km-long predominantly gravel bar connecting the northern bedrock outcrop with the Lubec Spit is the largest bar (Fig. 11.8d). It has the form of an intertidal tombolo, with a flat, sandy upper surface and coarse gravel clasts (often covered with algae) lining the margins. The predominantly sandy bar extending landward from the eastern bedrock outcrop clearly shows the influence of wave refraction in the opposite orientations of algae and drag marks on the east and west sides of the feature.

Sand, mixed with subordinate-dominant quantities of gravel, is the most texturally abundant sediment type in the Lubec Embayment. A mixed sand-gravel flat (Figs. 11.2 and 11.7) lies seaward of most of the Lubec Spit and continues north of the study area. The surficial sediments are extremely heterogeneous, and algal-covered pebbles and cobbles are common (Fig. 11.8e). Small depressions exist in places, and small ripples are locally common. Very crude stratification of sand pods and gravel layers occurs in the subsurface, but as with the gravel-dominated environments, there is little distinctive and continuous stratification.

On the seaward opening of the tidal inlet, an ebb tidal delta occurs (Figs. 11.2 and 11.8f). It is composed of distinct channel margin linear bars of sand and a main body of mixed gravel and sand (Figs. 11.2 and 11.8f). Discrete ebb and flood channels bound it and are bordered by the linear bars. The tidal creek (ebb channel) makes two 90° bends around the landform, which stands 1.5 m above it. Historical aerial photographs show that the bend in the creek has increased as the spit tip and inlet have migrated to the south (right in the image).

The Lubec Spit is a large mixed sand and gravel deposit, with a mantle of wind-blown sand less than a meter thick (Fig. 11.7). The spit crest ranges from 6.6 to 6.2 m in height from proximal to distal ends, and widens to more than 30 m where spit recurvatures occur. The grain size has modes at 2.25 and −3.4 phi, but surficial sediment is generally very spatially and temporally heterogeneous (Fig. 11.9a–c). Alternating sand and gravel beds form layers within the steep beachface, but these have limited spatial continuity (<1 m). A high frequency (500 mHz) Ground Penetrating Radar line along the upper beach failed to record any coherent reflectors within the beach or spit recurvatures. In the winter, the beach is often covered with ice and an ice foot develops during extremely cold periods (Fig. 11.9c).

Well-sorted fine-medium sand dominates the lower, outer flat (sand flat, Figs. 11.2 and 11.7) and tidal channel that crosses the flat. It represents 25 % of the study area and its surface is covered with numerous ripples and other bedforms created by time-varying current and wave directions (Fig. 11.9d). Cobbles are relatively rare here, but those that occur usually contain attached macroalgae (most commonly *Fucus vesiculosis*).

Muddy sediment dominates in the salt marsh and back-barrier lagoons, but mud is also an important constituent of the mixed mud-gravel flat at the southern margin of the embayment (mixed gravel-sand-mud flat, Figs. 11.2 and 11.7). Here, a veneer of algal-covered cobbles rests over a thin (<40 cm) deposit of muddy sand (Fig. 11.9e). The gravel-sand-mud flat forms a shallow basin, with modern mud resting uncomfortably over glacial-marine muddy sediment. On one edge of the flat, a freshwater peat deposit with a 15 cm high scarp crops out (Fig. 11.9f).

Fig. 11.9 Photographs of environments within the Lubec Embayment: (**a**) Lubec Spit following a spring high tide. Note the concentration of algae, all of which have attached gravel clasts, at the base of the sand dunes; (**b**) Lubec Spit a few days after a spring high tide. The algal wrackline borders the dune edge as in Fig. 11.9a, but wind-blown sand covers gravel on the beach. This is very near Fig. 11.9a; (**c**) Lubec Spit during a winter freeze. This is approximately the same position as Fig. 11.9a, b. Note the thick algae deposits beneath the ice and the well developed ice foot at the base of the beach; (**d**) Sand Flat showing complex bedform assemblage near the mean low water line (Modified from Walsh 1988; (**e**) Mixed mud and gravel flat viewed from the freshwater heath. The substrate is soft mud mixed with cobbles and pebbles. Algal-covered cobbles are very abundant on the flat. Note the shipwreck in the background. This is the position of the "Basin" from 1830 (Fig. 11.4c); (**f**) eroding freshwater peat deposit with a 10 cm high scarp (see Fig. 11.6 for location of peat)

11.4.3 Algal Transport of Clasts

The algal-covered clast transport experiment involved: (1) outer and topographically lower tidal flat locations (stations 1–3, elevation MLW-1.5 m); (2) mid-tidal flat locations (stations 4–9, elevation 1.5–3.0 m above MLW); (3) upper tidal flat locations (stations 10–14, elevation 3.0–4.0 m above MLW); and (4) back-barrier

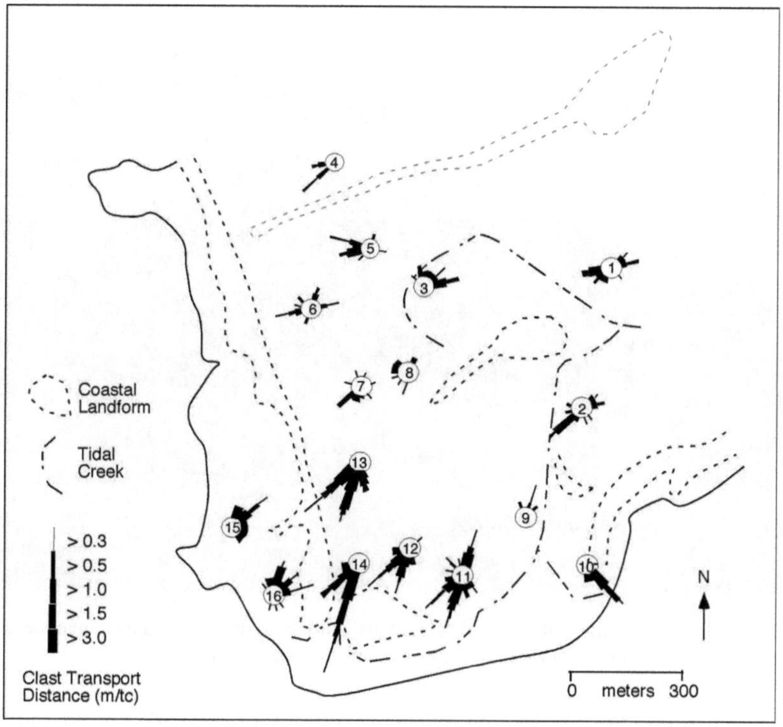

Fig. 11.10 Location of the algal-clast movement experiment and direction and distance of clast movement. The tidal creeks and landform outlines of Lubec and Quoddy Spits are included for spatial reference (From Walsh 1988)

locations (stations 15, 16, elevation >4.0 m) (Fig. 11.10). Of the 98 algal-colonized clasts deployed between June 22–24, 1986 88 % were observed during the July 9–10 survey, 81 % during the August 20–22 survey and 31 % in the final October 25 check. Because of the poor recovery rate in October, comments are confined to the June–August observations. All clast movements were monitored following an ebb tide.

All low tidal flat stations (1–3) were on sandy substrates (Figs. 11.6 and 11.10). Clast movement here was more bi-directional than at higher elevations, with some clasts moving generally landward and others seaward. Algal-covered clasts in the area were few, but most had flood-oriented fronds when observed following ebb tide. Clasts from stations 1 to 2 moved in a net landward direction; those from station 3 went seaward. The mean distance traveled over the study was 0.5 m/tidal cycle (m/tc).

Stations on the mid-tidal flat were on mixed sand and gravel substrates (Figs. 11.6 and 11.10), though station 8 was on a small sand/gravel bar. Clasts averaged a short transport distance of 0.1 m/tc with a high degree of directional variability. Stations 4 and 5, on opposite sides of the intertidal tombolo, showed the probable effects of wave refraction, with clasts converging on the bar. The net motion of clasts from all stations, except station 9, was generally towards land. Station 9 was located near a tidal creek, and though movement was limited, clasts went in a seaward direction in the apparently ebb-dominated channel.

Fig. 11.11 Structures informally termed "algal bars" collect in the flood tidal channel near the tip of the spit. *Arrows* point in the direction of transport, towards the spit tip

The high-tidal flat stations were on sand (stations 11, 13, 14) or sand and gravel substrates (stations 10, 12, 14) (Figs. 11.6 and 11.10). All were near the tidal inlet and displayed strong net landward (towards inlet) movement of clasts with an average rate of 0.6 m/tc). All but station 11 showed unidirectional movement. Clasts from several stations, and all from station 14, collected into "algal bars. These bars possess up top 15 cm of relief and contain a large quantity of pebbles and cobbles attached to the algae (largely *Fucus vesiculosis*). The algal bars migrate towards the tip of the Lubec Spit where the algae and associated sediment add to the growth of the spit at its terminus (Fig. 11.11).

The back-barrier stations (15, 16) were on sandy substrates (Figs. 11.6 and 11.10) and most clast movement was toward the Lubec Spit. The average transport rate was 0.2 m/tc.

11.5 Discussion

11.5.1 Shoreline Changes

Contrary to expectations (Masselink and Short 1993), both the Lubec and Quoddy Spits have been very dynamic since the eighteenth century (Fig. 11.4). The Quoddy Spit grew southwest since the earliest map depictions and may still be growing, though only slightly. The Lubec Spit changed its orientation 180° in the same time

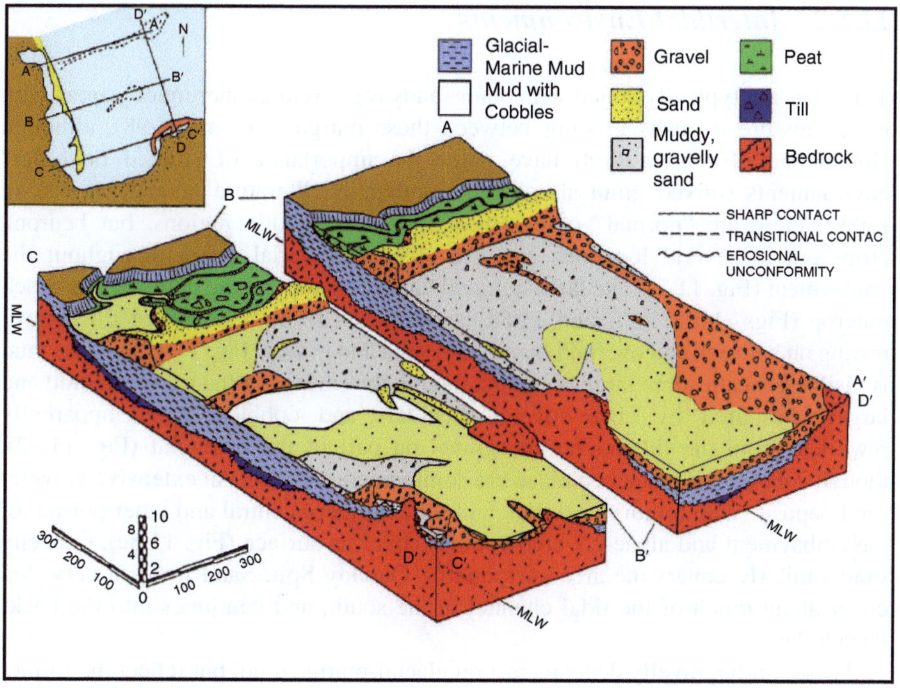

Fig. 11.12 Block diagram of tidal flat, spits and associated salt marsh. The *vertical* and *horizontal* scales are in meters. Note how thin the Holocene tidal flat sediments are above the Pleistocene material (From Walsh 1988). The main tidal channel runs directly on glacial-marine, muddy sediment

interval, broke up twice into barrier islands and spits and migrated almost half a kilometer landward. Its growth has permitted infilling of its backbarrier region by a salt marsh, and deflected the drainage from its backbarrier region (Figs. 11.4 and 11.8f). In the sense of Orford et al. (1996), the Lubec Spit appears to have transitioned from a "consolidated" domain in the eighteenth century through a "breakdown" domain and is now in a "reformation" phase, whereas the Quoddy Spit may be in an "inception" domain of slow growth. Although Orford et al. (1996) implied no evolutionary sequence in domains, their terminology aptly describes the historic dynamics of the spit systems in Lubec.

All barrier spit growth is towards the southeast corner of this embayment, towards a raised freshwater peat bog. The bog margin is retreating at about 1.0 m/year (Mansfield 2012), and the U.S. Army Corps of Engineers built a stone revetment along a small part of its length in 2012 to safeguard the road across the top. It is not possible to evaluate historic changes to the tidal flat because there are no prior maps of this environment. A block diagram (Fig. 11.12) shows that the Holocene sediment is generally less than 1.0 m thick except beneath the barrier and salt marsh. Residual freshwater peat crops out in the central flat (Fig. 11.9f), demonstrating the formerly greater extent of that environment.

11.5.2 Intertidal Environments

Tidal flats are typically zoned with outer sandy regions and inner muddy areas with mixed texture deposits grading between these margins (Eisma 1998), although Holland and Elmore (2008) have noted the importance of "mixed tidal flat" environments (mixed grain sizes) and summarized literature on the subject. In Lubec, there are "normal" outer sandy and inner muddy regions, but bedrock crops out in several locations and gravel-sized material exists throughout the embayment (Fig. 11.7). The largest gravel flat occurs around the outermost bedrock outcrop (Figs. 11.2, 11.7, and 11.8a) and appears to be a reworked till deposit resting on bedrock and overlain by modern sand and gravel (Fig. 11.12). Dense mud is visible beneath some large boulders, though the surface of the till is armored and largely obscured by algae-covered boulders and cobbles. Sand, apparently reworked from the till, covers the gravel on part of the outer flat (Fig. 11.12), though cobbles with attached algae are common on the sand. An extensive gravelly sand deposit (with minor mud) dominates the northern central and inner portion of the embayment and algae-covered cobbles cover its surface (Fig. 11.8e). Gravelly sand similarly covers the area surrounding Quoddy Spit. Sand forms a very thin cover along much of the tidal channel in the south, and continues into the back-barrier flat.

These sandy/gravelly deposits rest on glacial-marine mud, but reflect the former position of historic barriers (Fig. 11.4) and are likely reworked remnants of those barriers. The small freshwater peat outcrop should be older than the gravel deposits, but it is an erosional remnant with overlying materials removed (Fig. 11.9f). Scattered throughout the embayment are sand and gravel bars (Fig. 11.2). The only predominantly muddy areas are along the southern margin of the embayment where glacial-marine mud lies close the modern sea floor and crops out at the base of the fresh water peat (Fig. 11.7). Even here, cobbles and boulders with attached algae are common (Fig. 11.9e).

Coarse-grained flats such as Lubec are not well described in the literature except from high latitude regions where ice processes dominate (Dionne 1988). Lubec, Maine is on a paraglacial coastline and glacial effects linger in the form of eroding, raised gravel beach deposits and intertidal till and glacial-marine mud materials. Just as wave-dominated beaches elsewhere recycle material during a general transgression (Boyd et al. 1987), so here in Lubec, and likely other paraglacial localities, sand and gravel traverse the tidal flat as they connect former and present locations of beaches and bluffs.

11.5.3 Sediment Source(s)

The primary source of sediment to the beaches of the embayment is from erosion of sand and gravel, along with glacial-marine mud, bluffs along the margin and from reworking remnant till deposits in the intertidal and shallow subtidal region. Bluffs of peat, glacial-marine mud and regressive sand and gravel are all eroding along the southern margins of the embayment (Fig. 11.3), but the peat is presumably being oxidized and/or dispersed into the marine environment, while the mud is probably collecting in the back-barrier salt marsh. The peat bluff is retreating at up to 1.0 m/year (Mansfield 2012), and the linear shape of the margin suggests that all bluffs are retreating at roughly equal rates.

Remnant till deposits crop out around bedrock outcrops in the intertidal region, and drag marks on the flat and the tracer experiment indicate that some of this material is moving towards the spits. The raised gravel flat and sand and gravel bars also suggest that reworking of flat deposits remains an active process.

Carbonate sediment was not examined specifically, but qualitative observations indicate that shell fragments are concentrated in patches and are unlikely to comprise more than 5 % of flat and beach sediment.

11.5.4 Transport Mechanism(s)

The most remarkable aspects of the sand and gravel barriers in Lubec Embayment are their rapid historic breakdown and reorganization in a location with relatively low wave energy. The current growth of the Lubec Spit at more than a meter per year is also notable in an area with such a small fetch. Growth of the beach since the 1907 recurvature (Fig. 11.6) has required approximately 1,100 m^3/year for more than a century. Waves clearly deliver and move some sediment to the beach, especially sand (Fig. 11.8). The <20 cm waves, with only an hour or so each day they can effectively influence the spit, cannot move cobble-sized clasts along Lubec Spit, however. There is also no size sorting along the Lubec Spit to suggest wave-driven longshore transport.

The Quoddy Spit, on the other hand, does display size sorting along its length (Fig. 11.8b, c), with coarsest gravel near the eroding bluff source. Quoddy Spit is more exposed to waves. It directly faces the dominant wind direction and its position adjacent to the outer flat means that the water offshore is deeper than at Lubec, and larger waves have an opportunity to strike the beach longer. Most of the large clasts, and many attached smaller sand and gravel grains on Quoddy Spit are attached to algae, and wracks of algae and stones cover the spit (Fig. 11.8b, c).

Shaler (1895) was one of the first observers to see the potential for algal transport of clasts, though few others have continued a study of this process. In one early paper Emery and Tschudy (1941) considered both onshore and offshore transport of cobbles by kelp and discussed how important this mechanism was for both the deep sea and coast. Ben Avraham (1971) thought that algae introduced onto sandy Cape Cod (USA) beaches could transform them into gravel strands by algal transport of larger clasts. Kudrass (1974) and Gilbert (1984) were the first to begin quantitative studies, noting that the algae had to be three times the mass of the stone to float it, but that both floating of smaller rocks and dragging of larger ones brought material largely towards land. More recently Garden and Smith (2011) found that 27 % of all seaweed on a New Zealand beach was still attached to gravel.

Despite the unanimity of the few authors who noted how significant algal transport of clasts to beaches is, no study has developed a sediment budget involving algal transport. Although we do not yet have the means to do this in Lubec Embayment, clearly algal transport occurs there and could be a significant component of the sediment budget. Furthermore, algal transport appears to be an important mechanism for moving large clasts across tidal flats (and potentially shallow subtidal regions) from one depleted bluff source or unstable barrier position to another. This mechanism has not been discussed for other areas (Boyd et al. 1987), but may possibly be even more important in a regime with larger waves. Though lacking large waves, in Lubec, algal transport benefits from the powerful tidal currents in this macrotidal setting.

Ice transport is another important process in northern regions (Dionne 1984) and though not quantified in Lubec, was observed to carry blocks of tidal flat sediment onto the beaches and salt marsh (Walsh 1988; Wood et al. 1989). Ice may also be important in facilitating algal transport by pushing boulders and large clasts around on former barrier sites and till localities and, thus, releasing trapped clasts buried by larger rocks and abetting subsequent algal transport.

11.6 Conclusions

In a fetch-limited, macrotidal embayment gravel spits have undergone growth and rapid dynamical changes through the historic period. The tidal flat is modified from a "normal" textural zoning from outer to inner flat by reworking of older glacial and post-glacial deposits possibly by ice and certainly through transport of gravel clasts by attached algae. Gravel-sized clasts with attached algae abound all over the tidal flat and line the tip of Lubec Spit, which is still lengthening rapidly. Ice is also a factor in this north temperate location and ice may annually move large stones around, freeing up algae attached clasts to later transport.

References

Antony EJ, Orford JD (2002) Between wave and tide dominated coasts: the middle ground revisited. In: Cooper JAG, Jackson D (eds) Proceedings, Ireland. J Coast Res Spec Issue No. 36, pp 8–15

Bastin ES, Williams HS (1914) Description of the Eastport quadrangle. U.S. Geological Survey Atlas of the United States, Eastport Folio No. 192, 25 p

Ben Avraham Z (1971) Accumulation of stones on beaches by Codium fragil. Limnol Oceanogr 16(3):553–554

Boyd R, Bowen AJ, Hall RK (1987) An evolutionary model for transgressive sedimentation on the eastern shore of Nova Scotia. In: FitzGerald DM, Rosen P (eds) Glaciated coasts. Academic Press, New York, pp 88–114
Coastal Engineering Research Center (1984) Shore protection manual, vol 1. U.S. Army Corps of Engineers, Vicksburg
Colby GN (1881) Atlas of Washington County, Maine. George N. Colby & Company, Houlton and Machias, 1:42,000
Dionne JC (1984) An estimate of ice-drifted sediments based on the mud content of the ice cover at Montmagny, middle St. Lawrence Estuary. Mar Geol 57(1–4):149–166
Dionne JC (1988) Characteristic features of modern tidal flats in cold regions. In: de Boer PL, van Gelder A, Nio SD (eds) Tide-influenced sedimentary environments and facies. Reidel, Dordrecht, pp 301–332
Dorion CC, Balco GA, Kaplan MR, Kreutz KJ, Wright JD, Borns HW (2001) In: Weddle TK, Retelle MJ (eds) Deglacial history and relative sea-level changes, northern New England and adjacent Canada. Geological Society of America Special Paper 351, pp 215–242
Eisma D (1998) Intertidal deposits: river mouths, tidal flats, and lagoons. CRC Press, Boca Raton, 525 p
Emery KO, Tschudy RH (1941) Transportation of rock by kelp. Bull Geol Soc Am 52(6):855–862
FitzGerald DM, Van Heteren S (1999) Classification of paraglaccial barrier systems: coastal New England, USA. Sedimentology 46:1083–1108
Folk RL (1974) Petrology of sedimentary rocks. Hemphill Publishing Company, Austin, 182 p
Garden CJ, Smith AM (2011) The role of kelp in sediment transport: observations from southeast New Zealand. Mar Geol 281(1):35–42
Gilbert R (1984) The movement of gravel by the alga *Fucus vesiculosus* (L) on an intertidal Arctic flat. J Sediment Petrol 54(2):463–468
Hill H, Kelley JT, Belknap DF, Dickson SM (2004) The Effects of storms and storm-generated currents on the sand beaches in southern Maine. Mar Geol 210(1–4):149–168
Holland KT, Elmore PA (2008) A review of heterogeneous sediments in coastal environments. Earth Sci Rev 89(3–4):116–134
Kelley JT, Belknap DF, Claesson S (2010) Drowned coastal deposits with associated archeological remains from a sea-level "slowstand": northwestern Gulf of Maine, USA. Geology 38(8):695–698
Kudrass HR (1974) Experimental study of nearshore transportation of pebbles with attached algae. Mar Geol 16:M9–M12
Mansfield ME (2012) The critical leading edge of Gulf of Maine salt marshes-interface with freshwater wetlands. University of Maine Masters thesis, Orono, Maine, 152 p
Masselink G, Short AD (1993) The effect of tide range on beach morphodynamics and morphology: a conceptual beach model. J Coast Res 9(3):785–800
Orford JD, Carter RWG, Jennings SC (1996) Control domains and morphological phases in gravel-dominated coastal barriers of Nova Scotia. J Coast Res 12(3):589–604
Orford JD, Jennings SC, Pethick J (2003) Extreme storm effect on gravel-dominated barriers. In: Davis RA (ed) Proceedings on the International Conference on Coastal Sediments 2003. (Corpus Christi, TX), CD-ROM published by World Scientific Publishing Corp. and East Meets West Productions, p 12
Orford JD, Forbes DL, Jennings SC (2012) Organisational controls, typologies and time scales of paraglacial gravel-dominated coastal systems. Geomorphology 48(1–3):51–85
Osberg PH, Hussey AM, Boone GM (1985) Bedrock geologic map of Maine. Augusta, Maine, Maine Geological Survey, scale 1: 500,000, 1 sheet
Putnam R, Stone J, Titcomb S (1785) Township survey No. VIII including Moose Island. Machias, Maine, Registry of Deeps, scale 1:50,000, 1 sheet
Shaler NS (1895) Sea and land, features of coasts and oceans with special reference to the life of man. Smith, Elder and Co., London, 252 p
Short A (1991) Macro-meso tidal beach morphodynamics-an overview. J Coast Res 7(2):417–436

Smith DC, Borns HW, Anderson RS (1989) Relative sea-level changes measured by historic records and structures in coastal Maine. In: Anderson WA, Borns HW (eds) Neotectonics of Maine: studies in seismicity, crustal warping and sea-level change, vol 40. Maine Geological Survey Bulletin, Augusta, pp 127–137

U.S. Coast and Geodetic Survey (1862) Preliminary chart of Eastport Harbor. Washington, DC: Maine Historical Society, 1:40,000, 1 sheet

U.S. Geological Survey (1908) Eastport, Maine topographic map. Washington, DC, 1:62,500, 1 sheet

Wailing HF (1861) Map of Washington County, Maine. Lubec, Maine, Lubec Town Library, 1:117,000, 1 sheet

Walsh JA (1988) Sedimentology and late Holocene evolution of the Lubec Embayment. Orono, Maine. University of Maine, Masters thesis, 434 p

Wood M, Kelley JT, Belknap DF (1989) Patterns of sediment accumulation in the tidal marshes of Maine. Estuaries 12(4):237–246

Chapter 12
Sandy Spits and Their Mathematical Modeling

Magnus Larson, Jaime Palalane, and Hans Hanson

Abstract A mathematical model of spit growth was derived based on the sand conservation equation, where the boundary fluxes of sand were specified as input and specific assumptions were made about the cross-sectional shape of the evolving sandy barrier. Analytical solutions were developed to the governing equation to primarily simulate spit elongation under unrestricted and restricted growth, although the time evolution of the spit cross-sectional area was also described for some cases. The solutions were validated with data from the field and the laboratory, including field data on unrestricted spit growth from Sweden and United States, and laboratory data on restricted spit growth from the Coastal and Hydraulics Laboratory in the United States. The laboratory data allowed for validating analytical solutions of spit growth involving varying spit cross-sectional area and active profile height, as well as the effect of sand transport through an inlet that erodes the tip of the spit. Two field cases constituting more complex sand transport conditions at inlets, where different morphological features interacted with the spit evolution, were also investigated, but numerical approaches were employed to solve the governing equations. These two cases included spit growth at Fire Island Inlet on Long Island in the United States, and at Chilaw Inlet in Sri Lanka. Overall, the mathematical model of spit growth reproduced the data well, although some calibration of relevant parameters and coefficients were typically needed.

M. Larson (✉) • H. Hanson
Water Resources Engineering, Lund University, Box 118, 22100 Lund, Sweden
e-mail: Magnus.Larson@tvrl.lth.se; Hans.Hanson@tvrl.lth.se

J. Palalane
Water Resources Engineering, Lund University, Box 118, 22100 Lund, Sweden

Eduardo Mondlane University, C.P. 257 Maputo, Mozambique
e-mail: Jaime.Palalane@tvrl.lth.se

G. Randazzo et al. (eds.), *Sand and Gravel Spits*, Coastal Research Library 12,
DOI 10.1007/978-3-319-13716-2_12

12.1 Introduction

One of the first definitions of a spit was provided by Johnson (1919) as "an embankment (that) has its distal end terminating in open water". Even earlier than that, Wheeler (1902) pointed out the major characteristics of a spit: (1) predominant drift in one direction; (2) shape of a long narrow bank; (3) commencing at some salient point; (4) running in a direction parallel to the general coastline trend; and (5) leaving a protected bay behind as it expands. For the present study, the more specific definition given by Kraus (1999) is suitable where he defined spits as "organized surface-piercing accumulations of sediment that grow by transport directed from a landmass or sediment source toward a water body". In the geomorphology literature a spit has often been classified as a type of barrier (Shephard 1952) and Otvos (2012) presents a comprehensive discussion of coastal barriers.

Spits typically consist of sand or gravel and are commonly occurring morphological features at inlets, river mouths, and the down-drift ends of a barrier islands (Schwartz 1972). Thus, they may form at the ocean, lake, or bay side of inlets, entrances, and river mouths (Kraus and Seabergh 2002). Apart from the scientific interest in spits and their evolution, engineers have often studied spits with regard to their penetration into river mouths or inlets, restricting the flow rate and possibly even causing closure of the inlet (river mouth; *e.g.*, Tanaka et al. 1996). A reduced inlet opening could imply problems for navigation as sediment causes shoaling of the channel. Furthermore, a smaller inlet area because of spit growth results in less water exchange between the bay (lagoon) and the ocean (or lake), which may cause water quality problems.

The spit growth can occur under a predominant longshore transport driven by wave-generated mean currents with little interference from other currents (*e.g.*, tidal or wind-driven currents), denoted unrestricted growth (Kraus 1999). If other currents, often directed perpendicular to the main spit growth direction (*i.e.*, the direction of the longshore current bringing the sediment alongshore), are sufficiently strong to influence the sediment transport and spit development, the spit growth is considered to be restricted. In the case of restricted spit growth, a balance could be obtained between the longshore sediment transport and the transport by other currents, so that the sediment supply to the spit equals the sediment removal and a dynamic equilibrium prevails with a constant spit length.

The present study focuses on how to mathematically model spit evolution for an increased understanding of their behavior and the governing factors. Also, by developing such models quantitative predictions can be made that will be of use in practical investigations. After a brief review of the physical processes governing spit growth, a general model for spit growth is formulated based on the sediment conservation equation, following the work by Kraus (1999). The model is only applied to describe spit growth in the direction of the main shoreline trend, and not cases where the spit is curving. A number of analytical solutions to the governing equations are derived for schematized, yet realistic cases. These solutions primarily encompass the spit elongation including both restricted and unrestricted growth, but also the case of a varying spit width is treated. Solutions are presented for spit growth into increasing water depth and into an inlet where the offshore sediment transport is changing under a reduced cross-sectional area. The analytical solutions are compared with data from the field and laboratory. Two field cases are also briefly discussed where spit evolution is modeled in a more complex setting of an inlet with different morphological features and sediment transport paths. The paper ends with concluding remarks concerning the usefulness of the model developed and the solutions obtained.

12.2 Physical Processes of Spit Growth

Spits have often been classified with regard to their shape, which is closely linked to the formative processes and geomorphological constraints to their development. Again, Johnson (1919) was a pioneer in formulating a classification system for spits, identifying the following types: (1) linear, (2) recurved, (3) compound, recurved, (4) complex, and (5) serpentine (the latter type not very common and not discussed here). Although the division into the different types was mainly based on the spit configuration, Johnson (1919) provided explanations regarding their formation as a result of the processes acting and how they varied in time and space.

A linear spit develops (in the direction of the main coastline trend), if the sediment transport due to the longshore current (Q_L) is dominant compared to other transport mechanisms. The presence of tidal or wind-induced currents may induce transport perpendicular to the main coastline trend (Q_C), making the spit recurve. Johnson (1919) only considered such currents as agents in the formation of recurved spits, but subsequent research has shown that this kind of shape can also occur due to wave refraction at the tip of the spit or variation in the direction of the incoming waves (Evans 1942; Petersen et al. 2008).

If the transport mechanisms affecting spit growth vary in time compound, recurved spits can evolve, as periods with dominant linear growth ($Q_L \gg Q_C$) is followed by periods when the spit recurves (Q_L and Q_C of similar magnitude). Variations in Q_L that affect spit growth can occur on many time scales, from single storm events, over seasonal fluctuations, and to long-term climatic changes (Allard et al. 2008). Complex spits experience compound, recurved development at the same time as secondary spits form, growing independently from the main spit, typically on the bay side.

Kraus (1999) tabulated the most important parameters controlling spit geometry and evolution together with the governing processes acting on different time scales (short and long term). The parameters discussed were length, elongation speed, width, elevation, depth of closure, tendency to recurve, and presence of overwash fans. The five first of these parameters will be included in the mathematical model of spit growth presented here.

The spit length refers to the distance from its origin (which may be arbitrarily defined) to the tip. This length could refer to a specific time, but also to a situation where a quasi-equilibrium state has been attained where Q_L and Q_C are in balance. Thus, the longshore sediment transport rate, and its temporal and spatial variation, is a decisive factor for the length, both over the short- and long-term. If the spit penetrates an inlet, Q_C and its dependencies will be important, as well as any geological controls present (Larson et al. 2002a). The elongation speed corresponds to the change in spit length per unit time and is related to Q_L, the spit geometry (*e.g.*, width, elevation, depth of closure), and the presence of any currents generating a transport Q_C.

The geometry of the spit is often described in terms of the cross-sectional shape, which in turn is characterized by the width (often taken at the mean sea level), elevation (to the crest of the subaerial part of the spit), and the depth of closure. This shape is to a large extent determined by cross-shore sediment movement initiated by the waves, but also on the tidal range and the general bathymetry of the area where the spit is developing. The maximum elevation is typically a function of the wave runup, which is proportional to the square-root of the wave height and to the wave period directly. For gravel beaches, the grain size will influence the wave runup markedly, since percolation of water into the beach becomes important. The width of the spit is also closely related to the wave runup, whereas the depth of closure is primarily determined by the wave height, period, and grain size (Hallermeier 1981).

12.3 Mathematical Modeling of Spit Growth

12.3.1 General Formulation

The derivation of a mathematical model for spit evolution will to a large degree follow the work by Kraus (1999), also discussed in Kraus and Seabergh (2002). However, the model will be generalized to describe varying spit width and will employ different formulations for the influence of spit cross-sectional shape, active profile height (*i.e.*, berm height plus depth of closure), and inlet sediment transport. Analytical solutions to the model will be developed for a range of situations, necessitating simplified descriptions of initial, boundary, and forcing conditions. For more complex conditions, numerical solutions can easily be obtained using the present model.

The basic governing equation for the spit evolution may be written,

$$\frac{\partial}{\partial t}\left(\int_0^{x_s} A_s dx\right) = Q_{in} - Q_{out} \tag{12.1}$$

where A_s is the cross-sectional area of the spit at a distance x from the origin, x_s the location of tip of the spit, Q_{in} the longshore sand transport at the updrift end of the spit ($x=0$), Q_{out} the transport at the downdrift end of the spit ($x=x_s$), and t time. Equation 12.1 is obtained by employing the sand volume conservation equation for the entire spit with Q_{in} and Q_{out} being the boundary fluxes and the time derivative of the integral represents the changes in spit volume. The selection of the initial location of the spit, corresponding to $x=0$, is arbitrary, although closely related to the specification of Q_{in}. For unrestricted spit growth there would be no transport at the tip of the spit, that is, $Q_{out}=0$, and all sediment supplied to the spit would result in growth, either through elongation or widening of the spit. In the case of restricted spit growth, for example, at an inlet or river mouth where cross-shore flows may erode the tip of the spit, Q_{out} would differ from zero and typically vary with time.

In the general case, the cross-sectional area of the spit would vary with time and for constant wave and water level conditions approach an equilibrium value ($A_s=A_e$). For a growing spit, sand is supplied by the longshore transport, but the build-up of the cross-sectional shape is due to the cross-shore transport. The cross-shore response is often faster than the longshore response, so a reasonable assumption may be that $A_s=A_e$, implying that the variation in time of A_s is not described. This assumption simplifies the solution of Eq. 12.1 and avoids the need to find an equation for the build-up of the spit cross section. Here, both the cases where $A_s=A_e$ and A_s varies with time will be studied.

The particular cross-sectional shape of the spit may also be of interest, especially since existing data often report the spit plan form with regard to mean se level. Thus, assumptions have to be made about the spit shape and the active profile height, that is, the vertical distance over which the sand transport occurs. Kraus (1999) assumed that the active height (D) is the sum of the berm elevation (B) and the depth of closure (D_c), and the same assumption is employed here. In his model, Kraus (1999) estimated A_s as a constant spit width (W_s) times D ($= B+D_c$), implying a rectangular shape over the active height. Other spit cross-sectional shapes, however, may be more realistic, although in the present type of model, emphasizing an analytical approach, the difference between various assumed shapes is only a constant form factor. In the case of a trapezoidal cross section, with a linearly sloping beach at the landward and seaward side, the cross-sectional area is,

$$A_s = DW_s\left(1 + K_\beta \frac{D_c - B}{2W_s}\right) \tag{12.2}$$

where $K_\beta = 1/\tan\beta_s + 1/\tan\beta_l$ with β_s and β_l being the slope on the seaward and landward side, respectively. In Eq. 12.2, $A_s = \psi D W_s$ where ψ is a constant that depends on the particular cross-sectional shape of the spit. Thus, in comparison with data, the shape will not have any influence on the general trend concerning, for example, spit elongation, but with regard to the magnitude of change and the estimated Q_{in} from the data it will be of importance. This aspect will be discussed more in connection with the data comparison.

12.3.2 Unrestricted Growth

Unrestricted spit growth implies no transport at the tip ($Q_{out} = 0$), but all the incoming sand will benefit the growth. Furthermore, if $A_s = A_e$, Eq. 12.1 may be simplified to:

$$\frac{d}{dt}(A_e x_s) = Q_{in} \tag{12.3}$$

Assuming that Q_{in} is constant, the following solution is obtained for the spit elongation,

$$x_s = \frac{Q_{in}}{A_e} t \tag{12.4}$$

where $x_s = 0$ when $t = 0$. Thus, for a constant influx of sediment to the spit, a linear growth rate is expected, as previously discussed by Kraus (1999). If Q_{in} is allowed to vary in time, Eq. 12.3 can still be solved analytically for simple expression of this variation. Kraus (1999) presented the solution for a sinusoidal variation in Q_{in}. The speed of propagation of the spit is given by $dx_s/dt = Q_{in}/A_e$ and if this speed is known from field observations, the mean longshore transport rate feeding the spit may be estimated as $Q_{in} = (dx_s/dt)A_e$.

12.3.3 Time-Varying Cross-Sectional Spit Shape

As previously pointed out, cross-shore transport is the main factor for building the cross-sectional shape of the spit, implying that A_s varies with time. A detailed model of this build-up, will not allow for analytical treatment and here a simple formulation will be applied that captures the main features of the process. The basic assumption is that at a particular location x, a portion of Q_{in} is used to build up the cross section towards its equilibrium value (A_e) and the remaining portion is transported down-drift. At any given time, A_s/A_e is assumed to attain a ratio corresponding to Q/Q_{in}, where Q is the local transport rate, giving:

$$A_s = A_e \frac{Q}{Q_{in}} \tag{12.5}$$

In order to close the problem, it is necessary to specify how Q varies with x, and a general, heuristic expression is,

$$\frac{Q}{Q_{in}} = \left(1 - \frac{x}{x_s}\right)^m \tag{12.6}$$

where m is a constant (>0). Equation 12.6 yields $Q = Q_{in}$ for $x = 0$ and $Q = 0$ for $x = x_s$, with the value on the power m determining the shape of the variation between these two end points. If $m = 0$, then $Q = Q_{in}$, implying that $A_s = A_e$ for all x at all times, and Eq. 12.4 describes the spit elongation. For all other values on m, the spit elongation will be larger for a specific Q_{in}, since the cross section will not be built up to equilibrium instantaneously, and more sand will be transported downdrift towards the tip.

Substituting Eqs. 12.5 and 12.6 into Eq. 12.1, still letting $Q_{out} = 0$, results in the following,

$$\frac{d}{dt}(A_e x_s) = (m+1)Q_{in} \tag{12.7}$$

which for Q_{in} being a constant yields:

$$x_s = (m+1)\frac{Q_{in}}{A_e}t \tag{12.8}$$

The time variation in cross-sectional area at a particular location x is obtained from Eqs. 12.3, 12.6, and 12.8 as,

$$A_s = A_e\left(1 - \frac{A_e x}{(m+1)Q_{in}t}\right)^m \tag{12.9}$$

where $0 < x < (m+1)Q_{in}t/A_e$.

12.3.4 Increasing Active Profile Height

Similar to Kraus (1999), the effect of an increase in D with x from a constant value D_o was investigated, but using a slightly different formulation. Kraus (1999) assumed linearly increasing depth with distance from the spit starting at an active height D_o, whereas in the present study the following expression was used,

$$D = D_o + \Delta D(1 - \exp(-\alpha x)) \tag{12.10}$$

where ΔD is the increase in D occurring at large distances from the spit and α a rate coefficient quantifying the approach towards the final depth $D_o + \Delta D$. Substituting Eq. 12.10 into Eq. 12.1 and using Eq. 12.2 for introducing D yields an equation that can be solved analytically. It is also possible to employ a varying spit cross-sectional shape (*i.e.*, W_s), but only for $m = 1$. The solution for this case is given by,

$$\frac{1}{2}x_s\left(1 + \frac{\Delta D}{D_o}\right) + \frac{\Delta D}{\alpha D_o}\left(\frac{1}{\alpha x_s}(1 - \exp(-\alpha x_s)) - 1\right) = \frac{Q_{in}}{A_e^o}t \tag{12.11}$$

where A_e^o is the equilibrium cross-sectional area defined based on D_o and x_s is given as an implicit function of t. Again, Eq. 12.11 is valid for unrestricted spit growth.

For small values on x_s, Eq. 12.11 may be approximated by $x_s = 2Q_{in}t/A_e^o$, which corresponds to Eq. 12.8 for $m=1$ and using an equilibrium area based on D_o. For the other asymptotic case when x_s becomes large, Eq. 12.11 reduces to $x_s = 2Q_{in}t/A_e^\infty$, which again is obtained from Eq. 12.8 for $m=1$, but using an equilibrium area (A_e^∞) based on $D_\infty = D_o + \Delta D$. Thus, Eq. 12.11 produces linear growth rates for the spit elongation for small and large values on x_s.

12.3.5 Restricted Growth

At an inlet (or river mouth) it is expected that Q_{out} is different from zero (see Eq. 12.1) and that the tip of the spit will suffer from erosion due to this transport. As the spit grows across the inlet, a reduction in cross-sectional area of the inlet (A_i) occurs that typically implies a larger velocity in the inlet and a higher Q_{out}. A state may be attained when $Q_{in} = Q_{out}$, which implies no change in the spit shape and equilibrium prevails. If no such equilibrium is possible, the spit may grow across the inlet and close it (*e.g.*, Ranasinghe and Pattiaratchi 1999; Larson et al. 2009).

In order to model spit growth at an inlet, including a possible equilibrium state, the variation in Q_{out} as the inlet cross section is reduced has to be specified. Kraus (1999) assumed that the spit evolved from a specific location x_o towards the middle of the inlet channel (located at x_c), and when $x = x_c$, $Q_{out} = Q_{in}$ and equilibrium would occur. For $x = x_o$, $Q_{out} = 0$, and between x_o and x_c the transport through the inlet would grow linearly. Although producing a correct behavior, the formulation for Q_{out} is somewhat *ad hoc* and x_o is difficult to estimate for a particular case.

Here, a different expression for Q_{out} was employed based on simple considerations of the physics. It is assumed that the transport through the inlet can be described by $Q_{out} = C_t u_i^3 x_i / g$ (*e.g.*, Watanabe et al. 1991; Kraus 1998; Larson et al. 2011), where u_i is the velocity through the inlet, x_i the initial inlet width (before spit penetration), C_t an empirical transport coefficient, and g acceleration due to gravity. Furthermore, assuming a rectangular inlet cross section and a volume flow through the inlet that remains approximately constant as the spit grows yields the following expression,

$$Q_{out} = \frac{Q_o}{(1 - x_s/x_i)^2} \tag{12.12}$$

where Q_o is the transport through the inlet before the spit starts affecting the flow ($x_s = 0$). It is apparent from Eq. 12.12 that inlet closure cannot be described since $Q_{out} \to \infty$ as $x_s \to x_i$, and at some point $Q_{out} = Q_{in}$, implying equilibrium. In reality, the assumption of a constant flow will be violated as the spit grows through the inlet; however, to model the effect on the inlet flow in a more sophisticated manner requires a numerical approach.

Substituting Eq. 12.12 into Eq. 12.1 yields:

$$\frac{\partial}{\partial t}\left(\int_0^{x_s} A_s dx\right) = Q_{in} - \frac{Q_o}{(1 - x_s/x_i)^2} \tag{12.13}$$

For a cross-sectional area not varying in time ($A_s = A_e$), the solution to Eq. 12.13 is,

$$\sqrt{\delta}\,\mathrm{arctanh}\left(\frac{\sqrt{\delta}x_s}{(1-\delta)x_i - x_s}\right) + \frac{x_s}{x_i} = \frac{Q_{in}}{A_e x_i}t \tag{12.14}$$

where $\delta = Q_o/Q_{in}$ and x_s is given as an implicit function of t. At equilibrium, Eq. 12.13 gives,

$$x_{se} = x_i\left(1 - \sqrt{\delta}\right) \tag{12.15}$$

with x_{se} being the spit length at equilibrium, where the spit starts growing from one side of the inlet. For small values on the relative spit length (*i.e.*, x_s/x_i), Eq. 12.14 may be simplified to yield an explicit expression,

$$x_s = \frac{1}{2}x_i\left(1 + \frac{t}{t_o}\right)\left(1 - \sqrt{1 - \frac{4(1-\delta)t/t_o}{(1 + t/t_o)^2}}\right) \tag{12.16}$$

where $t_o = A_e x_i/Q_{in}$, which corresponds to the time it would take for the spit to propagate across the inlet assuming unrestricted growth. For small t/t_o Eq. 12.16 results in $x_s = x_i(1-\delta)t/t_o$.

Figure 12.1 illustrates in non-dimensional quantities the solution given by Eq. 12.14 for different values on δ. The following quantities were introduced: $x_s' = x_s/x_i$, $t' = t/t_o$, and δ ($= Q_o/Q_{in}$). The approximate solution (Eq. 12.16) is also given and for small x_s' the agreement is quite good. In fact, for most values on δ, Eq. 12.16 is accurate up to more than 90 % of the equilibrium value (x_{se}'). The final approach to the equilibrium spit length occurs asymptotically in time and in Fig. 12.1 the calculation was stopped when $x_s' = 0.99x_{se}' = 0.99\left(1 - \sqrt{\delta}\right)$.

Equation 12.14 is the solution for $A_s = A_e$, that is, $m = 0$ in Eq. 12.6. However, Eq. 12.13 can be solved for any value on m and the only difference is that the right-hand side of Eq. 12.14 should be multiplied by $m + 1$. Equation 12.6 was formulated based on the assumption of unrestricted growth, and if the tip of the spit is eroding because of Q_{out} a more suitable formulation would be $Q = Q_{out} + (Q_{in} - Q_{out})(1 - x/x_s)^m$. This expression yields Q_{in} for $x = 0$ and Q_{out} for $x = x_s$, and reduces to Eq. 12.6 for $m = 0$. If Q_{out} is constant, the solution is the same as Eq. 12.14, but the right-hand side should be multiplied with $(m+1)/(m\varepsilon + 1)$, where $\varepsilon = Q_{out}/Q_{in}$. If Q_{out} is allowed to vary with x_s/x_i as given by Eq. 12.12, an analytical solution is still possible, but it is rather extensive and will not be reported here.

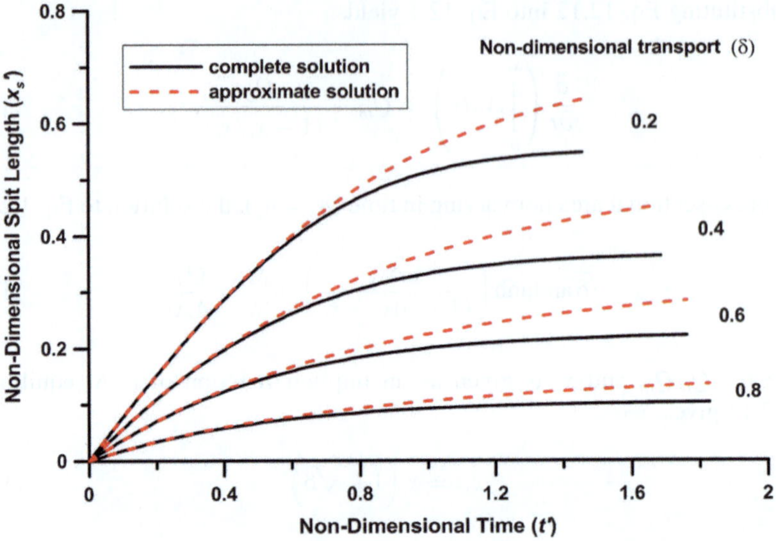

Fig. 12.1 Analytical solution in non-dimensional form for spit elongation into an inlet, where transport rate increases with reduced inlet width (complete and approximate solution shown)

12.3.6 Coupling to Inlet Processes

Restricted spit growth typically occurs at a river mouth or inlet where several other morphological features are present, for example bars and shoals (FitzGerald 1988; FitzGerald et al. 2000). The interactions between the spit and these features are complex, but many times necessary to describe. Even if such a description is greatly schematized, analytical solutions are in general not possible, but numerical approaches must be taken to simulate the evolution of the spit and the features.

Kraus (2000) presented a schematized approach to simulate the sediment transfer at an inlet by using the analogy between a specific morphological unit (feature) and a reservoir. Each such unit (reservoir) has a certain maximum capacity to store sediment corresponding to its equilibrium volume (V_{ME}), and the rate at which sediment is transported downdrift (Q_D) from the unit is directly proportional to the ratio between the amount of sediment stored in the unit (V_M) and V_{ME}. Thus, the governing equations for the evolution of the unit are, respectively, a sediment volume conservation equation and a sediment transfer equation,

$$\frac{dV_M}{dt} = Q_S - Q_D \tag{12.17}$$

$$Q_D = Q_S \frac{V_M}{V_{ME}} \tag{12.18}$$

where Q_S is the sediment transport into the unit (could be transport coming out from another unit or the beach located updrift).

For each morphological unit modeled a set of equations as those given by Eqs. 12.17 and 12.18 is employed, where couplings between the units are obtained through the sediment transport rates (Q_S and Q_D). The most updrift located unit will typically be fed by the longshore sediment transport and the most downdrift located unit will supply the downdrift beach with sediment through its output sediment transport (*i.e.*, bypassing). Often the units are connected in series and sediment is passed through them from the updrift to the downdrift side (Larson et al. 2002b), but more complex couplings are possible (Larson et al. 2006). In Eq. 12.18 the equilibrium volume of the morphological unit must be specified, which is often based on empirical relationships (*e.g.*, Walton and Adams 1976).

Kraus (2000) in his first modeling efforts identified three morphological units, namely the ebb shoal, bypassing bar, and attachment bar. Subsequently, other units have been included in such modeling, including the flood shoal, inlet channel, and spits. The main difficulties in applying the reservoir model is the schematization of the inlet into morphological units and defining the pathways of sediment transport between them. Typically, extensive observations and data from the site are required to set up the model and to validate it.

In a spit modeling context, when solving Eq. 12.1 to obtain the spit growth Q_{in} and Q_{out} must be specified. The transport Q_{in} may be given directly by the longshore sediment transport (Q_L), but often only a portion of Q_L may feed the spit and the other portion goes to offshore shoals and bars. Also, Q_{out} may be given by the cross-shore transport in the channel (Q_C), although other sources of sediment typically contribute to Q_C than the erosion of the spit.

12.4 Case Studies

A number of case studies were performed to test the analytical solutions developed, both for unrestricted and restricted growth. Three field cases from the literature focused on unrestricted growth, namely Skanör-Falsterbo Peninsula in Sweden and San Bernardo River Mouth and Corpus Christi North Beach in Texas, United States. In all these cases pronounced spits could grow unrestricted in the direction of the main longshore transport. The laboratory studies on spit growth carried out at the US Army Engineer Research and Development Center (ERDC) were employed to validate the solutions for a time-varying spit cross-sectional area and active profile height, as well as for inlet sediment transport eroding the tip of the spit.

Furthermore, two case studies from the field are presented that involve more complex conditions of spit growth, requiring numerical solutions, and where the spit growth model is a part of a larger model simulating coastal evolution. The first case focused on modeling the evolution of the spit at Fire Island Inlet on Long Island (Hoan et al. 2011). Spit elongation was simulated for conditions where the input longshore transport varied in time as well as for inlet conditions involving structures and dredging. The second case is from Chilaw Inlet on the western coast of Sri Lanka, where a reservoir-type model was developed to simulate spit elongation (Larson et al. 2009).

12.4.1 Skanör-Falsterbo Peninsula, Sweden

A pronounced spit, known as Badreveln Spit, has been steadily growing in the northern part of the Skanör-Falsterbo Peninsula in the very southwest corner of Sweden since the latter half of the nineteenth century (Blomgren and Hanson 2000; Larson and Hanson 2013). The spit started to form after a harbor was constructed to the south that caused a redirection of the transport and new growth path for the spit. A relict spit exists behind Badreveln spit that indicates the transport direction and spit evolution prior to the harbor construction.

Data on the evolution of the Badreveln Spit was presented by Blomgren and Hanson (2000) and its elongation was modeled by Hoan et al. (2011) using the analytical solution presented by Kraus (1999) for unrestricted growth. Here, the data were first extended using images from Google Earth that encompassed the years 2007, 2009, 2010, and 2012; thus, the studied period of spit growth was from 1916 to 2012, where the start of the period was selected based on the first clear documentation of the spit location.

The procedure employed to investigate how well analytical solutions could reproduce observed data was similar for all case studies. For each case, some parameters in the solution was estimated to yield an optimum fit between measured and calculated values. In order to assess the validity of the solutions, the general trend of the calculations was compared with the measured trend. Also, the optimum parameter values were, if possible, compared with observations at the site to judge how realistic the values were.

Figure 12.2 illustrates how well Eq. 12.8 could describe the elongation of Badreveln Spit. In spite of some fluctuations, the spit elongation rather well follows a linear trend with a best-fit line indicating an elongation rate of about 28 m/year ($k = (m+1)Q_{in}/A_e$). The figure also includes lines corresponding to rates that are 20 % more and 20 % less than the best-fit rate to indicate the sensitivity of the elongation to the variation in k (provides a quantitative estimate of how uncertainties in k will affect x_s). Overall the agreement between the analytical solution and the measurements is good, although there is a tendency to overestimate k during the first 50 years and underestimate it during the last 50 years. This could be related to the reduction in sediment transport that the harbor initially caused, before the deposition updrift the harbor was sufficient to allow for bypassing of material to feed the spit.

Hoan et al. (2011) estimated that the width of the spit was about 70 m based on aerial photos and Hanson and Larson (1993) observed an active profile height of about 5 m from beach profile surveys (see also Larson and Hanson 2013). Thus, setting $m = 0$ and assuming a rectangular spit cross-sectional shape (*i.e.*, $\psi = 1$; see Eq. 12.2) yields a Q_{in} of about 10,000 m^3/year. However, for another shape, for example, trapezoidal, Q_{in} would be about 25,000 m^3/year after suitable assumptions about side slopes ($\psi = 2.5$). There is limited information on the net sediment transport along the Skanör-Falsterbo Peninsula, but Larson and Hanson (2013) estimated it to be about 40,000 m^3/year south of the harbor from numerical

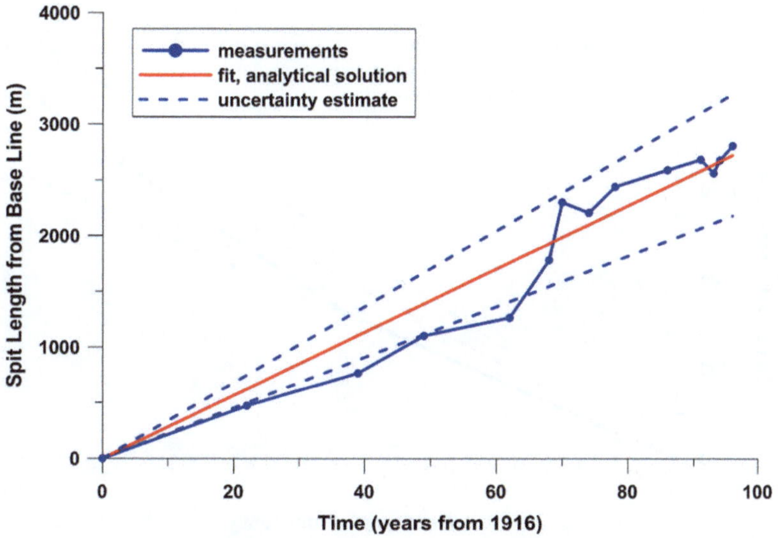

Fig. 12.2 Fit of analytical solution for unrestricted growth to data on spit elongation from Skanör-Falsterbo Peninsula in Sweden

modeling, and about half of this transport would pass the harbor and feed the spit. Thus, the value on Q_{in} obtained from the analytical solution is in agreement with the data existing from the site.

12.4.2 San Bernard River Mouth, Texas

San Bernard River mouth is located in Texas on the Gulf of Mexico coast (see Kraus and Lin 2002). The river is a small stream with limited flow and sediment transport. East-west directed sediment transport along the coast makes the mouth subject to spit growth and closure. Dredging of the river mouth has been carried out periodically since the 1940s to satisfy navigational requirements and to prevent closure.

Kraus and Lin (2002) presented data obtained from aerial photos on the spit length as well as on the width of the spit and the width of the river mouth (see also Kraus and Seabergh 2002). These data showed that the spit grew linearly in the westward direction with an almost constant width of 300 m. The time period covered by the aerial photos was from April 1989 to July 2001. In the present analysis the length of the spit was simply referenced to the location on the first photo. Beach profiles around the river mouth were also surveyed at three occasions with a distance between transects of about 30 m. The spit displayed an approximately trapezoidal shape, except close to the tip, with a maximum elevation of 1.5 m above mean low tide.

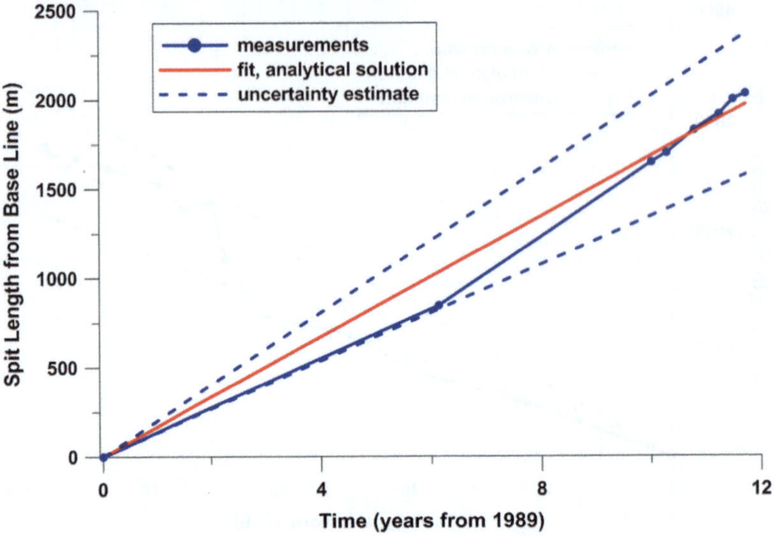

Fig. 12.3 Fit of analytical solution for unrestricted growth to data on spit elongation from San Bernard River Mouth, Texas

Figure 12.3 illustrates how well the analytical solution for unrestricted spit growth (Eq. 12.8) could be fitted to the data on spit elongation from San Bernard River mouth. By employing the solution for unrestricted growth it was assumed that the erosion of the spit due to the river flow is negligible, which is in accordance with the observed weak flow in San Bernard River. The best fit gave an elongation rate of $k = 168$ m/year for the time period studied. The effect of varying k is also shown in the figure by plotting the spit evolution for k-values that are 20 % more and 20 % less than the best-fit k. Although the data indicate somewhat more rapid spit growth during the latter part of the studied period, a linear fit seems to work well.

The profile data obtained by Kraus and Lin (2002) yielded an estimate of the active profile height of about 6 m, which would yield a transport rate feeding spit growth of $Q_{in} = 300{,}000$ m^3/year for a rectangular spit shape. However, a detailed analysis by Kraus and Lin (2002) of the spit and river mouth bathymetry showed a limited depth in the mouth that would indicate that a lower active profile height contributing to spit growth; that is, sediment transported at a depth below the river mouth depth would bypass the mouth and not contribute to the spit growth. Such an interpretation would give $D = 3$ m and $Q_{in} = 150{,}000$ m^3/year (rectangular spit shape). For this value on D, B and D_c would be approximately the same and $\psi = 1$ (Eq. 12.2). The value on Q_{in} is in agreement with the estimate of Kraus and Lin (2002) provided based on their measurements and also in the range of previous studies.

12.4.3 Corpus Christi North Beach, Texas

Corpus Christi Beach is a bay shore located on the western side of Corpus Christi Bay (Kraus 1999), which is connected to the Gulf of Mexico mainly through Aransas Pass. The beach is approximately oriented north-south and is presently contained by a jetty at its southern end and a groin at its northern end. During the time of observation, however, there was no groin at the northern end and sediment could be transported towards the north without any interruption.

In October 1977 the US Army Corps of Engineers started to construct an artificial beach consisting of close to 400,000 m^3 of silt overlaid by 300,000 m^3 of sand (mean diameter 0.4 mm). Longshore sediment transport to the north immediately started to build a spit at the northern end that grew towards the opening to Nueces Bay (Kraus 1999). By 1982 the spit had extended more than 600 m and started to impinge on the causeway passing the opening to Nueces Bay. Kraus (1999; see also Kraus and Seabergh 2002) digitized aerial photos to document the spit location at specific times, including the spit length and the spit width. The period of analysis extended from 1977 to 1984, where the spit growth during the last few years was affected by the causeway, as observed through an increase in width. During the major part of the spit growth, the width remained fairly constant, being about 20 m, whereas towards the end of the investigated period when $x_s > 500$ m the width increased to about 30 m close to the tip.

Figure 12.4 displays the fit of the analytical solution for unrestricted spit growth (Eq. 12.8) to the data on spit elongation from Corpus Christi North Beach. The optimum value on k was estimated to 80 m/year, and lines corresponding to 20 % more and 20 % less than the best-fit k are given as an indication of the sensitivity towards k. A linear fit describes the spit elongation well, although there is a tendency for the rate to decrease towards the end of the period, probably related to the impingement on the causeway and the spreading out of the spit. Beach profile surveys were conducted in order to monitor the beach fill and from that Kraus (1999) estimated and active profile height of 3.5 m. Thus, for a rectangular spit shape Q_{in} becomes about 5,500 m^3/year, whereas if a trapezoidal shape is assumed based on the surveyed profile shape, $\psi = 2.0$, implying $Q_{in} = 11{,}000$ m^3/year. Kieslich and Brunt (1989) calculated the longshore sediment transport to be about 9,000 m^3/year based on repetitive profile surveys and Kraus (1999) obtained $Q_{in} = 8{,}500$ m^3/year using his analytical model for spit growth.

12.4.4 ERDC Physical Model Study

No data sets from the field were available to validate the analytical solutions developed for varying spit cross-sectional area (*i.e.*, width) and active profile height, as well as for the influence of a current (river or inlet) on spit erosion. However, one data set from the laboratory discussed by Kraus (1999) and Kraus and

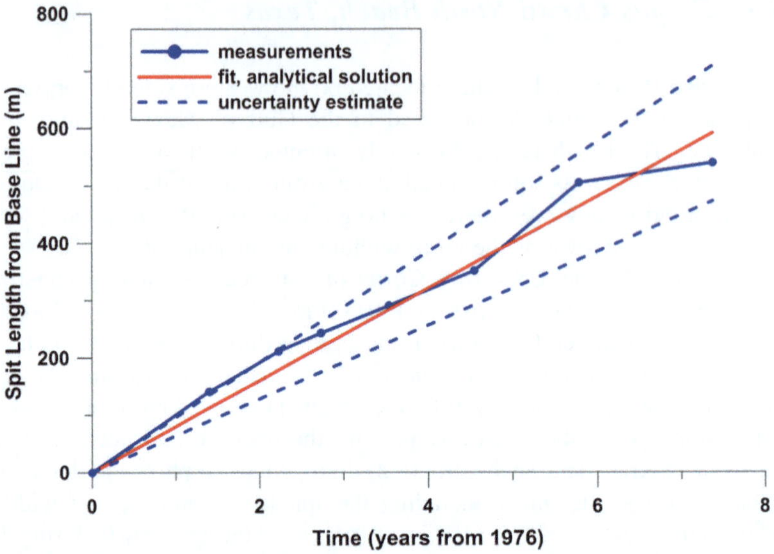

Fig. 12.4 Fit of analytical solution for unrestricted growth to data on spit elongation from Corpus Christi North Beach, Texas

Seabergh (2002) was found suitable for this purpose, although the documentation of the experiment is rather limited. For the present purpose enough information could be extracted from the existing papers to perform a satisfactory validation of the analytical solutions developed.

In the late 1990s the Coastal and Hydraulics Laboratory of the US Army Engineer Research and Development Center (ERDC) constructed a physical model to study processes around inlets (Seabergh 1999). The model consisted of a 46 m wide, 99 m long, and 0.6 m deep concrete basin with parallel depth contours in the offshore that followed an equilibrium profile (Dean 1977). An inlet was built on the bay side of the model having an opening width of about 2.5 m and a maximum water depth in the inlet just over 0.15 m. The basin was equipped with a wave generator on the ocean side, and tidal flows could be induced through a pumping system.

One experimental study carried out in the model encompassed spit evolution and its interaction with the inlet flow (Kraus and Seabergh 2002). Sand with a median grain size of 0.13 mm was placed in the model to build a beach updrift the inlet and waves approaching at an angle of 20° created a longshore transport that built a spit penetrating into the inlet. In total 22 experimental cases were carried out with different wave conditions and tidal flows. Measurements were performed of the spit properties (*e.g.*, length and width) during the experiment. Kraus (1999) presented some cases pertaining to the growth of the spit width towards equilibrium at selected locations along the spit and spit elongation for the cases of no current in the inlet and a steady flood current in the inlet. These data will be employed here to validate the analytical solutions.

The experimental case from Kraus (1999) displaying the development of the spit width towards equilibrium at different locations was first utilized for comparison with Eq. 12.9. This equation was developed for unrestricted growth, although it may also be employed for the case of a constant transport rate at the inlet (Q_{out}), if the solution is multiplied with an appropriate factor as discussed above. The data were given in terms of the spit width (W_s), whereas the analytical solution includes the cross-sectional area A_s. Thus, a conversion based on the spit cross-sectional shape is needed (*e.g.*, Eq. 12.2 for a trapezoidal shape). However, if the active profile height is not changing, as well as the shape factor ψ, then A_s (and A_e) may be replaced directly by W_s (and W_e) in Eq. 12.9.

When comparing Eq. 12.9 with the data some parameters have to be set that are not known *a priori*, including m and the factor $Q_{in}/(D\psi)$. These values were calibrated through a least-square fit with the data. Furthermore, W_e needs to be specified as well, which was done based on observations, yielding a value of 0.65 m. Since some effects are expected regarding spit growth from the inlet flow, the data was placed in two groups that displayed similar growth behavior and for which the net supply of sediment ($Q_{in}-Q_{out}$) would be different. In reality, this net supply would gradually decrease with time, but in order to simplify the data comparison, this grouping was applied. In fact, data from a location close to the tip of the spit could have been yet another group, but these data were very limited and thus not included. The first group of data included the measurements of W_s as a function of time (t) at the locations $x = 0.3$, 0.6, and 0.9 m downdrift the inlet side from which the spit growth started, whereas for the second group $x = 1.2$, 1.5, and 1.8 m.

Figures 12.5 and 12.6 illustrate the obtained agreement with the data for the first and second group, respectively. The best fit was obtained for $m = 1$ for both groups, whereas the optimum value on the transport parameter $Q_{in}/(D\psi)$ was 0.018 and 0.010 for the first and second group, respectively (note: Q_{in} in this expression represents the net supply to the spit). The agreement obtained indicates that the simple approach to model the growth of the spit cross-sectional area, resulting in Eq. 12.9, works satisfactorily. The general trend in the measured evolution of W_s is well reproduced together with the magnitude, although the latter relies mainly on the calibration procedure. The transport was smaller for the second group, as expected, since the reduced inlet cross section would have increased Q_{out} (and reduced the net supply of sediment to the spit), which should be of greater importance at the more downdrift locations included in group two.

Another set of data presented by Kraus (1999) encompassed spit elongation into an inlet for no tidal current. This corresponds to unrestricted spit growth, but since the water depth varied as the spit penetrated into the inlet, the linear behavior predicted by Eq. 12.8 failed to describe the observed response. Thus, Eq. 12.11 describing the spit evolution for an active profile height increasing exponentially towards a maximum value was employed. The solution is valid for $m = 1$, which corresponds to the best-fit value obtained when the time evolution of W_s was studied.

In the least-square fitting of Eq. 12.11 to the data, three parameters have to be specified, namely α, Q_{in}/A_e^o, and $\Delta D/D_o$. As previously discussed, for small values on x_s, the solution to Eq. 12.11 reduces to $x_s = 2Q_{in}t/A_e^o$, whereas for large values

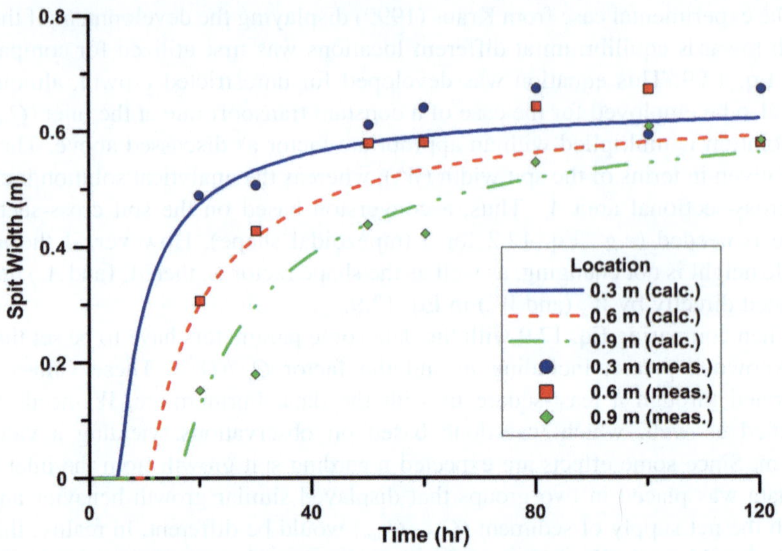

Fig. 12.5 Measured and calculated evolution of the spit width at different locations from the ERDC experiment for the first group of data

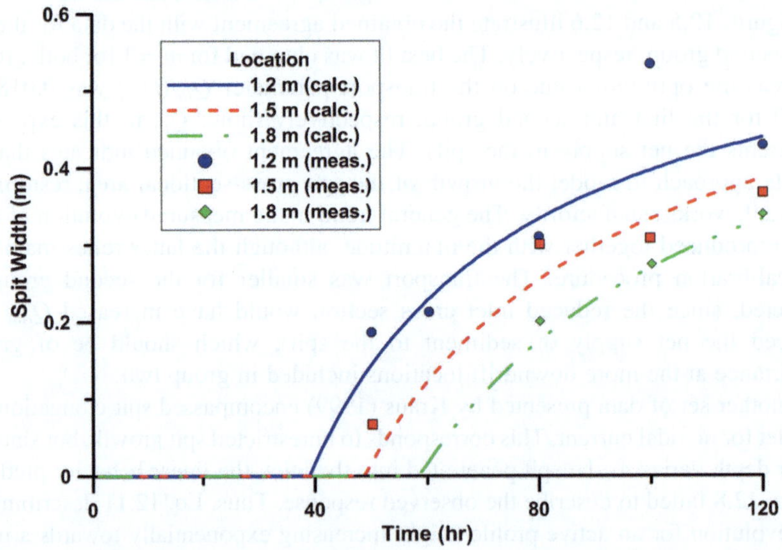

Fig. 12.6 Measured and calculated evolution of the spit width at different locations from the ERDC experiment for the second group of data

on x_s it becomes $x_s = 2Q_{in}t/A_e^{\infty}$. Viewing the data presented by Kraus (1999), both the earlier and latter phase of spit growth can be approximated by straight lines, although with widely different slopes. The best-fit linear slope is $k_o = 4.5$ for small x_s and $k_{\infty} = 0.45$ for large x_s. For constant Q_{in}, the ratio k_o / k_{∞} is equal to $1 + \Delta D/D_o$, which for the present case indicates $\Delta D/D_o = 9$. This value seems high, yielding a quite low value for D_o, but it will still be used here since detailed information about the inlet cross-sectional bathymetry is lacking.

Using $\Delta D/D_o = 9$ in the fitting procedure, optimum values on the two remaining parameters became $\alpha = 0.8$ and $Q_{in}/A_e^o = 0.043$. The least-square fitting was done utilizing a Newton-Raphson technique to solve for x_s for specific t, since Eq. 12.11 is implicit. Figure 12.7 illustrates the agreement between the analytical solution and the measured data. The analytical solution catches the trend of spit elongation well, even though it underpredicts a bit during the initial phase and overpredicts during the final phase. The optimum value on the transport parameter for the present case was similar to the optimum value obtained for the spit width modeling, after correction with regard to the geometry.

The final case studied from the ERDC experiment presented by Kraus (1999) involved spit elongation under the influence of an inlet current. The analytical solution given by Eq. 12.14 was employed for this case and $m = 1$ was assumed. In Eq. 12.14, there are two parameters to assign, namely δ and Q_{in}/A_e. However, at equilibrium $x_{se} = x_i\left(1 - \sqrt{\delta}\right)$, so if x_{se} and x_i are known δ can be calculated. In the present case, x_{se} was estimated to be about 2.1 m and x_i close to 2.5 m, implying $\delta = 0.04$. Using this value on δ, Eq. 12.14 was least-square fitted to the data yielding an optimum value on $Q_{in}/A_e = 0.019$. A Newton-Raphson technique was also used

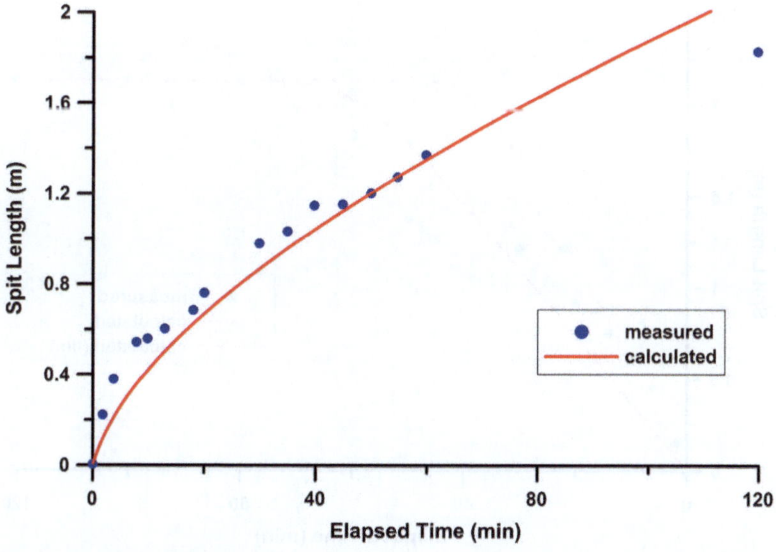

Fig. 12.7 Measured and calculated spit elongation from the ERDC experiment for a case with penetration into an inlet without flow

in this case to obtain x_s for a specific t, since Eq. 12.14 is implicit as well. Figure 12.8 shows the agreement between calculated and measured spit elongation. Although the agreement is good initially and when x_s approaches equilibrium, the measured evolution exhibits a somewhat different behavior during the intermediate period of spit growth.

An alternative, more direct way to estimate Q_{in}/A_e would be to use the asymptotic solution ($i.e., x_s = x_i(1 - \delta)t/t_o$) for small x_s to get an optimum estimate of this parameter. This estimate would then be used in the complete solution. The dashed line in Fig. 12.8 shows the result of this method. A better agreement is obtained initially, but for intermediate values the discrepancy with the measurements become even larger than when all points were included in the fit. Overall, the comparison between the analytical solution and the measurements indicate that the formulation used for Q_{out} seems to work initially, but as x_s becomes a large portion of x_i, the formulation becomes a poor description of how Q_{out} varies with $x_i - x_s$. This is expected since the assumption of a maintained flow is not suitable when a large portion of the inlet is closed. Instead a more detailed model for the inlet flow is needed, but such a model would not facilitate an analytical approach.

12.4.5 Fire Island Inlet, Long Island

Fire Island Inlet (FI) is located on the south shore of Long Island and is one of several inlets that contribute to the water exchange between the Atlantic Ocean and the Great South Bay (see Fig. 12.9). The inlet is located at the most western point of

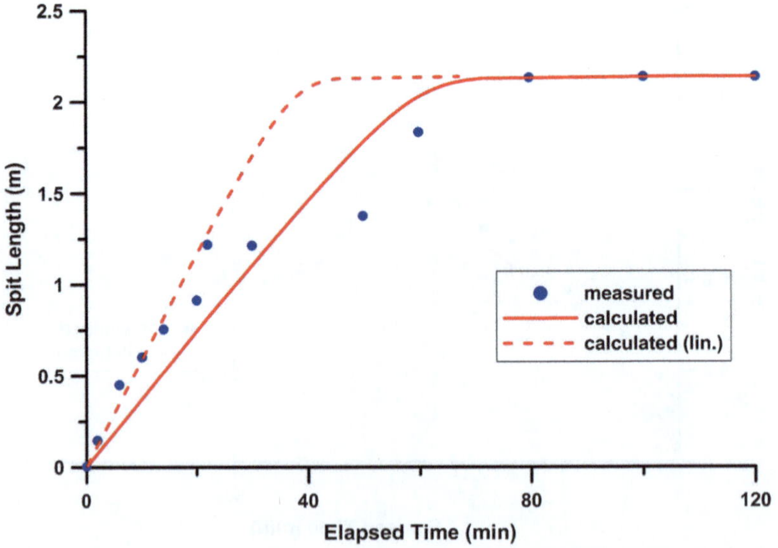

Fig. 12.8 Measured and calculated spit elongation from the ERDC experiment for a case with penetration into an inlet with flow

Fig. 12.9 Map over Long Island and the study area around Fire Island Inlet (Courtesy of U.S. Army Corps of Engineers)

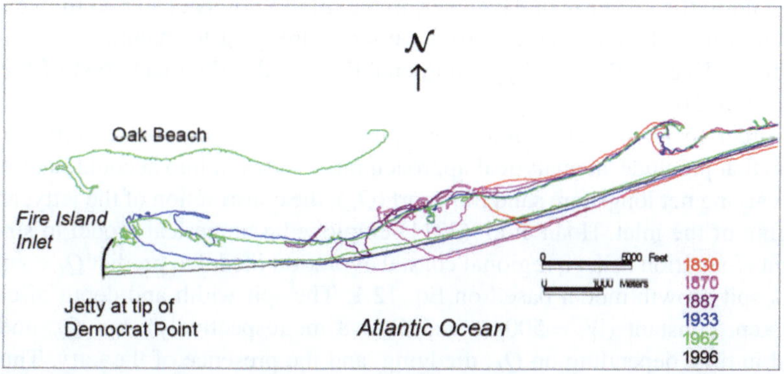

Fig. 12.10 Recorded spit growth at Fire Island Inlet (Courtesy of U.S. Army Corps of Engineers)

a sandy barrier island, which also includes two other major inlets, namely Shinnecock Inlet (SI) and Moriches Inlet (MI). The net longshore transport rate along this barrier island is to the southwest with an average annual rate gradually increasing from zero at Montauk Point to about 200,000 m^3/year at FI, with some local variation at SI and MI (Kana 1995; Rosati et al. 1999; Larson et al. 2002b).

FI has existed since the early 1700s (Smith et al. 1999) and the western tip of the barrier island has grown in the direction of the net longshore transport, creating a spit, forcing the inlet to move towards the west. Figure 12.10 illustrates the growth of the spit from the early 1800s based on data from Kana (1995) and Kraus and

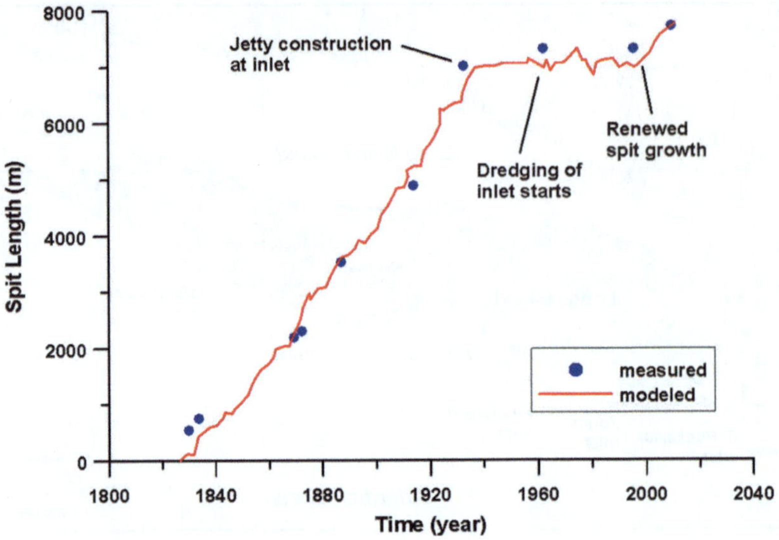

Fig. 12.11 Modeled and measured spit elongation at Fire Island Inlet

Seabergh (2002). In 1939 a jetty was constructed to prevent the movement of the inlet, halting the spit growth. However, in the mid-1950s sand started to bypass the jetty and fill in the channel, requiring periodic dredging to maintain navigational conditions. Eventually, this bypassing enabled spit development west of the jetty (Hoan et al. 2011).

In order to realistically model the spit elongation a number of complicating factors that preclude an analytical approach must be taken into account, including a time-varying net longshore sand transport (Q_L), the construction of the jetty, and the dredging of the inlet. Hoan et al. (2011) employed a numerical model to simulate the spit elongation using a regional coastal evolution model to predict Q_L combined with a spit growth model based on Eq. 12.1. The spit width and depth of closure were kept constant ($W_e = 500$ m and $D_c = 8$ m, respectively), but Q_{in} and Q_{out} varied in time depending on Q_L, dredging, and the presence of the jetty. Thus, the spit was subject to restricted growth with the inlet sediment transport and dredging contributing to Q_{out}. The sand transport in the inlet was calculated with the formula proposed by Watanabe et al. (1991), similar to the expression used to arrive at Eq. 12.12. The coastal evolution model used for simulating Q_L is described in detail in Hoan et al. (2011; see also Larson et al., 2002b), and here only the spit growth will be discussed.

Figure 12.11 illustrates the result of modeling the spit growth on the eastern side of FI. The location of the spit in 1825 was arbitrarily set to $x_s = 0$. The measurements clearly show the influence of the jetty and the dredging, which significantly affected the spit elongation rate (the measurement from 2010 was added by Hoan et al. 2011, based on a satellite image). Several decades after the construction of the jetty spit growth ceased completely, initially due to the blocking effect of the jetty, but after about a decade dredging was needed as well to prevent further spit growth. After the implementation of proper boundary and forcing conditions, the model yields good agreement with the measurements. Note that during the time period when the spit was free to grow (*i.e.*, before the jetty construction), the FI moved accordingly, leading to an elongation rate that was close to constant, similar to unrestricted growth described by Eq. 12.8.

12.4.6 Chilaw Inlet, Sri Lanka

Chilaw Inlet (CI) is located on the west coast of Sri Lanka about 120 km north of the capital Colombo. During the summer monsoon (May to December) strong waves produce a net longshore sand transport to the north that tends to close many inlets on the west coast, including CI, since the tidal range is too small and the river flow too low to keep the inlets open. The closure of CI occurs through spit growth across the inlet, typically starting during the month of May (Wikramanayake and Pattiaratchi 1999). Towards the end of July the inlet mouth is normally artificially opened to allow for safe navigation; the lagoon serves as anchorage for a large number of fishermen (Baranasuriya 2001; Ranasinghe and Pattiaratchi 1997). In October or November, the flow in the nearby Deduru River will be strong enough to flush away the spit and the inlet becomes open again.

Figure 12.12 provides an aerial photo of CI, a schematic description of the most important morphological units around the inlet connected with the spit growth, and a view of CI looking south. At the start of the monsoon (May), increased wave action initiates a marked longshore sediment transport towards the north, passing a sandstone reef before reaching CI. Also, on top of the reef, a sand dune is normally formed during calm condition that is eroded away by the monsoon waves, contributing to the northward sediment flow. When the sediment reaches CI, two pathways are possible at the tip of the reef: sediment will move (1) around the tip of the sandstone reef and initiate a growth of the sand spit across the inlet or (2) seaward and be deposited on the ebb shoal. Wave refraction, breaking, and diffraction occur at the inlet because of the ebb shoal and the sheltering sandstone reef, causing onshore movement of sand from the shoal to the spit. Thus, a combination of cross-shore and longshore transport causes the spit growth and eventually inlet closure (Wikramanayake and Pattiaratchi 1999; Ranasinghe et al. 1999).

Figure 12.13 illustrates in a schematized manner the sediment transport pathways and morphological units identified at CI (Larson et al. 2009). The rate Q_{IN} is the longshore sediment transport from the beaches south of the reef and Q_D is the transport from the dune on top of the reef. The transports Q_{IN} and Q_D yield the sediment flow to the reef beach that will determine the outflow from the this beach (Q_R) When the transport Q_R reaches the tip of the reef it is divided into two transport components: (1) the first component, caused by diffraction behind the tip, generates an easterly transport across the inlet, initiating the spit growth, and (2) the second component generates a transport in the north-east direction to the ebb shoal. The

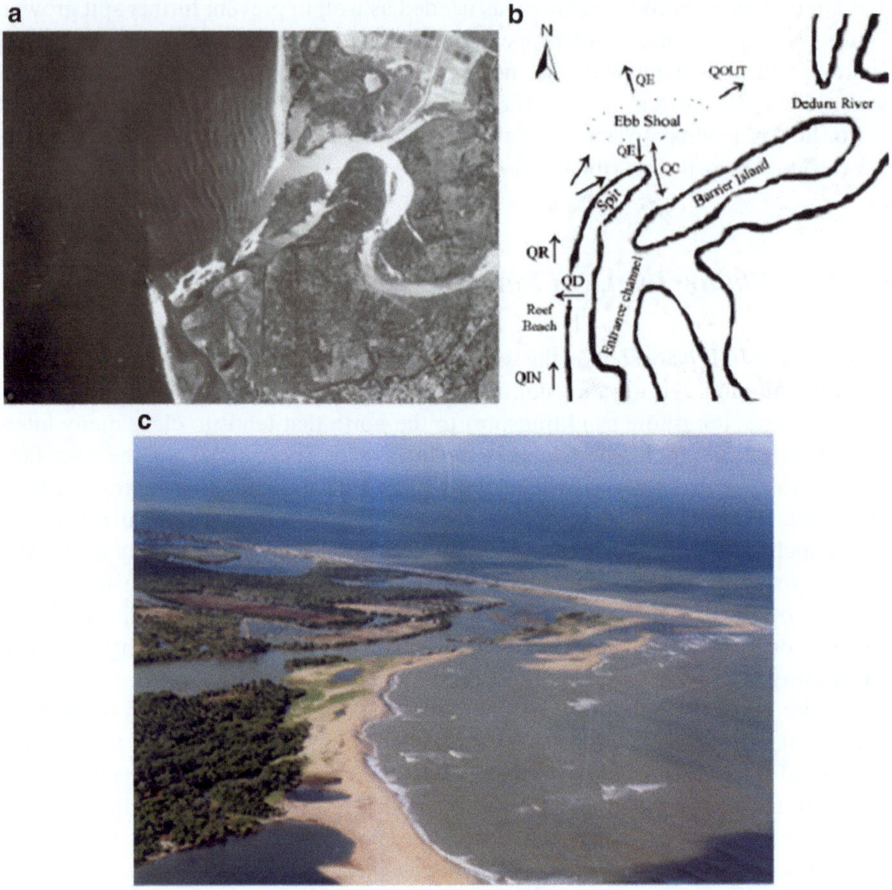

Fig. 12.12 Spit growth at Chilaw Inlet on the western coast of Sri Lanka; (**a**) aerial photo of the inlet, (**b**) schematized description of morphological units and sediment transport pathways, and (**c**) photo of the inlet looking south

ebb shoal connects the sediment pathways from the reef beach, spit, and the down-drift side of the inlet. Sediment comes to the ebb shoal from Q_R or from the spit (Q_C) when the inlet current is strong and erodes the tip of the spit.

Sediment leaves the ebb shoal through sediment bypassing (Q_{OUT}) or through Q_E, which is the wave-related cross-shore transport. The reservoir model by Kraus (2000) was used to describe the ebb shoal as well as the reef beach (see Eqs. 12.17 and 12.18). The cross-shore transport (Q_E) may either be onshore or offshore directed, depending on the wave and sediment conditions. In the modeling, a heuristic formula for Q_E was employed where the rate depends on the dimensionless fall speed (see Larson et al. 2009). The spit loses sand through the transport generated by the inlet flow (Q_C), where the material moves both towards the lagoon and the ebb shoal due to the tidal flows. The CERC formula was employed to

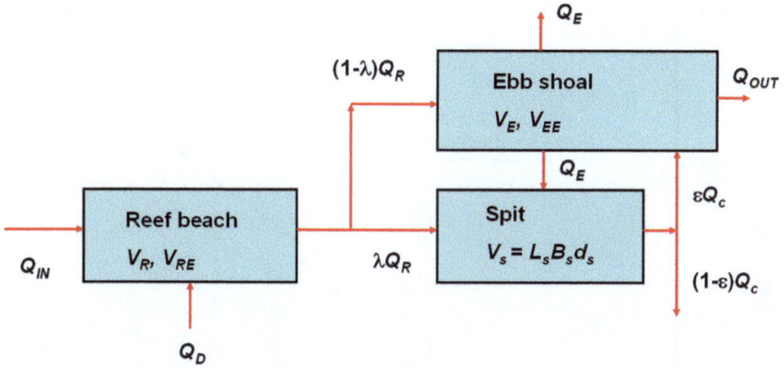

Fig. 12.13 Schematized description of morphological units and sediment transport pathways at Chilaw Inlet

calculate Q_{IN} from a time series of measured waves, representative for the monsoon period. The erosion of the dune on top of the reef beach (Q_D) was estimated from the formula proposed by Larson et al. (2004).

The evolution of the three morphological units (the reef beach, ebb shoal, and spit) is governed by the following sand conservation equations,

$$\frac{dV_R}{dt} = Q_{IN} + Q_D - Q_R \tag{12.19}$$

$$\frac{dV_E}{dt} = (1 - \lambda)Q_R + \varepsilon Q_C - Q_E - Q_{OUT} \tag{12.20}$$

$$\frac{dV_S}{dt} = \lambda Q_R + Q_E - Q_C \tag{12.21}$$

where V_R is the sediment volume stored along the reef beach, t time, V_E is the volume in the ebb shoal, ε determines the relative amount of sediment that is deposited on the ebb shoal, λ a coefficient determining how much of Q_R that is transported to the spit, and V_S the volume of sediment stored in the spit. The output of sediment from the reef beach, Q_R, and the ebb shoal, Q_{OUT}, is modeled using the reservoir concept,

$$Q_R = (Q_{IN} + Q_D)\frac{V_R}{V_{RE}} \tag{12.22}$$

$$Q_{OUT} = ((1 - \lambda)Q_R + \varepsilon Q_C)\frac{V_E}{V_{EE}} \tag{12.23}$$

where V_{RE} and V_{EE} is the equilibrium volume of the reef beach and ebb shoal, respectively.

In the spit modeling, a constant Q_C was assumed that is smaller than the sediment inflow to the spit ($Q_S = \lambda Q_R + Q_E > Q_C$). This implies that the channel

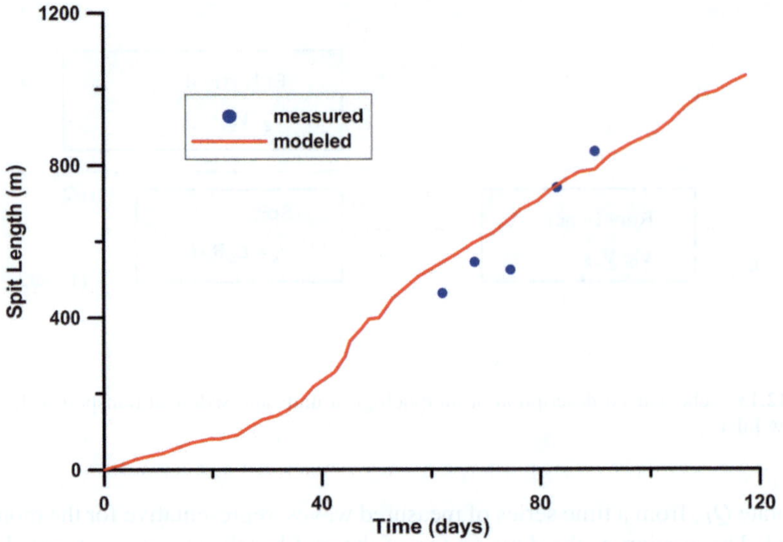

Fig. 12.14 Modeled and measured spit elongation at Chilaw Inlet

maintains its transport capacity Q_C and that the excess transport $Q_S - Q_C$ will produce the spit growth. The growth of the spit may be computed from:

$$\frac{dx_s}{dt} = \frac{Q_S - Q_C}{W_s D} \tag{12.24}$$

where Eq. 12.21 was used assuming $V_S = x_s W_s D$ (W_s and D are taken as constants).

A field campaign was carried out from the 20th of May to the 30th of July in 2006 to survey the topography of the spit, sand bar, and part of the barrier island (Angelbratt and Kluft 2007). The surveys were performed using both a leveling instrument and a total station. Seven field days yielded sufficient data for detailed analysis of the spit growth, namely the dates 060528, 060611, 060702, 060708, 060715, 060723, and 060730 (year, month, and day). Based on these field data, the width of the spit was set to $W_s = 40$ m in the modeling.

Equations 12.19, 12.20, 12.21, 12.22, 12.23, and 12.24 was solved simultaneously using a numerical approach to simulate the spit elongation taking into account the interaction with the other morphological units. Figure 12.14 illustrates the agreement between the simulated and measured spit elongation for the month of July when a clear spit emerged. After 90 days (the 30th of July; the simulation was arbitrarily started on May 1st), a spit length of about 800 m was obtained in the model, which corresponds well to the measured spit length. From observations of the morphological response, it seems like the spit grows at a slower rate early in the monsoon, during May and June, and later, in July, the growth accelerates. The basic model shows this behavior, and during the first 50 days the spit growth is retarded, after which the spit grows at an almost constant speed. This speed is obtained when the reef beach has reached its equilibrium volume and all the sediment from the south is bypassed.

In order to solve the set of equations, a number of coefficients and quantities had to be assigned values, which was done based partly on existing empirical equations and partly on observations and data from the site. The equilibrium and initial volume of the reef beach were estimated to 70,000 m^3 and 10,000 m^3, respectively, whereas the ebb shoal equilibrium volume was computed to 8,000 m^3 according to the formula presented by Walton and Adams (1976). The ebb shoal was assumed to contain about 4,000 m^3 sand when the south-west monsoon starts. The transport rate Q_C was calculated to 200 m^3/day as a representative value using the transport formula developed by Camenen and Larson (2005). The two non-dimensional parameters was set to $\lambda = 0.55$ and $\varepsilon = 0.5$. The former value was determined through calibration, requiring that the correct spit length was obtained at the end of the measurement period, whereas the latter value was based on the assumption that the tidal flow was sinusoidal.

12.5 Concluding Remarks

The mathematical model of spit growth was employed to describe the observed spit evolution both in the laboratory and the field for a wide range of hydrodynamic and sedimentary conditions. In order to solve the governing equation together with the appropriate boundary and forcing conditions analytically, schematization and simplifications were necessary. However, in spite of these simplifications, observed spit growth could be well reproduced with these solutions. For the more complex field cases studied, where the interaction between the spit growth and adjacent morphological features had to be included in the modeling, the system of equations developed could not be solved analytically, but numerical techniques were required.

In general, the behavior and trend of the spit quantities simulated (*e.g.*, length, width) were satisfactorily described by the mathematical model; however, to obtain the correct magnitude of these quantities, the values of different parameters and coefficients had to be determined through calibration against data. Thus, in any application of the spit model developed in this study, in order to yield robust and reliable results, comparison with data is necessary for proper validation and selection of appropriate parameter and coefficient values.

The basic formulation of the model assumed a spit growth in the direction of the longshore sand transport and the main shoreline trend. No effort was made to include recurved, compound, and complex spits. In general, for simulating such spits, an approach is needed that resolves the two-dimensional transport field. Kraus (1999) modeled a recurved spit employing a heuristic description of the evolution by directly specifying the cross-shore movement of the spit in terms of location and the velocity at the tip. Such a model will reproduce the recurving behavior of a spit, but the pre-specification of the velocity is difficult.

Acknowledgements This work was partially funded by the Regional Sediment Management Program under the Inlet Geomorphologic Work Unit of the Coastal Inlets Research Program of the U.S. Army Engineer Research and Development Center, and partly by Sida/SAREC under grants SWE-2005-332 and SWE-2010-038. Support was also provided by Sida/SAREC under grant

2011-002102 for continued bilateral research cooperation between Swedish institutions and Eduardo Mondlane University in Mozambique, Program Integrated Water Resources Management – Quantitative and Qualitative Aspects of IWRM for Sustainable Development in Southern Mozambique.

References

Allard J, Bertin X, Chaumillon E, Pouget F (2008) Sand spit rhythmic development: a potential record of wave climate variations? Arçay Spit, western coast of France. Mar Geol 253:107–131

Angelbratt A, Kluft J (2007) Seasonal closure of Chilaw Inlet in Sri Lanka: physical processes and mathematical modeling. Unpublished M.S. report, Department of Water Resources Engineering, Lund University, Lund

Baranasuriya PW (2001) Hydrographic investigations for the design of an anchorage in a complex lagoon estuary: a spatial Odyssey. In: Proceedings of 42nd Australian Surveyors Congress, pp 1–12

Blomgren S, Hanson H (2000) Coastal geomorphology at the Falsterbo Peninsula, southern Sweden. J Coast Res 16(1):15–25

Camenen B, Larson M (2005) A bedload sediment transport formula for the nearshore. Estuar Coast Shelf Sci 63:249–260

Dean RG (1977) Equilibrium beach profiles: U.S. Atlantic and Gulf coasts. Ocean engineering technical report no. 12. Department of Civil Engineering and College of Marine Studies, University of Delaware, Newark

Evans OF (1942) The origin of spits, bars and related structures. J Geol 50:846–865

FitzGerald DM (1988) Shoreline erosional-depositional processes associated with tidal inlets. In: Aubrey DG, Weishar L (eds) Hydrodynamics and sediment dynamics of tidal inlets. Springer, Berlin, pp 186–225

FitzGerald DM, Kraus NC, Hands EB (2000) Natural mechanisms of sediment bypassing at tidal inlets. CHETN-IV-30. Coastal and Hydraulics Laboratory, US Army Engineer Research and Development Center, Vicksburg

Hallermeier RJ (1981) A profile zonation for seasonal sand beaches from wave climate. Coast Eng 4:253–277

Hanson H, Larson M (1993) Sand transport and coastal development at Skanör-Falsterbo. Report no. 3166, Department of water Resources engineering, Lund Institute of Technology, Lund University, Lund, Sweden (in Swedish)

Hoan LX, Hanson H, Larson M, Kato S (2011) A mathematical model of spit growth and barrier elongation: application to FireIsland Inlet (USA) and Badreveln Spit (Sweden). Estuar Coast Shelf Sci 93:468–477

Johnson DW (1919) Shore processes and shoreline development. Wiley, New York

Kana TW (1995) A mesoscale sediment budge for long island, New York. Mar Geol 126(1):87–110

Kieslich JM, Brunt DH (1989) Assessment of a two-layer beach fill at Corpus Christi Beach, TX. In: Proceedings of coastal zone 89, ASCE, New York, pp 3975–3984

Kraus NC (1998) Inlet cross-sectional area calculated by process-based model. In: Proceedings of 26th coastal engineering conference, ASCE, pp 3265–3278

Kraus NC (1999) Analytical model of spit evolution at inlets. In: Proceedings of coastal sediments 99, ASCE, Charleston, pp 1739–1754

Kraus NC (2000) Reservoir model of ebb-tidal shoal evolution and sand bypassing. J Waterw Port Coast Ocean Eng 126(6):305–313

Kraus NC, Lin L (2002) Coastal processes study of San Bernard River mouth Texas: stability and maintenance of mouth. Technical report TR-02-10. Coastal and Hydraulics Laboratory, US Army Engineer Research and Development Center, Vicksburg

Kraus NC, Seabergh WC (2002) Inlet spits and maintenance of navigation channels. CHETN-IV-44. Coastal and Hydraulics Laboratory, US Army Engineer Research and Development Center, Vicksburg

Larson M, Hanson H (2013) Coastal erosion and protection in Sweden. In: Pranzini E, Williams AT (eds) Coastal erosion and protection in Europe. Earthscan Ltd, Routledge, Oxon, pp 31–46

Larson M, Rosati JD, Kraus NC (2002a) Overview of regional coastal sediment processes and controls. CHETN-XIV-4. Coastal and Hydraulics Laboratory, US Army Engineer Research and Development Center, Vicksburg

Larson M, Kraus NC, Hanson H (2002b) Simulation of regional longshore sediment transport and coastal evolution – The Cascade model. In: Proceedings of 28th coastal engineering conference, ASCE, Wales, pp 2612–2624

Larson M, Erikson L, Hanson H (2004) An analytical model to predict dune erosion due to wave impact. Coast Eng 51:675–696

Larson M, Kraus NC, Connell KJ (2006) Modeling sediment storage and transfer for simulating regional coastal evolution. In: Proceedings of 30th coastal engineering conference, World Scientific, pp 3924–3936

Larson M, Wikramanayake N, Hanson H, Ranasinghe R (2009) Seasonal closure of Chilaw Inlet, Sri Lanka: physical processes and mathematical modeling. In: Proceedings of coastal dynamics 09, ASCE (on CD)

Larson M, Hanson H, Kraus NC, Hoan LX (2011) Analytical model for the evolution of coastal inlet cross-sectional area. In: Proceedings of coastal sediments 11, ASCE (on CD)

Otvos EG (2012) Coastal barriers – nomenclature, processes, and classification issues. Geomorphology 139–140:39–52

Petersen D, Deigaard R, Fredsoe J (2008) Modelling the morphology of sandy spits. Coast Eng 55(7–8):671–684

Ranasinghe R, Pattiaratchi C (1997) Investigation for Chilaw Anchorage: data report CTD measurements. January 1996. May 1996. WP 1296 RR, Centre for Water Research, The University of Western Australia, Australia

Ranasinghe R, Pattiaratchi C (1999) The seasonal closure of tidal inlets: Wilson Inlet – a case study. Coast Eng 37(1):37–56

Ranasinghe R, Pattiaratchi C, Masselink G (1999) A morphodynamic model to simulate the seasonal closure of tidal inlets. Coast Eng 37:1–36

Rosati JD, Gravens MB, Smith WG (1999) Regional sediment budget for Fire island to Montauk point, New York, USA. In: Proceedings of coastal sediments '99, ASCE, New York, pp 803–817

Schwartz ML (ed) (1972) Spits and bars. Dowden, Hutchinson & Ross, Stroudsburg

Seabergh WC (1999) Physical model for coastal inlet entrance studies. CHETN-IV-19. Coastal and Hydraulics Laboratory, US Army Engineer Research and Development Center, Vicksburg

Shephard F (1952) Revised nomenclature for depositional coastal features. Am Assoc Pet Geol Bull 36:1902–1912

Smith WG, Watson K, Rahoy D, Rasmussen C, Headland JR (1999) Historic geomorphology and dynamics of Fire Island, Moriches and Shinnecock Inlets, New York. In: Proceedings of coastal sediments '99, ASCE, New York, pp 1597–1612

Tanaka H, Takahashi F, Takahashi A (1996) Complete closure of the Nanakita River mouth in 1994. In: Proceedings of 25th coastal engineering conference, ASCE, Orlando, 4545–4556

Walton TL, Adams WD (1976) Capacity of inlet outer bars to store sand. In: Proceedings of 15th coastal engineering conference, ASCE, Honolulu, 1919–1937

Watanabe A, Shimizu T, Kondo K (1991) Field application of a numerical model of beach topography response. In: Proceedings of coastal sediments '91, ASCE, Seattle, pp 1814–1828

Wheeler WH (1902) The sea-coast. Longmans, Green, and Co, London

Wikramanayake N, Pattiaratchi C (1999) Seasonal changes in a tidal inlet located in a monsoon regime. In: Proceedings of coastal sediments '99, ASCE, New York, pp 1462–1477

Chapter 13
Spits on the French Atlantic and Channel Coasts: Morphological Behaviour and Present Management Policies

Hervé Regnauld, Stéphane Costa, and Jonathan Musereau

Abstract Spits are present all along the Channel and Atlantic coasts of France. They belong to different types, depending on their composition and genesis. Some of them were initiated when RSL stabilized about its present level, 2,500 years ago. Many others are much more recent and were created when major land-use changes took place in mainland watersheds. Soil erosion increased the fluvial sediment supply and many river mouths were stabilized. All these spits continue to evolve under variable degrees of human control (mainly on sediment delivery). This paper present three examples (Arçay, Talbert, Cayeux) which illustrate the main types of issues faced by coastal planners dealing with spits in western France.

13.1 Introduction

Spits are frequent features along the French Atlantic and Channel coasts: a large part of the coast lines follow an east-west trend and the dominant wave directions are from the west (Fig. 13.1). Consequently, longshore drift is very active. There is a significant difference between the Channel and the Atlantic coasts. In the Channel, dominant waves are from the west, irrespective of the path of low pressure systems. Most of the storms are also from the west. From Ushant to Dunkirk (Fig. 13.1) drift-aligned features may be found almost at each river mouth. Exceptionally, in a shallow bay just east of a important cape, wave are refracted and a spit may grow from east to west, such as in Sables d'Or, along the Channel coast. On the Atlantic coasts of the Bay of Biscay, low pressure systems may have many different paths and waves may come from anywhere between South and North West. From Ile de Sein to the Loire estuary, the coast follows a northwest–southeast trend. Dominant waves vary between west and southwest. Drift-aligned features are not very frequent, though some of them are very long, such as in Quiberon. Between the

H. Regnauld (✉) • J. Musereau
University of Rennes, Rennes, France
e-mail: herve.regnauld@uhb.fr

S. Costa
University of Caen, Caen, France

G. Randazzo et al. (eds.), *Sand and Gravel Spits*, Coastal Research Library 12,
DOI 10.1007/978-3-319-13716-2_13

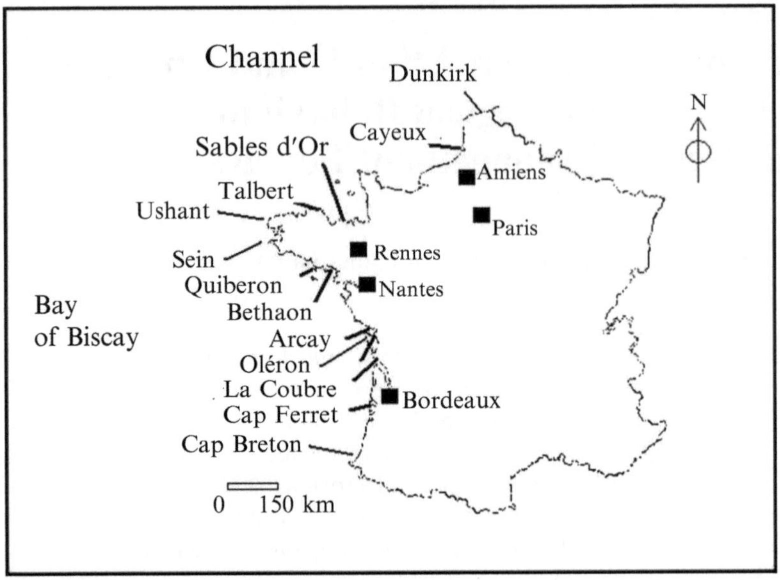

Fig. 13.1 Location of the main spits mentioned in the text

Loire estuary and the Gironde estuary (a north–south trending coast, with many bays), dominant waves are from the west but many islands and river mouth create a complicated pattern of local wave directions. South of the Gironde, the entire coast is drift-controlled (dominant waves from west/northwest) and the sediment from the estuary is moved south to the Pays Basque, where it is evacuated seaward by a very steep submarine canyon the head of which reaches almost to the coast : the Cap Breton Gouf (Gouf meaning canyon).

Most of these spits have a high present morphological variability and have been creating problems for local coastal communities for centuries. They could extend and temporarily close the entrance of a harbour (e.g. in Bethaon, Baie de la Vilaine, South Brittany, Traini 2009), they could produce sand banks and change the channels between an island and the main land, such as in Oléron (Bertin 2008). More recently with the rise of coastal tourism and recreational sailing, moving spits have been seen as problems because they make navigation difficult, such as in La Courbe (Bressolier 1979). Many of them were reinforced by dykes or wooden fences as early as 1832 in Oleron Island, (Musereau et al. 2007) and 1906 in Pointe d'Arçay spit (Verger et al. 1972) so that they could withstand storms without being breached. The general issue in their management is the input of sediment. The spits are not fed by enough material to remain in equilibrium with the present wind and waves conditions, though local situations are very different from one site to the other.

In a first part this paper will present three examples which may serve as archetypes. The first one is a perfectly natural feature. It is a 3 km-long gravel spit, the Sillon de Talbert which is located on the northern coast of Brittany and has been extensively studied during the last 50 years. The second one in the Vendée and Charente regions, Pointe d'Arçay is a "man-made" sandy spit which began to appear during the eighteenth century when the mouth of a river, Le Lay, was totally modified in order to create large polders. The third one is the Cayeux spit, at the Somme river mouth. It is a very good example of a natural feature the evolution of which is now entirely controlled by human activities.

In a second part the paper will address general issues about spit management from two different points of view: from one perspective they can be seen as protecting vulnerable low lands and if breached by storms floods cause impacts on human activities in the lowlands, from an alternative perspective spits are dynamic landforms and part of the geomorphological heritage and should be considered as mobile forms.

13.1.1 Spits as Mobile Features

The many spits along the French coasts are of different types. Some of them (Cap Ferret, Talbert for instance) are old features, dating at least from the late Holocene RSLR (Stephan 2011) and are believed to behave today in a similar way as they did in the past. Some are more recent and historical data shows they begun to grow when land use and coastal management modified heavily the sediment cells and delivered more material to the coast (e.g. Fleche de l'Aiguillon, de Verbois-Oleron island, d'Arçay...). Others display the opposite situation: human activity has interrupted the sediment input and they are, today, undergoing a deficit.

13.1.2 Sillon de Talbert Spit

The Sillon (sillon means spit and tombolo in French) is a gravel spit (Fig. 13.2) which extends about 3 km seaward and ends with a large hook. The Sillon de Talbert has been used for years by sea weed gatherers. They came on the spit with wheels carts, and later with small lorries at low tide and collected sea weed for a local factory. This spit has been extensively studied because any time a storm created a breach, sea weed gathering was interrupted. This happened in 1962 and in 1967. In 1974, rocky armour was installed on the seaward flank of the proximal end of the Sillon. This held the shoreline, but shifted the focus of erosion further along the spit A subsequent proposal was to enclose the gravel with a textile sheet (1987). A test installation on a short part (about 250 m long) of the center of the spit was unsuccessful since the spit continues to retreat. The test zone was then extended towards the spit hook but was not more efficient. So, the device was abandoned.

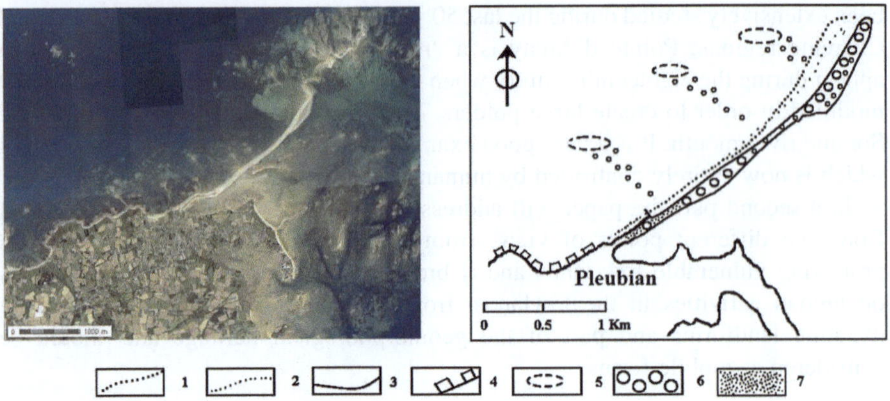

Fig. 13.2 Synthetic geomorphological behaviour of Sillon de Talbert spit. *1* Coastline 1775; *2* coastline by 1837; *3* coastline by 2010; *4* low cliff with periglacial head furnishing sand; *5* intertidal skerries furnishing gravel; *6* gravel ridge accumulation (with local areas of retreat); *7* sand accumulation (with local areas of retreat)

In order to understand the spit behaviour studies have been undertaken as early as 1980 by the French Geological Survey, then by Hallegouet et al. (1989), Regnauld et al. (1993), Pinot (1994), Morel (1995) and later by Stephan (2011).

The Sillon appears on a 1666 map and is also visible on maps in 1764, and 1789. The oldest air photos go back to 1930. As the maps are not very precise the measured movements have an unknown error margin. Between 1775 and 2000 the Sillon was retreating via roll over at a rate of 0.5–1 m a year. During the same period it lengthened by 200–250 m. From 1930 to 2000, (for which a higher precision is achieved) the retreat rate is highly variable. Stephan (2011) showed that it is above 4 m a year between 2006 and 2008 but was only 0.25 m a year between 1998 and 2006.

An important point is that this spit is not mainly fed by longshore drift. Some parts of the low cliffs which are located updrift are covered with periglacial head deposits and they produce some sediment (caption 8 of Fig. 13.2), some gravel and mainly sands which accumulate at the base of the spit. This source has been very active during historical times but is almost completely spent today and the most important present source sites are the skerries (small eroding topographic highs) which are located in the intertidal zone (caption 9 of Fig. 13.2) and produce short trailing spits in the direction of the Sillon. This Sillon is much more a swash-aligned feature than a drift-aligned feature.

For this reason, by the year 2000 the French "Conservatoire du Littoral", which is in charge of managing large parts of the coasts decided to let the Sillon retreat and behave "naturally". In 2006 it was designed as a natural preserve. The only heavy interventions are undertaken when a storm creates a large breach in the spit. The breaches are artificially sealed.

13.2 Pointe d'Arçay Spit (Fig. 13.3) and Belle Henriette Lagoon

The Le Lay river mouth was initially a delta-like feature. The river had several arms and was able to shift from to another if a large storm took place during high waters season and high spring tide. The local population decided to channel the river into one single arm, so that the rest of the area could be turned into polders, locally named "prises". The almost immediate consequence was that the abandoned delta eroded very quickly (Cooper and Pilkey 2004). A spit began to grow from the eroded delta sediment. It is not mentioned on Cassini's map (1767) but was already 4 km long in 1850. The lagoon which is between the spit and the main land is called "Belle Henriette". By the beginning of the twentieth century the entire former delta material was dispersed. The sediment input is considerably reduced, although not totally absent, as some sand is moved from updrift areas. Now the spit is cannibalising itself (Orford et al. 2002) and the retreat rate between 1880 and 1905 is 1 m/year. The storms are often able to breach it more frequently. The polders were flooded in 1906 and a 600 m long dyke was built. The spit was breached once more in 1911 during the peak of a spring tide. The army quickly built a new dyke, which was destroyed in 1928, then rebuilt. Another flood took place in 1930, and again in 1957. During that time (from 1906 to 1960) the distal part of the spit extended by some 2 km (Galichon 1984).

At the beginning of the 1970s the risk of breaching posed such a risk to the polders and back-barrier villages that groynes were built across the beach in order to stop the

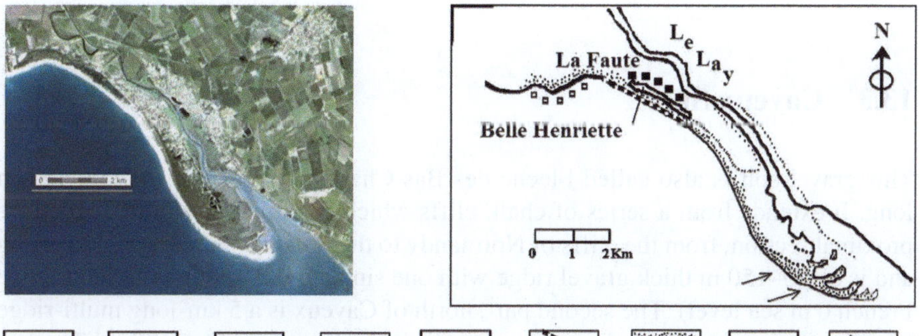

Fig. 13.3 Synthetic geomorphological behaviour of the Pointe d'Arçay spit. Caption 1–4 shows the different positions of the coastline with a variable error margin (in space and in time). *1* Coastline by 1824; *2* coastline by 1960; *3* coastlines by 2000; *4* coastline by 2010. *5* former river mouth material eroded after the river is diverted; *6* zone of cannibalisation. *7* Sand accumulation (with local areas of retreat); *8* artificial devices, mainly groynes; *9* present accumulation zones along the spit

drift. They also had the effect of preventing southward growth of the spit. The first scientific studies (Verger et al. 1972) presented a map, which at this moment is the only analysis of the spit behaviour: it does not show the successive positions of the coastline but the main wave patterns and sediment fluxes. With this map the scientists present a list of all the processes that control the spit behaviour. From that moment (this was an unexpected consequence of the map) the local people begin to accept the idea that the spit behaviour is characterised by a great variability. The amount of sand varies during the year and there are always one or two places which have lower resistance than others. It means a breach may always happen and not always in the same spot. Instead of building one single long dyke all along the spit the local authorities built a short dyke and set some groynes where they thought the breach would be the most dangerous. Storms of 1999 and 2008 produced floods through breaches located elsewhere and the present situation is alarming. Down drift of the groynes the retreat rate is 2–4 m/year (Musereau 2010). The present issue for coastal managers is to consider if such an artificial feature can or should be maintained for ever in its present condition as most of its sediment source is spent.

The debate is intense. The spit is an artificial feature because it was created when local people changed the course of the Le Lay river. But once it begun to grow, it has been behaving as a natural feature for about 100 years. When the first dykes were built they were always located along the proximal part. The distal part of the spit has never been managed, nothing has ever been built on it and it may be considered (according to Tounsi and Verger 2007) as a natural feature. It extended by 4,000 m between 1850 and 2000 and several scientists think it is one of the best examples of coastal mobility. They think it should be taken as a part of the French geomorphological heritage and should be considered as a laboratory for monitoring relative sea level rise impacts on mobile coastal features.

13.3 Cayeux Spit

This gravel spit is also called Flèche des Bas Champs (Fig. 13.4) is about 16 km long. It extends from a series of chalk cliffs which contain many flint layers. The proximal section, from the cliffs of Normandy to the village of Cayeux is 7 km long and is a 110–150 m thick gravel ridge with one single crest, about 8 m high (above French 0 m sea level). The second part, north of Cayeux is a 5 km-long multi-ridge complex. It may comprise as much as seven ridges for a width of 600 m. The last part is about 4 km long and is a highly mobile accumulation of one single gravel ridge overlying sand banks (Costa 1997; Dolique 1998; Dolique and Anthony 1999). According to ancient maps (the first one dating back to 1754) this spit has not changed considerably in overall morphology. The accuracy of the maps is, however, difficult to assess, especially at the root of the spit, where it leaves the cliff (Ault harbour), therefore no precise conclusion may be drawn. It seems that the mouth of the river Somme has had many different arms (or channels) and that one of them used to flow along the spit, to the south, during the nineteenth century.

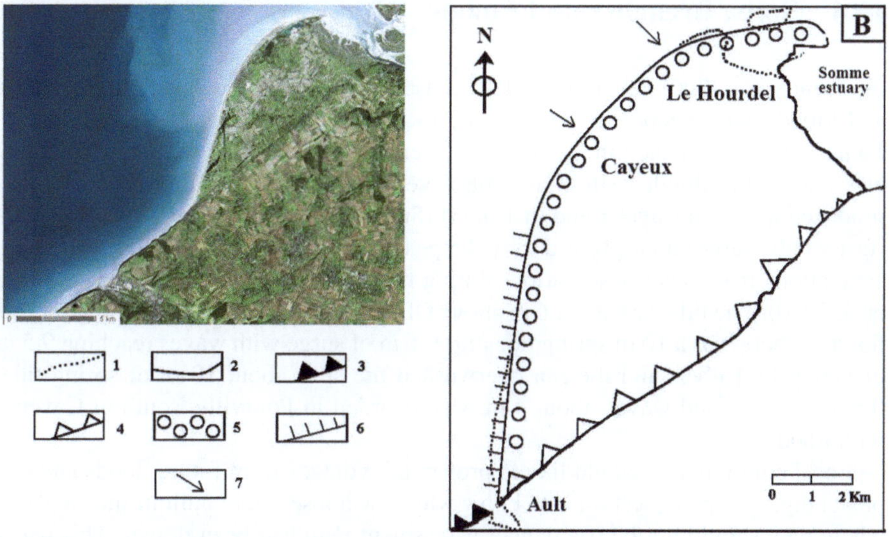

Fig. 13.4 Synthetic geomorphological behaviour of Cayeux spit. *1* Coastline of by 1745; *2* coastline by 2000; *3* high cliff retreating and producing flint gravel; *4* abandoned cliff; *5* gravel ridge accumulation (with local areas of retreat); *6* artificial devices, mainly groynes; *7* present accumulation zones along the spit

The spit began to grow about 2,500 years ago (Broquet and Beun 1981) and was fed by the erosion of the updrift cliffs which produced 20–30,000 m^3 of gravel per year. During the nineteenth and twentieth centuries many harbours (Fécamp, Dieppe) and jetties were built outside estuaries along the cliffs and at two nuclear plants in Paluel and Penly. Longshore drift has been interrupted and today, only 2,000–3,000 m^3 of sediment reaches the spit. The consequences were easy to forecast and gravel extraction was forbidden and groynes were built in an effort to sustain the spit. A first series of 90 groynes, each one 90 m long and separated from the next one by 90 m was erected between 1964 and 1984. Gravels taken from the updrift area of the nuclear plants were brought to the spit (estimated to be about 780,000 m^3 between 1984 and 1994). After a storm in 1990, 35 more groynes were built and about 24 more are considered today, though the decision is not taken yet.

13.3.1 Spits and Coastal Management

The three above-mentioned spits are good examples of the variety of situations coastal managers have to cope with. The first issue deals with short term events and is about flood risks. The second one has a longer scope and deals with coastal behaviour, storminess changes and RSLR.

13.4 Spits, Breaches and Floods

In recent years (2008, 2010 and 2011) coastal floods have destroyed many houses and killed many people in France. The most spectacular event (Storm Xynthia, January 2010) happened in the region of Vendée Charente and the Pointe d'Arçay was one of the flooded sites. In 2008 a very strong storm (called Johanna) has produced floods in Cayeux and in Talbert (Suanez 2010). The peak of these storms (fast winds, surge) took place during the peak of a high spring tide. The following estimations from local observations during the storm fit well with models (Caspar et al. 2010). The tide was about 6 m above OD sea level in Arçay and the region was flooded; there was a 10 m spring tide plus 0.6 m of surge with waves reaching 9.5 m of height in Talbert and the sea overwashed the spit; about 10 m of spring tide, 0.8 m of surge and waves about 4 m was recorded in Pourville, south of Cayeux, with floods.

Local communities would like to protect the villages from future floods and spit breaching. An open question is to know why new houses were built in such a place, where they should not be if a danger assessment map had been drawn. This paper will not consider this point as it is still being debated between political entities in France.

It will concentrate on old houses and old villages which have been built legally but which are now exposed to floods.

In order to protect buildings behind spits in France two main types of management operations are considered. Most of the time the first solution is to fill the breaches in the spit so that water cannot flow into the land ward direction to the low lands (the "Bas Champs" in Cayeux, the "Prises" in Arçay). This requires an availability of sediment ready to be used. In Arçay, as the source is almost totally depleted, sand is taken from the nearby sea floor, between the mainland and the island of Ré. This area, the Pertuis Breton has strong tidal currents which move sand banks from the northwest to the south. The sand which is extracted from the sea floor is expected to be replaced by new sand banks, although the different estimations about the quantity of sand transport vary by a factor of 3 (LCHF 1987; Bertin 2008). In Talbert (the only spit which is not currently underfed) the gravel used to fill the breaches is taken mainly from overwash on the landward side of the spit. In Cayeux it is brought from the up drift area.

The other common approach is to erect dykes and there is much debate concerning where to build them. Should they follow the entire spit or should they be limited to definite areas? The cheapest solution is to build the shortest dykes and therefore local planers have to decide which areas are going to be let unprotected. This issue is specially difficult in Arçay, where many camping sites are scattered along the spit. Tourism is a very important part of the local economy as this part of Vendée and Charente has more sun hours (over 2,200 per year) than any other site along the French Channel and Atlantic coasts (Musereau and Regnauld 2009). The dyke (caption 15 on Fig. 13.2) was built along the village and some land, especially a former lagoon, called "Belle Henriette" was left as a flooding basin, supposed to collect waters in case of surge and storm during the peak of high spring tides. In Talbert there is no tourist building on the spit. The tourist places are in the vicinity of the Pleubian village and they are not threatened by floods. In Cayeux floods may affect fields, small industrial activities and houses. This is why the spit is constantly reinforced and when a very strong storm is forecast local authorities are on the alert and have some quantity of gravel ready to be used in the event of a breach.

These management policies are costly and it is difficult to forecast if they will be able to cope with any change in relative sea level and storminess patterns.

13.5 Spit and Global Change

Today some parts of the spits do accumulate material. The distal part of Arçay has extended at a rate of several meters per year during the last 10 years. In Cayeux two areas are growing. One is just north of Cayeux and between 1961 and 2000 the coastline was prograding seaward between 4 and 7 m/year. In le Hourdel the rate is about 3.4 m/year, though some very localised parts are losing material (estimated at 2.8 m/year south of Le Hourdel and 1 m/year south of Cayeux). In Talbert the Sillon is not accumulating a lot but is able to whistand storms with a 20 year return time (Stephan et al. 2010).

There is much speculation about the pattern of storm intensity and frequency along the Atlantic and Channel coasts of France. Recent PhDs (Jouan 2005; Mastapaud 2011; Schoenenwald 2013) have extensively studied the storm record and have not found any clear trend and many papers do not fully agree (Dalla Marta 2009; Dupuis et al. 2006; Tabeaud and Bredif 2013). On the Atlantic coasts, in the Charente region the strongest known storm was in 1957 (since 1905) and this has not yet been surpassed. The February 15th 1957 storm was characterised by fast winds (above 24 m/s on average) and the tide level reached 6.27 m. The second most intense storm was on February 28th 2010. Average wind speed was around 23 m/s and the tide level was 5.96 m. There is no clear trend to signify that storm intensity is increasing. The frequency of events is another problem.

The sensitivity of these spits to storms is highly related to the frequency of events. On the Talbert spit Stephan (2011) has shown that the spit is resilient enough to cope with most of the marine surges and submersions. The same is probably true for other spits or gravel features (Costa et al. 2004), though there is no precise data to fix a threshold, after which the feature would be destroyed, or severely hit. They are vulnerable in case of violent storms the peak of which happens during a high spring tide. Other storms do not pose significant risks to human activity. The question shifts from storm frequency in general to the combination of storm frequency, spring tide height and surges coinciding at each local site (Cooper et al. 2004; Regnauld et al. 2004). Moreover, Ferreira et al. (2006) have demonstrated that although one single storm may not be a problem, that storm groups can be. A storm group is a series of storms following each other or a single period of time with more than three consecutive (or nearly consecutive) days with a strong gradient pressure. If a storm happens shortly after the previous one the spit has not enough time to re organise itself and is far more vulnerable. This leads to the hypothesis that if low to medium intensity storms arrive in clusters they may be far more impacting than one single extreme storm. A recent inventory shows that these types of event were twice as frequent during the second half than the first half of the twentieth century (Musereau and Regnauld 2014). There is no study to demonstrate whether this is just a chance or if it is a real trend.

The Sillon du Talbert is probably the most interesting site to observe the impact of these possible meteorological changes on coastal evolution. It is the least human-impacted feature and there is almost no threat to any buildings. The decision taken by the French "Conservatoire du Littoral", to let it behave "naturally" is certainly a very good one, even if it has not been without difficulties (Stephan 2011). The distal part of Pointe d'Arçay could also be allowed to behave "naturally", but there is a demand to secure the inhabited areas at the base of the spit. In Cayeux the scale of development poses a severe limitation on efforts to allow the spit to behave naturally.

13.6 Conclusion

The study of the behaviour of these three spits illustrates different points. First of all, sediment supply is a crucial factor in spit behavior. The source sites are probably the most important sites to survey regularly. The second point is about sensitivity to storm and surges. A spit must be understood as a mobile feature. It may roll over itself landward, it may extend along shore or change from one behaviour to the other. This has not been easily accepted by local communities. The very recent storms which have induced large floods have helped scientists to convince local communities that they should regard spits as mobile features. The non-intervention policy of the "Conservatoire" is also having the same effect: coastlines are going to change with "global change", but at locally variable rates and patterns. The behavior of each spit is very site-dependent and each of them must be considered individually.

References

Bertin X (2008) Morpho dynamique séculaire, architecture interne et modélisation d'un systéme baie-embouchure tidale: le Pertuis de Maumusson et la Baie de Marennes Oléron. PhD Université de La Rochelle, pp 1–198

Bressolier C (1979) Processus d'action dynamiques et réponses sédimentaires à la pointe de la Courbe. Mem Lab Geomorph EPHE 27:1–157

Broquet P, Beun N (1981) La sédimentation holocène dans les Bas-Champs de Cayeux (Somme). Evolution des lignes de rivage et du réseau hydrographique. Ann Soc Géol Nord, Lille, t.C, pp 31–41

Caspar R, Costa S, Lebreton P, Letortu P (2010) les submersions de tempêtes de la nuit du 10 au 11 Mars 2008 sur la côte d'Albatre. Norois 215:115–132

Cooper JAG, Pilkey OH (2004) Longshore drift, trapped in an expected universe. J Sed Res 74:599–606

Cooper JAG, Jackson DWT, Navas F, McKenna J, Malvarez G (2004) Identifying storm impacts on an embayed, high energy coastline: western Ireland. Mar Geol 210:261–280

Costa S (1997) Dynamique sédimentaire et risque naturel: l'impact des aménagements, des variations du niveau marin et des modifications climatiques entre la baie de Seine et la baie de Somme. PhD, Université de Paris 1, pp 1–376

Costa S, Cantat O, Pirazoli P, Lemaitre M, Delahaye D (2004) Vents forts et submersions marines en Manche Occidentale: analyse météo- marine sur la période historique récente? Actes XVII Colloque International de Climatologie, pp 277–280

Dalla Marta PM (2009) The return period of wind storms over Europe. Int J Climatol 29:437–459

Dolique F (1998) Dynamique morphosedimentaire et aménagements induits du littoral picard au sud de la Baie de Somme. PhD, Université de Dunkerque, pp 1–417

Dolique F, Anthony E (1999) Influence à moyen terme d'un estran macro tidal sur la stabilité d'un cordon de galets: la flèche de Cayeux (Picardie, France). Géomorphologie 1:23–38

Dupuis H, Michel D, Sottolichio A (2006) Wave climate evolution in the Bay of Biscay over two decades. J Mar Syst 63:105–114

Ferreira O, Garcia T, Matias A, Taborda R, Dias JA (2006) The role of storm groups in the erosion of sandy coasts. Earth Surf Proc Land 31:1058–1060

Galichon P (1984) Hydrodynamique sédimentaire des flèches littorales sableuses: cas de la point d'Arçay (Vendée). PhD Université de Paris 11, pp 1–234

Hallegouet B, Bouroules MA, Guennoc P (1989) Introduction des géotextiles dans les ouvrages de protection du littoral contre l'érosion marine. Bull Centre Géomorph Caen 36:227–230

Jouan D (2005) Evolution de la variabilité de la fréquence et de l'intensité des tempêtes en Europe du Nord Ouest. PhD, University Rennes 2, UMR CNRS 6554, pp 1–341

LCHF (Laboratoire Central Hydraulique de France) (1987) Catalogue sédimentologique des cotes françaises, Collection de la Direction des Etudes et Recherches d'Electricité de France, pp 1–559

Mastapaud A (2011) Impacts des tempêtes sur la morphodynamique du profil côtier en milieu macrotidal. La baie de Somme, Picardie. PhD Université du Littoral Côte d'Opale, pp 1–321

Morel V (1995) Impact des actions anthropiques sur les cordons de galets. Hommes Terres Nord 1–2:58–64

Musereau J (2010) Les tempêtes et le recul du littoral français. Mise au point et application d'un indice d'érosion dans la Zone des Pertuis Charentais (France). Approche de la gestion des cordons littoraux : mise au point et application d'un indice d'érosion (Zone des Pertuis Charentais, France). PhD dissertation, University Rennes 2

Musereau J, Regnauld H (2009) Coastal artificialisation and public policies; the example of the beach of Marennes (Seudre estuary, France). Open Geogr J 2:16–24

Musereau J, Regnauld H (2014) Storms impact on morphodynamics of human controlled coastal features in western France: the prevailing role of local management practices. J Coast Conserv 18(5):538–550

Musereau J, Regnauld H, Planchon O (2007) Vulnérabilité aux tempêtes des dunes littorales: développement d'un modèle de prédiction du dommage à travers l'exemple de Saint Trojan (Ile d'Oléron, France). Ann l'Assoc Int Climatologie 4:1–22

Orford JD, Forbes DL, Jennings SC (2002) Organisation controls, typologies and time scale of paraglacial gravel-dominated coastal systems. Geomorphology 48:51–85

Pinot JP (1994) Fixer le plan ou gérer le profil: l'exemple du Sillon du Talberv. Cah Nantais 41–42:307–316

Regnauld H, Carter RWG, Monnier O, McKenna J (1993) Modélisation de l'Evolution d'une Flèche Littorale et relation avec une Elevation du Niveau Marin: Exemple du Sillon du Talbert, Bretagne, France. Gaïa Rev Geocienc 6:59–64

Regnauld H, Pirazzoli PA, Morvan G, Ruz M (2004) Impacts of storms and evolution of the coastline. Mar Geol 210(1–4):325–337

Schoenenwald N (2013) Les tempêtes en France et dans les îles Britanniques: des aléas aux événements. PhD dissertation, University of Paris 1 Panthéon-Sorbonne

Stephan P (2011) Les flèches de galets de Bretagne: morpho-dynamique passée, présente et prévisible. PhD, Université de Bretagne Occidentale, pp 1–558

Stephan P, Suanez S, Fichaut B (2010) Franchissement et migration des cordons de galets par roll over. Impact de la tempête du 10 mars 2008 dans l'évolution récente du Sillon de Talbert. Norois 215:59–77

Suanez S (2010) Impact morphogéniques des tempêtes. Norois Spec Issue 215:1–146

Tabeaud M, Bredif H (2013) Aux grands maux faut-il toujours préférer les grands remèdes? Bull l'Assoc Géographes Fr 1:24–35

Tounsi I, Verger F (2007) Etude dia chronique de la point d'Arcay (Vendée) d'après les images aériennes et spatiales. Photo Interprétation 43(4):13–16 and 41–42

Traini C (2009) L'estuaire de la Vilaine : évolution naturelle et anthropisation. PhD Université de Bretagne Sud, pp 1–162

Verger F, Auphan E, Moniot C (1972) La carte géomorphologique conçue comme un modèle dynamique. Mém Doc Serv Doc Cartographie Géogr 12:223–263, éd. du CNRS, 2 pl., 1 carte hors-texte

Chapter 14
The Sand Spits of the Rhône River Delta: Formation, Dynamics, Sediment Budgets and Management

François Sabatier and Edward Anthony

Abstract This chapter describes the morphology and dynamics of the three sand spits flanking the Rhône river delta: Gracieuse, Beauduc and Espiguette spits. The formation and distal extension of these spits (respectively +5, +10–15, +10–15 m/year) results from active wave-induced longshore transport that generates sand migration downdrift from erosional updrift sectors. These erosional sectors are systematically associated with former delta distributary mouths that are being reworked by waves. In other words, the sand brought down to the sea by the Rhône participates in the construction of sand spits only following mouth abandonment. Finally, as a result of the curvature of these spits, the overall aeolian regime induces sand migration seaward in the southern sectors of the spits near the transition zone between eroding and accreting sectors. Understanding these functional aspects of spit formation and growth is important in terms of management of the Rhône delta shoreline and its sediment budget by public authorities.

14.1 Introduction

The strong influence of waves in the Mediterranean has led to the development of deltas characterised by spit development. Among these, the Rhône delta, the third largest in the Mediterranean, after the Po and the Nile deltas, exhibits three large spits, Gracieuse, Beauduc and Espiguette (Fig. 14.1) and two much smaller delta-mouth spits, the Roustan spits, which are not further considered in this review. These three wave-formed spits are relatively recent features of the delta, having been formed respectively in 1892, 1711, and between the eighteenth and nineteenth centuries (Vella et al. 2005; Sabatier et al. 2006; Rey et al. 2009). Following the classic spit inception model of Zenkovitch (1967), these spits are associated with an updrift erosion zone, but no analysis of their morphodynamics has as yet been

F. Sabatier (✉) • E. Anthony
Aix Marseille Université, CEREGE UMR7330 CNRS, Marseille, France
e-mail: sabatier@cerege.fr; anthony@cerege.fr

G. Randazzo et al. (eds.), *Sand and Gravel Spits*, Coastal Research Library 12,
DOI 10.1007/978-3-319-13716-2_14

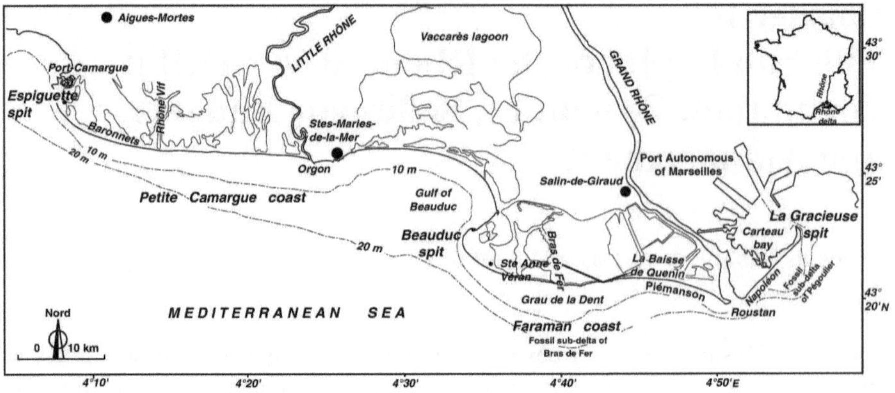

Fig. 14.1 The Rhône delta and names cited in the text

conducted. Understanding the way these spits evolve, is however, an increasingly necessary requirement for the sustainable management perspectives of the Rhône delta shoreline. These features, represent still relatively pristine natural areas, are potential sand sources for the renourishment of eroding shoreline sectors, and good indicators of the meso-scale organisation of the coastal sediment budget of the Rhône delta and its sediment cells, with important implications for future sediment husbandry in a context of rising sea level (Brunel and Sabatier 2009; Sabatier et al. 2005, 2009b). We will successively describe the morphology of these spits, and the genesis and dynamics of their terrestrial and submarine segments, followed by a conceptual model of the way they function.

14.2 Location and Description of Spits

14.2.1 Gracieuse Spit

The rectilinear Gracieuse spit, situated at the eastern extremity of the Rhône delta shoreline, is stretching towards the NE, under a dominant drift driven by waves from the SW (Fig. 14.2). The annual rate of growth of the spit is close to 5 m, a rate inferior to that of the other two spits. The terrestrial accretionary part of the spit is relatively flat, lacks proper dunes and vegetation, and is subject to regular marine submersion. This part of the spit is subject to northerly winds that can blow over a fetch of several hundreds of metres (<1 km), resulting in the seasonal formation of barkhans bordering the shoreline. The accumulation sector of the Gracieuse spit corresponds to a zone of marked shoreline curvature extending over 2–3 km. In the updrift zone, oblique and or crescentics bars are regularly observed, and these rapidly change into longshore bars the crest of the deepest of which lies about 2 m below the mean water level. The slope of the active part of the cross-shore

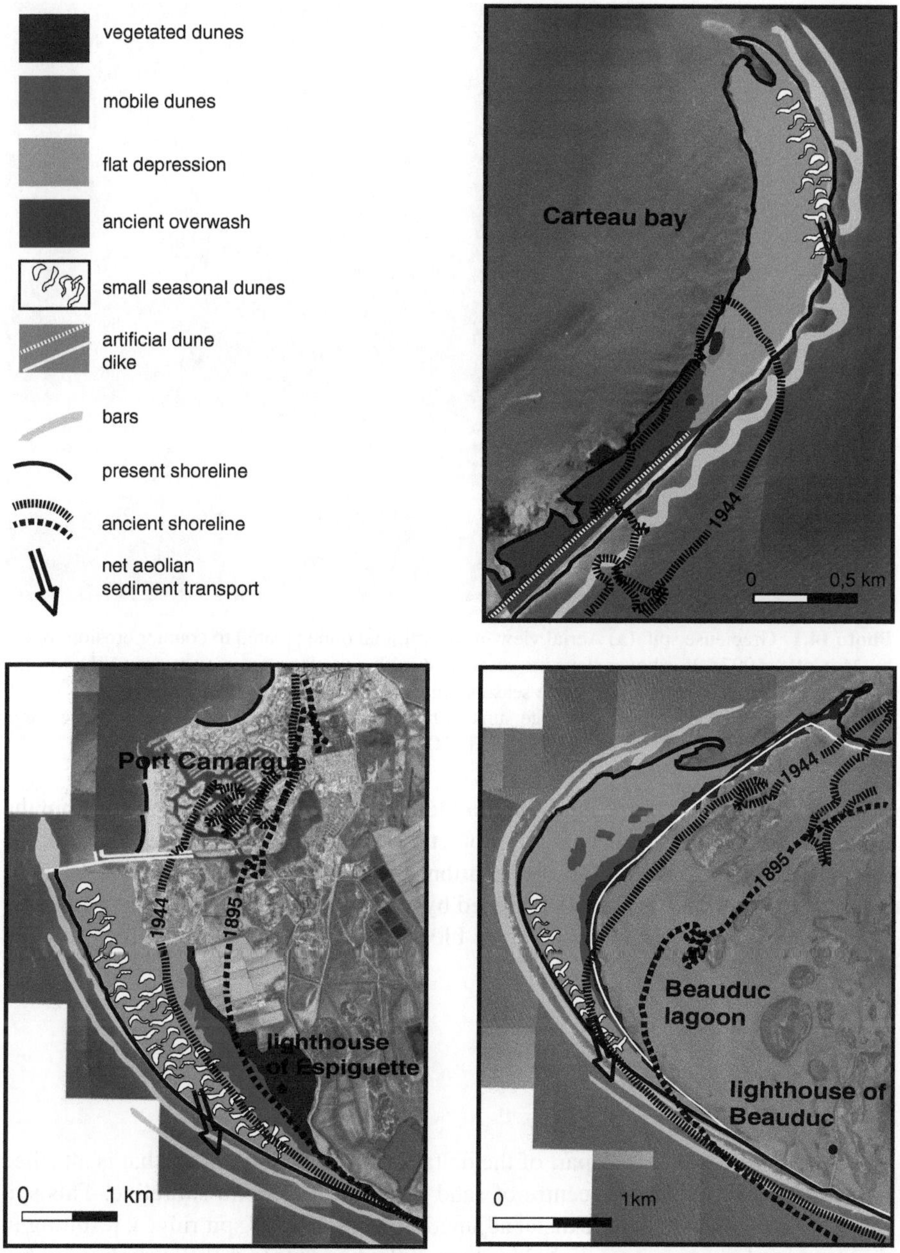

Fig. 14.2 Morphology of the Rhône delta spits

profile, which goes down to about −10 to −12 m, is about 5–6 %. The distal extremity of the spit exhibits one to two recurves under the effect of waves generated by NW (Mistral) winds. Gracieuse spit protects the access to the Port

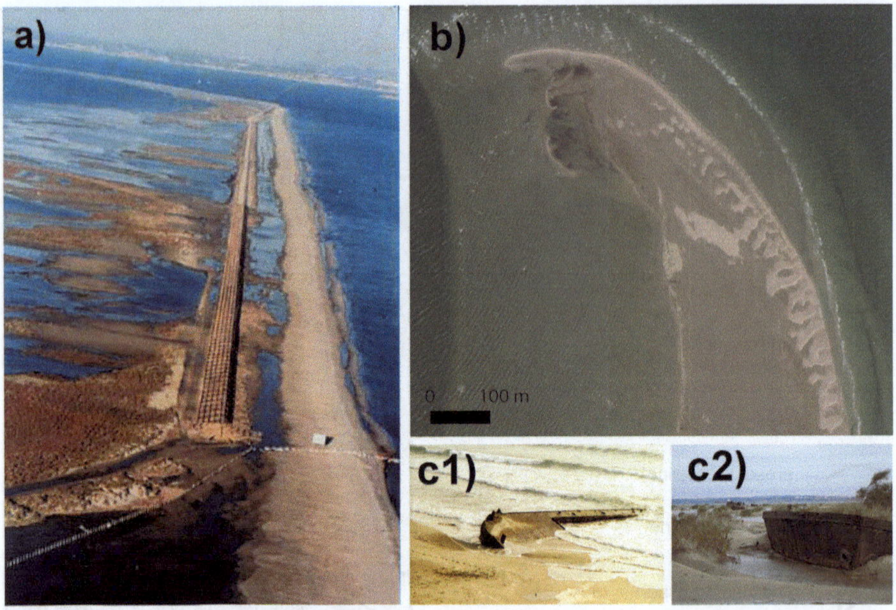

Photo 14.1 Gracieuse spit. (**a**) Aerial view of the artificial dune created to counter erosion (photo by M. Cotte, 1997). Washover deposits associated with spit shoreline retreat are visible on the backshore of the spit. (**b**) Small mobile seasonal dunes on the foreshore migrating towards the SE (Courtesy of IGG 2005). (**c**) *c1* A barge sunk in the 1970s on the shoreline. *c2* The same barge completely infilled with sand and now located 300 m offshore

of Marseille, the third largest port in the Mediterranean, from storm waves from the SE. In order to reinforce this natural protective role, the eroding central sector of the spit was consolidated through the creation of an artificial dune in 1992, whereas accretion at the tip of the spit is impeded by old barges that are sunk and ultimately become filled with sand. These barges block the longshore drift and therefore act like groynes (Photo 14.1).

14.2.2 Beauduc Spit

Beauduc spit, in the central part of the delta, exhibits a recurved tip that is attached to Beauduc bay, a net depocentre of sand on the Rhône delta shoreline. This spit differs from the others by the distal adjunction of successive spit ridges, resulting in a pronounced concave-seaward plan shape linked updrift to a still accretionary but much narrower proximal zone (Fig. 14.2). Over the last 10 years, the spit has been extending to the NE at a mean rate of 8 m/year and the distal accumulation sector is now 5 km long. Beyond a zone of barkhanoid dunes lining the shoreline, the central part of Beauduc spit exhibits a vast flat depression bounded by a continuous 3–4 m-high foredune ridge located at a distance of up to 500 m from the shoreline near the

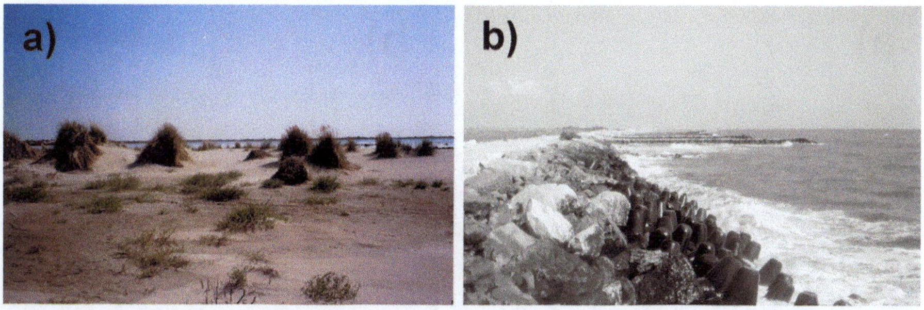

Photo 14.2 Beauduc spit. (**a**) Eroding dune barrier in the transition zone between accreting and retreating sectors. (**b**) Rock armouring on the Faraman coast. The beaches have practically disappeared in this eroding sector

distal tip. This vegetated dune flanks a dyke constructed in 1972 in order to protect the back-spit lagoon from marine submersion during storms (Beauduc lagoon). Beauduc spit is under the influence of longshore drift generated essentially by storm waves from the SE, but SW waves also contribute to growth of the spit at its distal sector. The submarine morphology generally comprises two bars in the southern sector of the spit and 3–4 bars in the northern sector. These bars are longshore features but the innermost bar is sometimes crescentic. In relation with the along-shore accumulation, the accretion zone down to depths of about −8 to −10 m shows a slope of nearly 4–5 %. Beauduc spit is sourced by erosion of the updrift Faraman sector of the Rhône delta. In order to combat erosion which had begun threatening the industrial exploitation of the salt-producing lagoon behind the spit, this part of the spit has been subject to various engineering interventions since the 1970s, and especially in the 1990s (Sabatier and Suanez 2003). At present, nearly 9 km of the spit shoreline in this eroding sector are equipped with groynes, a dyke and one breakwater (Photo 14.2).

14.2.3 Espiguette Spit

Espiguette spit, at the western extremity of the Rhône delta, has had its distal zone occupied since 1968 by the Port Camargue Marina and by a dyke aimed at blocking sand movement, but in its natural state, the spit displays recurves stretching N and NE (Fig. 14.2). Adjacent to the dyke, the spit shoreline progrades rapidly, by 10–15 m/year, nourished by strong drift generated by waves from the S and SE. The beach foreshore and backshore form a wide planar surface that tends to become infilled by small extremely mobile dunes, especially in its central part. Near the shoreline, aeolian dunes are formed seasonally and Espiguette spit is the only one of the three Rhône delta spits having truly vegetated dunes, with sparse colonies of marram grass and dense stands of pines. The spit submarine morphology is generally characterised by 2–3 longshore bars with the inner bar often exhibiting a cresentic morphology.

Photo 14.3 Espiguette spit. (**a**) Aerial view of hard defense structures on the Petite Camargue coast. Erosion is so important that the groynes are now cut off from the beach 15 years after their construction. (**b**) Mobile dunes in the central part of the spit. (**c**) Aeolian transport seaward during a Mistral event (Photo taken on the south side of the spit)

The slope (ca. 4–5 %) and depth (−8 to −10 m) of the distal spit zone are similar to those of Beauduc spit. Espiguette spit is sourced in sand by erosion of the Petite Camargue sector of the Rhône delta. This erosional zone, the backshore of which is also exploited by the salt-manufacturing industry and by farmers, has been consolidated by an impressive field of about 150 groynes over a length of nearly 20 km since the 1980s, with addition of dykes in places (Photo 14.3).

14.3 Spit Formation

14.3.1 Gracieuse Spit

The formation of Gracieuse spit commenced in 1892, in the wake of the artificial opening of the Roustan outlet (Fig. 14.3) through which the main Rhône channel now debouches. This opening marked the diversion of the liquid and solid discharge of the Rhône from the Pégoulier outlet resulting in its erosion (Fig. 14.4). Sand derived from this erosion was redistributed by longshore drift towards the NW, progressively forming Gracieuse spit. Between 1895 and 1944, the Pégoulier outlet retreat by 1,500 m, while Gracieuse spit extended by 1,400 m over the same period. At the same time, the beaches of Piémanson and Napoleon on either side of this mouth prograded respectively by 1,500 et 1,700 m. Beginning in the 1940s, the shoreline morphology, which is similar to that of the present day, developed from a

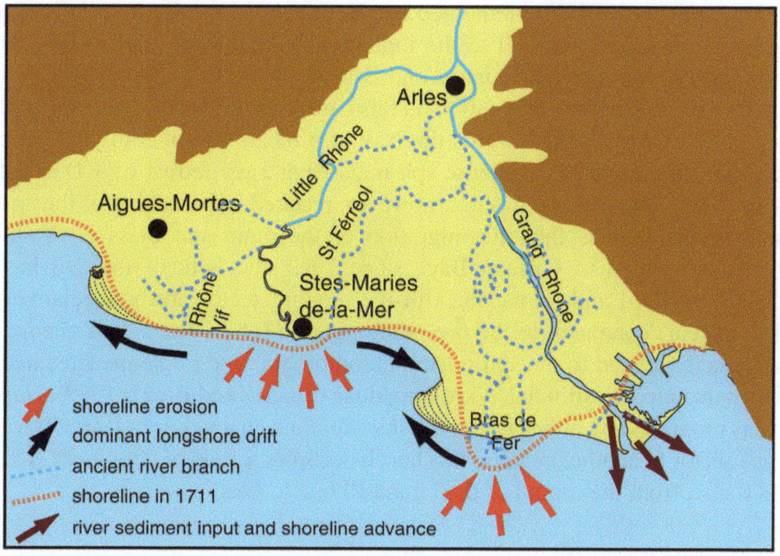

Fig. 14.3 Shoreline change on the Rhône delta between the eighteenth century and the present day

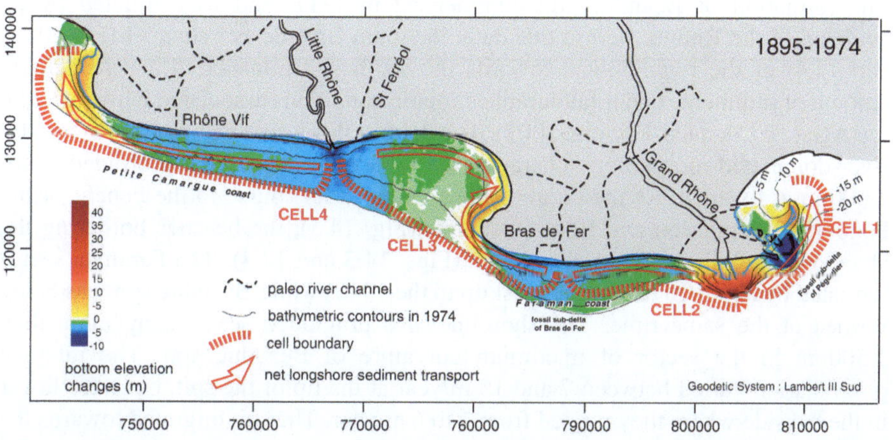

Fig. 14.4 Organisation of littoral drift cells on the Rhone delta shoreline (Modified after Sabatier and Suanez 2003, and Sabatier et al. 2009b)

flattened sinusoïdal plan shape to a rectilinear plan shape, while the spit grew by another 1,200 m at a mean rate of 19 m/year. This evolution occurred in two stages: (i) up to the mid 1960s, spit lengthening occurred towards the NE; (ii) since the 1970s, accretion has been affected by sunken barges at the tip of the spit, which act such as groynes. The latter block longshore drift and favour the diffraction of waves from the SW in such a way that the tip of the spit is now extending NNW.

These interventions were implemented by the Port of Marseille in order to prevent silting up of an access channel to the industrial sector of the port in Fos-sur-Mer (Suanez and Provansal 1998; Suanez and Bruzzi 1999).

The change in shoreline orientation on this part of the delta coast is associated with a hinge point that migrates towards the NW at a mean rate of 3 m/year. In consequence, the central part of Gracieuse spit underwent a net retreat of 400 m between 1944 and 1989. This erosion has not, however, resulted in breaching of the spit. The dominant effect here is that the migration of the main spit mass is resulting in significant deposition in Carteau Bay. Since 1992, the bulk of the spit has been artificially stabilised and retreat rates in the eroding sector are now relatively low (about 2 m/year) compared to the preceding period (3–6 m/year). This change is, of course, directly related to the spit rehabilitation programme implemented since 1988 and which included an artificial, 3 km-long dune (Suanez and Bruzzi 1999). Napoleon beach has prograded by about 120 m but the shoreline changes are extremely variable from one sector to another because this beach occupies a zone of temporary storage of sand in transit from the mouth of the Grand Rhône to Gracieuse spit.

14.3.2 Beauduc Spit

The formation of Beauduc spit commenced in 1711, and was initiated by an avulsion of the Rhône. Before this date, the main Bras de Fer channel (Figs. 14.3 and 14.4) of the river flowed towards the south and transported a considerable amount of sediments under favourable climatic conditions associated with the Little Ice Age (e.g., Arnaud-Fassetta 2003). The delta at this time exhibited an essentially river-dominated morphology characterised by a river mouth that protruded about 2,000 m at its apex. As the Grand Rhône changed its course to the benefit of the Pégoulier, Piémanson and Roustan outlets (Fig. 14.3), the beaches bordering the abandoned mouth were rapidly eroded (Figs. 14.3 and 14.4). The Faraman sector retreated by about 10 m/year, at least up to the 1940s while Beauduc spit was being formed at the same time. The shoreline also prograded seaward by more than 1,500 m in the sector of maximum curvature of Beauduc spit. The rates of progradation varied between 7 and 15 m/year at the tip of the spit, but were lower in the bayside where they ranged from 1 to 6 m/year. The spit migrated towards the NW, undergoing a process wherein the accretion sectors of the southern flank ended up being eroded to source the north-facing flank, in a process akin to cannibalisation. The spit recurves at the tip grow alongshore but never diverge from the mainland shore and finally end up welding to the latter.

The transition between the accreting sink zone and the eroding source zone is presently situated just south of the Veran lighthouse (Fig. 14.2). This limit evolves slowly in time, migrating in the direction of longshore drift. Up to the 1950s, the entire eroding Faraman coast experienced differing retreat rates on either side of the Grau de la Dent outlet. To the east, retreat attains 2–4 m/year, but at about 1989/ 1990 the shoreline became stable. This change occurred in the wake of the emplacement in 1987 of rock defences by the salt-manufacturing companies the Compagnie des Salins du Midi and Salines de l'Est. This shoreline stabilisation operation worked out fairly well (Suanez and Bruzzi 1999), but some poorly protected sectors started retreating again in 1992. West of the Grau de la Dent outlet, shoreline retreat in the sectors of Sainte Anne and Beauduc lagoons varied between 2 and 10 m/year over this period. The strongest erosion has been recorded at Veran where the beach has retreated by 650 m since 1895. Defense works constructed in this sector in 1990 have also stopped shoreline retreat. However, the 1997 storm and the damages it caused have left this part of the delta coast vulnerable to erosion (Sabatier et al. 2009b). Following this storm, a seawall was constructed, reinforced by rock revetments. The seawall was, however, progressively weakened at its base by wave reflection, resulting in deepening of the toe of the beach profile (Samat et al. 2005). Much of this wall has now collapsed.

14.3.3 Espiguette Spit

The formation of Espiguette spit started between the sixteenth and seventeenth centuries, probably at about 1530, the year a presently abandoned former branch of the Rhône (the Rhône Vif) changed its course to flow towards the Grau d'Orgon outlet (Fig. 14.3). Starting from this time, the Petite Camargue coast became subject to erosion, releasing sand for the construction of Espiguette spit by SE storm waves. As in the case of Beauduc spit, the history of Espiguette spit is marked by constant progradation. It is in this sector that the Rhône delta has prograded most in recent times (+1,500 m during the twentieth century). This accretion became all the more important following the construction of a dyke in 1970 at the distal tip of the spit to prevent it from blocking the entrance to Port Camargue Marina and to prevent harbour silting. Since the end of the 1990s, this dyke is, however, progressively bypassed by sand (Fig. 14.2).

The transition zone between the prograding spit and the eroding shoreline of Petite Camargue is now naturally migrating westwards at a rate of 6 m/year but the migration rate has significantly increased following the various stages of emplacement of rock revetments. We have here the classic downdrift erosion following the updrift implantation of groynes. From the mouth of the Little Rhône branch to the south of Espiguette spit, retreat rates progressively decrease, ranging from 4 to 1 m/year. The shoreline bordering the mouth of the Little Rhône is the sector of the Rhône delta that has retreated most since 1895. This retreat exceeds 1,000 m and is characterised by rates ranging from 3.5 to 10 m/year, depending on sector. It is worth noting, however, that retreat has slowed down since 1980, a consequence, as in other areas of the delta, of the construction of groynes in 1975. However, the efficiency of these structures remains doubtful (Suanez and Sabatier 1999). Rip currents generated at the flanks of these groynes and a spacing too large between groynes have been identified as two factors responsible for on-going erosion (Sabatier 2001).

14.3.4 Drift Cells and Sediment Budgets

A comparison of the Rhône delta shoreface bathymetric maps between 1895 and 1974 reveals a pattern of nearshore erosion and accretion echoed by the shoreline, especially in the sector of the present Grand Rhône mouth (Fig. 14.4). On the eroded shoreface, the relict sediments of the prodelta lobes of Little Rhône-St Férreol, Bras de Fer and Pégoulier are reworked shoreward to feed growth of Espiguette, Beauduc and Gracieuse spits (Sabatier and Suanez 2003). This organisation reflects a classic drift cell structure wherein updrift shoreline erosion sources downdrift spit growth (May and Tanner 1973; Swift 1976).

The easternmost cell (CELL1) is located east of the Grand Rhône, where the prevailing sediment transport is eastward. It is bounded on its western side by the mouth of the Grand Rhône, and on its eastern side by Gracieuse spit. This cell is thus fed by the erosion of the fossil delta lobe of Pégoulier (Suanez et al. 1998). Meanwhile, the present-day sand inputs of the Rhône (active source) have decreased (Sabatier et al. 2009b). The second cell (CELL2) also shows a general direction of transport from west to east, and extends from Ste. Anne lagoon as far as Grau de Roustan. Although shoreline retreat in this zone contributes to the supply of sand to Piémanson beach, part of the eroded material is stored in the vicinity of the Roustan prodelta, considered as an important sediment sink (Suanez et al. 1998). The limit between cells 2 and 3, located generally near Ste. Anne lagoon, is explained by a reversal of the longshore drift as a result of refraction of swell on the fossil Bras de Fer delta lobe (Sabatier 2001). The third cell 3 (CELL3) extends from Ste. Anne lagoon in the east as far as the Little Rhône in the west. This cell is made up of a central part subject to accumulation (Beauduc spit and Beauduc bay), fed in the east by beach retreat and erosion of the fossil Bras de Fer lobe, and in the west by beach erosion in the Saintes-Maries-de-la-Mer sector. Thus, drift convergence associated with the two eroded sectors results in an important sediment sink in the very central sector of the delta shoreline. Sediment supply to Beauduc bay via the Little Rhône is extremely limited, being dependent on the very low sand supply of this branch (Arnaud-Fassetta 2003). To the west of the Little Rhône, the fourth cell (CELL4) is defined by a dominant sediment transport directed towards Espiguette spit. As seen on the Faraman coast, the opposite sediment transport directions in cells 3 and 4 can be explained by divergence of the longshore drift related to wave refraction on the prodelta of the Little Rhône (Sabatier 2001). The sector in erosion from the Little Rhône to Baronnets releases sand for the spit.

The sediment budget of the shoreface discussed here concerns the western and eastern sectors of the Grand Rhône mouth and between 0 and the -10 m depth contour. The sediment budget of cell 1 (Gracieuse spit) involved erosion of the relict Pégoulier prodelta lobe to the tune of about 380,000 ($\pm$40,000) $m^3.year^{-1}$ and spit accumulation of about 41,000 ($\pm$50,000) $m^3.year^{-1}$ between 1895 and 1974. Thus, assuming that erosion releases sand for accumulation generated by longshore drift, we can infer that the mean long-term sediment contribution of the Grand

Rhône river to the littoral drift is of the order of about 30,000 (±90,000) $m^3.year^{-1}$ between 1895 and 1974 which is very low compared to the accumulation at the mouth 2,490,000 (±250,000) $m^3.year^{-1}$.

Along the western part of the Roustan coast, the shoreface, less affected by the river sediment input, yields a net sediment budget slightly in erosion (−280,000 ± 310,000 $m^3.year^{-1}$) or in equilibrium if we consider the error margin. This equilibrium budget implies longshore sediment redistribution between erosional and accretional sectors. Beauduc spit shows a net gain of 600,000 $m^3.year^{-1}$ whereas erosion of the fossil Bras de Fer lobe is of the order of about 510,000 $m^3.year^{-1}$. This slightly positive budget suggests alongshore redistribution of eroded sediments and/or opened boundaries between cells 2 and 3. In contrast, the Espiguette spit budget is slightly negative (−120,000 $m^3.year^{-1}$). This value testifies to the extremely low, if not nil, sand supply by the Little Rhône river branch to the cell 4, as well as potential offshore losses or to the benefit of the neighbouring cell 3.

14.4 Shoreface Sediment Transport

Downdrift of the spits, the drift rates calculated from bathymtric charts (1895–1974/1988) show values that are extremely high at the world scale and compared to other Mediterranean deltas. These values are 410,000, 610,000 and 860,000 $m^3.year^{-1}$ for Gracieuse, Beauduc and Espiguette spits, respectively. These very high rates are explained by: (1) the obliquity of storm waves relative to the eroding sectors (incident angles are in the range of 40–20° in the breaker zone), and (2) the fine sand size (D_{50} of about 0.16 mm in the breaker zone). Both of these conditions are conducive to large drift volumes, but these values may also beover-estimated as a result of the bathymetric differencing method used to calculate them. Finally, it is worth noting that there is a link between spit size and the annual drift volume.

In accreting shoreline sectors, numerical simulations indicate that drift is exerted in depths that progressively decrease in the dominant transport direction. Along Beauduc spit, for instance, drift is progressively limited from −5 to −2 m towards the distal tip. At the same time, the width of the breaker zone (equated with the width of the barred zone) also diminishes in the same direction whereas the submarine slope between 0 and −10 m becomes steeper. In contrast, the depth of closure (Dc) increases in the same direction, from −6 to −7 m in the erosional zones, to −9 to −11 m in the most prograded sectors. The spits are affected by drift processes that lead to sand being stranded in the upper part of the profile, leading to subaerial spit widening but also to an increase in the submarine slope because of seaward profile translation into deeper water. The greater slope generates cross-shore sand transport by gravitational processes to depths larger than those affected by longshore drift.

14.5 Aeolian Dunes and Sand Transport

14.5.1 Wind Conditions and the Potential Aeolian Sediment Budget

The wind rose of the Rhône delta shoreline is dominated by offshore winds from the N and NW (Mistral and Tramontane winds) and by onshore winds from SSW to SE that generate storms. The potential aeolian transport shows an overall equilibrium between these two sectors (Sabatier et al. 2009a). However, if we take into account phases of beach submersion and rainfall that limit onshore aeolian sand transport during storms, the N and NW winds are more efficient in terms of transport. Along portions of eroding coast with an overall E-W orientation, the net resultant aeolian transport is thus hardly favourable to dune construction. In contrast, in the distal curved sectors of the spits, the situation is different because the change in orientation engendered by curvature favours significant aeolian transport mainly by winds from the W to N sector, the dry fetch of which can exceed several hundreds of metres. Shoreline orientation thus plays a primary role in source and sink zones on the subaerial portions of the Rhône delta spits. In all three cases, the tips of the spits are subject to a system of cross-shore aeolian sand transport from the inner shores to the central parts of the spit, and then to the southern, seaward flanks. These seaward flanks therefore serve as temporary storage zones for sand that is ultimately transported to the sea by Mistral and Tramontane winds. Using the regional aeolian sand drift potential concept of Fryberger (1979), a sand transport threshold with a wind speed of 6 m/s at an above-ground elevation of 10 m, and data from aeolian sand traps (Sabatier et al. 2007, 2009a), the beaches of Gracieuse, Beauduc and Espiguette spit are thus stripped, respectively of 12,000; 24,000 and 40,000 m^3/year of sand by aeolian deflation. This sand is not lost to the sea, as it is simply recycled into local swash-induced beach accretion (Photo 14.3b) or transported by longshore drift towards the distal spit extremities.

14.5.2 The Role of Minor Seasonal Dune Activity

The beaches of the three spits commonly comprise small seasonal dunes. These features are commonly barkhanoid, and extend over 10–50 m over a post-storm season berm (Photo 14.1b). These mobile dunes have a maximum height of about 2.5 m and are completely submerged and reworked by waves and swash but can be rapidly reformed (in a few days) by Mistral and Tramontane winds. Coalescent barkhans during long periods of low wave activity can form mini-transverse or linguoid dunes. These dunes are thus typical summer features but they can also develop in winter between two storms whenever the northerly winds prevail. Even though these dunes are aligned with spit curvature, their crest orientation is determined by Mistral and Tramontane winds which generate southward dune migration.

In detail, these barkhanes are most commonly located on shoreline segments oriented N340 to N30 along Gracieuse spit, and N125 to N 170 along Beauduc and Espiguette spits. Work remains to be done to gain a finer understand of the dynamics of these dunes, including aspects such as topographic steering along the steep spit flanks.

Under the influence of the diminution of the longshore transport capacity, the beaches of the Rhône delta spits are prograding rapidly, by up to 15–20 m/year. The small dunes accumulating on the beach berms constitute a permanent but recyclable background sand stock, ultimately reworked and transported seaward. The fine-tuned local wind dynamics involved in the inception and evolution of these dunes are not known but we presume that they become increasingly more mobile as they grow in altitude above the level of the berm into an airflow zone of less impeded transport. On the inner spit flanks, the dominant N winds are oblique to the shoreline and lead to constant deflation, thus limiting the upward growth of these dunes, leading to their relatively constantly identical elevation in time. Aeolian deflation creates a flat surface seaward of the dunes, over which sand is transport obliquely relative to the shore downwind. Where this aeolian fetch becomes sufficiently large, which is the case of Beauduc and especially Espiguette, aeolian accumulation can ensue and the dunes colonised and semi-stabilised by vegetation (Fig. 14.2).

14.6 Discussion: A Conceptual Model of Spit Morphodynamics

The large-scale behaviour of the Rhône delta shoreface allows us to draw up a conceptual model of spit inception and growth: (1) accumulation of sediment at the river mouth, with advance of adjacent beaches and little river sediment input to beaches farther from the mouth; (2) shifting of the mouth; (3) erosion of relict prodelta lobes with beach retreat, and; (4) longshore redistribution of sediment from the eroded beaches to build up the spits (Sabatier et al. 2006). This conceptual model suggests that there is a “time-shift” between river sediment supply to the sea and the build up of a spit. In the course of the recent history of the Rhône delta, the river channels have shifted many times to build up the present deltaic plain and shoreface morphology and spits. Nowadays, because the river channels are stabilised by dykes and other engineering structures, avulsion is no longer possible (assuming that the dykes can resist the strongest flood events). Under these conditions, and with the decreased input of fluvial load into the sea, it thus appears unlikely that river sediments can supply the beaches of the Rhône delta coast. Moreover, the relict prodelta lobes are sediment reservoirs that are gradually being depleted. The chronic erosion of the coastline, which is caused by a deficit in sediment, a very low fluvial sediment supply and the redistribution of river sediments, is thus likely to continue in the future. All relict prodelta lobes show the same trend: whenever natural avulsion or artificial channel diversion occurs and the river mouth becomes abandoned, the shoreline in the vicinity of the mouth erodes to feed adjacent spits.

At a relatively short timescale, the lengthening of the three Rhône delta spits is merely a reflection of longshore drift on this coast. Beauduc and Espiguette spits evince similar characteristics as far as shape, size and growth are concerned, both evolving essentially under the influence of SE storm waves, but their distal tips differ in behaviour. On Beauduc spit, the tip is characterised by recurves towards the shoreline, whereas on Espiguette spit, the natural tip shows rectilinear extension alongshore (before the construction of the marina). A probable reason for this contrast is the difference in fetch length between Beauduc bay and Aigues Mortes bay (Fig. 14.1). The first being larger (7.5 km) than the second (5 km), waves generated by N to NE winds are better formed and can therefore more readily generate the drift and refraction conditions that lead to recurves. Another explication may reside in slightly different wave rose conditions between the two sites, with Beauduc probably being subject to stronger SE waves because of wave refraction across the ancient mouth lobe of Bras de Fer. This difference certainly expresses the complexity of the processes involved in distal tip growth. Gracieuse spit differs from the other two spits in its relatively rectilinear shape. These shape differences are probably related to the influence of the local bathymetry on wave refraction.

In the cross-shore dimension, spit progradation is manifested by progressive seaward sand accumulation, resulting in a body the seaward slope of which increases towards its distal end. Over the beach, aeolian transport prevails within a long-term NNW-SSE sand circulation system driven by Tramontane and Mistral winds. The small seasonal dunes bordering the shoreline serve as temporary sand storage features. The overall aeolian regime induces sand migration seaward in the southern sectors of the spits near the transition zone between eroding and accreting segments. This sand is then remobilised by wave-driven longshore drift towards the distal spit extremities. This sediment transport cycle thus involves both longshore drift and wind-driven transport but the volumes mobilised longshore drift are 10–100 times larger than those generated by aeolian processes on the beach.

14.7 Conclusion

The shoreline dynamics of the Rhône delta over the last two centuries has been strongly characterised by the development of three spits – Gracieuse, Beauduc and Espiguette – reflecting the importance of wave-generated longshore drift in the western Mediterranean setting of the delta. The plan shape, size and morphology of these spits reflect local differences in exposure to waves, winds and wave-bathymetry interactions. The three spits clearly reflect the organisation of the shoreline sediment budget and sediment cells of the Rhone delta. Aeolian processes play a variable, but much less important role in the morphological evolution and sediment budgets of the spits. The spits not only offer protection to port and marina accesses but also constitute important sand stocks. The spits also offer relatively pristine natural areas. They are, however, characterised, like many spits, by updrift erosion that poses a constant threat to installations and activities. This has necessitated the implementation of various shoreline defence projects that have met with mixed success. The spits are potential sources of sand for the nourishment of eroding shoreline sectors, but maintaining their integrity in a context of overall waning sediment supply and rising sea level will be more and more of a challenge in the twenty-first century.

References

Arnaud-Fassetta G (2003) River channel changes in the Rhone Delta (France) since the end of the Little Ice Age: geomorphological adjustment to hydroclimatic change and natural resource management. Catena 51:141–172

Brunel C, Sabatier F (2009) Potential sea-level rise influences in controlling shoreline position for French Mediterranean Coast. Geomorphology 107:47–57

Fryberger SG (1979) Dune forms and wind regime. In: McKee ED (ed) A study of global sand seas, USGS Professional Paper 1052. USGS and NASA, Washington, DC, pp 137–169

May JP, Tanner WF (1973) The littoral power gradient and shoreline changes. In: Cosates DR (ed) Publications in geomorphology. State University of New York, Binghamton, pp 43–60

Rey T, Lefevre D, Vella C (2009) Deltaic plain development and environmental changes in the Petite Camargue, Rhone Delta, France, in the past 2000 years. Quatern Res 71(3):284–294

Sabatier F (2001) Fonctionnement et dynamiques morphosédimentaires du littoral du delta du Rhône. PhD. thesis, Université d'Aix-Marseille III

Sabatier F, Suanez S (2003) Evolution of the Rhône delta coast since the end of the 19th century. Géomorphol Relief Processus Environ 4:283–300

Sabatier F, Provansal M, Fleury T (2005) Discussion of: PASKOFF R (2004) Potential implications of sea-level rise for France. J Coast Res 20(2):424–434. J Coast Res 21(4):860–864

Sabatier F, Maillet G, Fleury J, Provansal M, Antonelli C, Suanez S, Vella C (2006) Sediment budget of the Rhône delta shoreface since the middle of the 19th century. Mar Geol 234:143–157

Sabatier F, Chaibi M, Chauvelon P (2007) Transport éolien par vent de mer et alimentation sédimentaire des dunes de Camargue. Méditerranée 108:83–90

Sabatier F, Anthony E, Héquette A, Suanez S, Musereau J, Ruz M-H, Régnauld H (2009a) Morphodynamics of beach/dune systems: examples from the coast of France. Géomorphol Reliefs Processus et Environ 1:3–22

Sabatier F, Samat O, Ullmann A, Suanez S (2009b) Connecting large-scale coastal behaviour with coastal management of the Rhone delta. Geomorphology 107:79–89

Samat O, Sabatier F, Lambert A (2005) Erosion of the sandy bottom in front of a seawall (Véran site, Gulf of Lions, Mediterranean coast). In: 5th international conference on coastal dynamics, American Society of Civil Engineering, Barcelona, 4–7 Apr 2005

Suanez S, Bruzzi C (1999) Shoreline management and its implications for coastal processes in the eastern part of the Rhône delta. J Coast Conserv 5(1):1–12

Suanez S, Bruzzi C, Arnoux-Chiavassa S (1998) Données récentes sur l'évolution des fonds marins dans le secteur oriental du delta du Rhône (plage Napoléon et flèche de la Gracieuse). Géomorphologie: Relief Processus Environ 4:291–312

Suanez S, Provansal M (1998) Large scale shoreline change, Rhône delta. J Coast Res 14:493–503

Suanez S, Sabatier F (1999) Eléments de réflexion pour une gestion plus cohérente d'un système anthropisé exemple du littoral du Delta du Rhône. Revue de Géographie de Lyon 71(1):7–25
Swift DJP (1976) Coastal sedimentation. In: Stlaney DJ, Swift DJP (eds) Marine sediment transport and environnent management. Wiley-Interscience, New York, pp 311–350
Vella C, Fleury J, Raccasi G, Provansal M, Sabatier F, Bourcier M (2005) Evolution of the Rhône delta plain in the Holocene. Mar Geol 222–223:235–265
Zenkovitch VP (1967) Processes of coastal development. Oliver and Boyd, Edinburgh/London

Chapter 15
Long-, Mid- and Short-Term Evolution of Coastal Gravel Spits of Brittany, France

Pierre Stéphan, Serge Suanez, and Bernard Fichaut

Abstract Gravel spits of Brittany have experimented a long morphosedimentary evolution over the last millenia. Based on analysis of several back-barrier holocene sediment stratigraphies, distinct phases of construction and barrier breakdown were recognized, indicating the role played by storminess and sediment supply during the late-holocene period. Over the last centuries and decades, a deficit of sediment budget affecting several gravel spits is highlighted. Therefore, actual coastal evolution of most of them is mainly dominated by cannibalization, landward retreat by rollover and complete destruction of the spits in some places. This coastal erosion is related to the lack of significant offshore sediment input or from the erosion of unconsolided cliffs. Locally, anthropogenic forcing have axacerbated the erosion processes by sediment minings and/or the construction of hard defense structures. For the swash-aligned gravel spits, frequency and magnitude of overwash processes is controlling the rate of landward retreat by rollover. This morphodynamic behaviour is illustrated by topo-morphological surveys realised between 2002 and 2012 on Sillon de Talbert spit which has experienced a complete crestal removal during the 10 March 2008 Johanna storm. Although this event have a 50–100 year occurrence, the barrier has exhibited a rapid crestal rebuilding by overtopping processes, illustrating the great resilience of the spit. Thus, coastal erosion management strategies mainly based on hard defense structures are gradually abandoned for new management policies based on soft operations. Nowadays, gravel spits of Brittany are also considered as a geological heritage and management plans are establishing by local authorities.

P. Stéphan (✉) • S. Suanez • B. Fichaut
LETG-Géomer-Brest, UMR 6554 CNRS, Institut Universitaire Européen de la Mer, Université de Brest, Technopôle Brest-Iroise, Place Nicolas Copernic, 29280 Plouzané, France
e-mail: pierre.stephan@univ-brest.fr; serge.suanez@univ-brest.fr; bernard.fichaut@univ-brest.fr

G. Randazzo et al. (eds.), *Sand and Gravel Spits*, Coastal Research Library 12,
DOI 10.1007/978-3-319-13716-2_15

15.1 General Setting

In Brittany, gravel beaches are located mainly on the northern and western coast (Fig. 15.1). Their construction is mainly due to the shoreward removing of periglacial deposits initially accumulated on the coastal shelf during the post-glacial marine transgression. The southern part of Brittany is characterized by sandy beaches and coastal dunes, except locally where a coarse material is provided by erosion of cliffs formed by Pleistocene deposits (head). Nowadays, unconsolidated cliffs are considered as the most significant source of coarse sediments in Brittany (Guilcher et al. 1957, 1990). In the Bay of Brest, the highly weathered of shale cliffs also contribute locally to feed the gravel barriers. In the Bay of Brest, the indentations of the jagged coastline were favorable to the construction of a numerous small-scale barriers and spits with a high morphological diversity (Fig. 15.2a–d). The length and the volume of gravel spits in the bay of Brest never exceed 700 m and 100.000 m^3 respectively. Nevertheless, the Sillon de Talbert spit studied in the north of Brittany forms the bigger accumulation reaching 3.2 km long and sediment volume estimated at 1.23×10^6 m^3 (Fig. 15.2e, f). Like most spits on the Channel coasts, formation of the Sillon de Talbert spit began at about 6,000 cal.year.BP

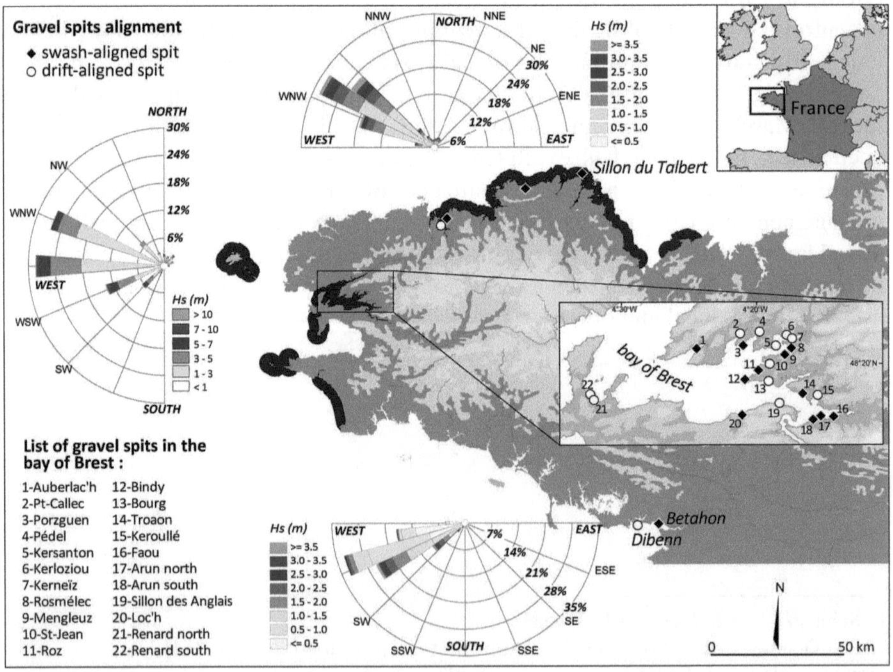

Fig. 15.1 Location maps of gravel beaches (shoreline in *grey color*) and spits (*points*) along the west coast of France. Wave climate of the northern, the southern and the western part of Brittany obtained by numerical run model over the period 1979–2002 (Source: LNHE-EDF and CETMEF-Brest laboratory)

Fig. 15.2 (**a**) Auberlac'h mid-bay spit, (**b**) Sillon des Anglais spit, (**c**) Rosmélec and Mengleuz opposite spits, (**d**) Loc'h spit, (**e**) Distal end of Sillon de Talbert spit, (**f**) Proximal dune section of Sillon de Talbert spit, (**g, h**) Removing of the frontal armor located on units 2 and 3 of the Sillon de Talbert spit in October 2004

when the relative sea-level rise decreased (Morzadec-Kerfourn 1995). This accumulation was enhanced by the presence of many rocky outcrops scattered on a wide seaward-inclined abrasion platform. A few hundred metres seaward from the spit, a peat deposit shows at the surface, in which a Final Neolithic archaeological site dating from around 2800 BC was uncovered. This human settlement, then located in the sheltered back-barrier area, bears witness to the considerable movement of the sediment accumulation due to the sea-level rise during the past millennia.

Gravel spits of Brittany belong to "composite gravel beach" type (Carter and Orford 1993; Jennings and Schulmeister 2002). The beachface is characterised by two distinct morphosedimentary parts separated by a break in slope at the mid tide level. The lower part of the beachface is dissipative with a low slope gradient (0.01 %), and takes the form of a sandy or rocky platform. The upper part of beachface is reflective and corresponds to the gravel barrier itself, mainly composed by sandy-gravel material and a pebble fraction dominating on the crest. Interstitial sandy fraction is important (around 50 %). Slopes values are between 5 and 17 % according to the different types of sediment assemblages that form the barriers. Gravel spits present generally a single ridge exceeding the high-water-spring-tide level of a few decimeters. Nevertheless, the crest is potentially affected by episodic overwash processes during storm events combined with high-tide level. Some spits present multiple ridge-crest morphology in their distal part related to longshore sediment supply. Behind the spits, back-barrier environments are most often characterised by salt-marshes infilled by sandy to peaty mud.

The coast of Brittany is considered with macrotidal environment, with a maximum tidal ranging from 10.85 m on the Sillon de Talbert, to 7.6 m in the Bay of Brest, and 6 m on the southern coast. Dominant swells mainly come from the west (Fig. 15.1) and show a notable increase during the winter. However, wave conditions vary locally depending on the shallow bathymetry. The spit of Sillon de Talbert appears to be affected by the most energetic swell conditions. The modal height (Hs) is between 1 and 1.5 m, and the modal period (Tpic) between 9 and 10 s. During storm events, the wave height can sometimes reach up to 9 m and the period up to 20 s. In these conditions, the Sillon de Talbert acts as a natural barrier against wave attack, offering a sheltered environment for the Bréhat Archipelago to the east and preventing flooding of low-lying Lanros peninsula to the south. The other spits of Brittany are located in more sheltered environments as shown the spits of the bay of Brest. The spits of Dibenn and Bétahon are protected by several shoals and islands. The spit of Linkin is also sheltered from dominant waves by a rocky promontory. Finally, the Bay of Brest is a fetch-limited environment and the waves height never exceed 1 m, whatever the wind conditions (Stéphan 2011a). According to their general orientation to incident waves, spits are swash- or drift-aligned (Fig. 15.1).

15.2 Long-Term (10^3 Years) Evolution of Gravel Spits of Brittany

The lithostratigraphy and biostratigraphy of several back-barrier sediment sequences were examined to reconstruct the late-holocene evolution of four gravel spits located in the bay of Brest. Benthic foraminifera assemblages were analysed to identify the exposure of the salt-marshes and to determine the phases of construction and breakdown of gravel spits. The characterization and the chronology of environmental conditions associated to sedimentary units were improved by using a foraminifera-based transfer function and 20 radiocarbon dates (Stéphan 2011a).

The results show that the stratigraphy of sediment infillings studied at Porzguen and Troaon salt-marshes are transgressives and reveal a similar general pattern of sedimentation. A basal well-humified peat containing reeds remnants (*Phragmites australis)* and detrital wood fragments overlays a weathered shale pre-Holocene surface on a maximum thickness of 1.7 m. The age of this basal unit is between 6,250 and 5,000 cal.year BP. Benthic foraminifera assemblages are dominated by *Trochammina inflata* and, secondarily by *Jadammina macrescens* and *Haplophragmoides wilberti* and are attributed by the transfer function to the upper limit of the tide levels. The basal peat units are overlayed by an organic-rich silty-clay 4 m thick containing root fragments of salt-marsh plants. Foraminifera assemblages are dominated by *Trochammina inflata* and *Jadammina macresens.* The contact between the basal peat and this overlying organic mud is gradational and reflect an increase of marine influence in a salt-marsh environment protected by a coastal barrier. At Porzguen marsh, the foraminifera assemblages are homogenous until the actual surface of the mash indicating no significant change of the back-barrier environment and a great stability of the gravel spit during the last millenia. At Troaon marsh, however, this unit contains two thin silty-sand horizons characterized by numerous shell fragments and dominated by *Haynesina germanica* and *Elphidium* species. These horizons reflect a change in the hydrodynamic conditions of the back-barrier environment related to the erosion of the initial gravel barrier. These coastal changes are not dated precisely but could be related to a phase of high storminess recognised around 3,000 cal.year BP in western part of France. Barriers breakdowns are identified in most late-holocene stratigraphies and correspond to a period of major climatic changes, known as the Bond Cold events (Bond et al. 1997).

At Arun marsh, the sediment infilling has a maximum thickness of 4 m. At the seaward marges of the salt-marsh, the back-barrier sediment sequence is characterised by a set of coarse washover fan deposits interstratified with a silty-clay layers dominated by *Jadammina macrescens* and *Trochammina inflata* assemblages. Two barrier erosion and breakdown phases associated to washover deposits are dated around 2,000–1,800 cal.year BP and 1,000 cal.year BP respectively. Here, long-term coastal changes are more important because the spit is exposed to more energetic wave conditions.

On the Loc'h spit, the sedimentary sequence is characterised by a coarse gravels unit corresponding to an initial gravel barrier. Two radiocarbon dates obtained from wood fragments located at the bottom and at the top of an underlying sandy-clay organic-rich layer give an age of AD 889–1023 and AD 1023–1157, respectively (Stéphan and Laforge 2013). Analysis of fossil foraminifera assemblages and use of the foraminifera-based transfer function demonstrates that the underlying deposit corresponds to a salt-marsh environment initially protected by an initial gravel barrier. The destruction of this barrier appears to take place from the eleventh to twelfth centuries and could be related to important variations in sediment supply from periglacial cliffs erosion, associated to extreme storm events. Storm deposits from the end of the early middle ages have been recognized by Giot (1998) and Devoy et al. (1996) in other coastal stratigraphies in the western part of Brittany. This period is obviously characterised by high storminess in the western France.

The back-barrier sediment sequences studied in the bay of Brest indicate that actual gravel spits have recorded a long morphosedimentary evolution over the last millenia. Distinct phases of construction and barrier breakdown were recognized, indicating a variability of storminess and sediment supplies during late-holocene period. However, morphogenic events are not synchronous because the high spatial variability of barrier exposition to the wind-waves in a fetch-limited environment. At Porzguen marsh, sediment sequence highlights a great stability of the gravel spit due to very sheltered conditions.

15.3 Historical Evolution of Gravel Spits

Ancient maps of the spits located in south Brittany which have been achieved at the beginning of the nineteenth century show a trend to landward retreat. In the bay of Brest, a landward migration is also suspected on swash-aligned gravel spits but the rates of migration are difficult to assess. On the Sillon de Talbert, however, coastal changes are more significant. On the first detailed maps drawn in 1666 and 1675, the gravel spit is represented as a welded barrier attached to the islands of the Olone Archipelago. A major morphological change can be seen on the Cassini's map, achieved in the second half of the eighteenth century. On that map, the Sillon de Talbert is deconnected from the Olone Archipelago by a 200–300 m long breach, transforming the barrier into a spit. This dislocation is dated in the early eighteenth century, probably due to the violent storm on 26 November 1703. This storm heavily hit the British Isles and western France and is supposed to have cost the lives of between 8,000 and 15,000 people. In England, this storm alone is thought to have killed 1,700 people and to have sunk 12 Royal Navy vessels. It is considered as the most violent storm of the past centuries (Lamb and Frydendahl 2005). The transformation of the initial welded barrier into a spit marked the beginning of a slow cannibalistic process throughout the nineteenth and twentieth centuries, which was accompanied by continuous landward retreat of the spit by rollover. The cannibalistic trend on the Sillon de Talbert is linked to a gradual change in its general direction in relation to incident waves, from a swash-aligned to a more drift-aligned position, leading to an increase in longshore drift. This sediment transport fed the distal end, which widened out through the formation of successive accretion ridges on the lee side of the tip. In return, the proximal end experienced a sediment deficit due to sediment loss by drift.

Over the last decades, the quantification of the recent mobility of spits in Brittany shows a sediment budget deficit (Fig. 15.3). This sediment depletion is

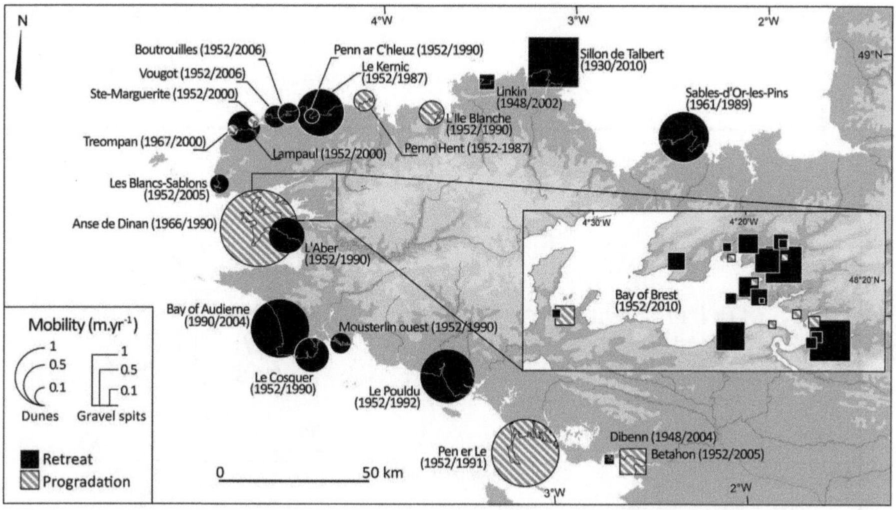

Fig. 15.3 Synthesis of data available on recent mobility of coastal barriers of Brittany (Stéphan 2011c)

illustrated by cannibalization (e.g. Dibenn spit, Fig. 15.4a), landward retreat by rollover (e.g. Sillon de Talbert spit, Fig. 15.4b) and, in some places, complete destruction of the gravel-dominated spits (e.g. Faou spit, Fig. 15.4c). Located on the southern part of Brittany, Betahon spit is the only case of recent progradation with a mean rate of seaward migration around 0.25 $m.year^{-1}$ (Fig. 15.3). The progradation is related to the construction of a foredune during 1980s by important sediment supply from the west of the bay. Six spits located in the inner parts of the bay of Brest are stable due to their very sheltered position and fetch-limited conditions. It is the case of Porzguen spit which also showed an exceptional stability during the last millennia. Seven spits have experienced a slow landward migration by rollover because of occasional impact of overwash events inducing, in this case, rates of retreat never exceed 0.2 $m.year^{-1}$. Seven spits show macro- and microscale cannibalization trends, reflecting scarcity of the longshore sediment supply. Four spits recorded a rapid rollover with rates of landward migration between 0.25 and 1 $m.year^{-1}$. The highest migration rates are recorded on the Sillon de Talbert (Fig. 15.4b). Finally, a case of complete destruction was observed on Faou spit during last decades. After a rapid landward retreat around 0.6 m/year between 1948 and 2004, two large breaches were opened during the 1970s leading to the gradual lowering of the crest and the sediment dispersion in the back-barrier zone by waves (Fig. 15.4c). Ultimately, the morphological evolutions measured over last decades reveal a scarcity of coarse grain sediment along the Brittany coastline.

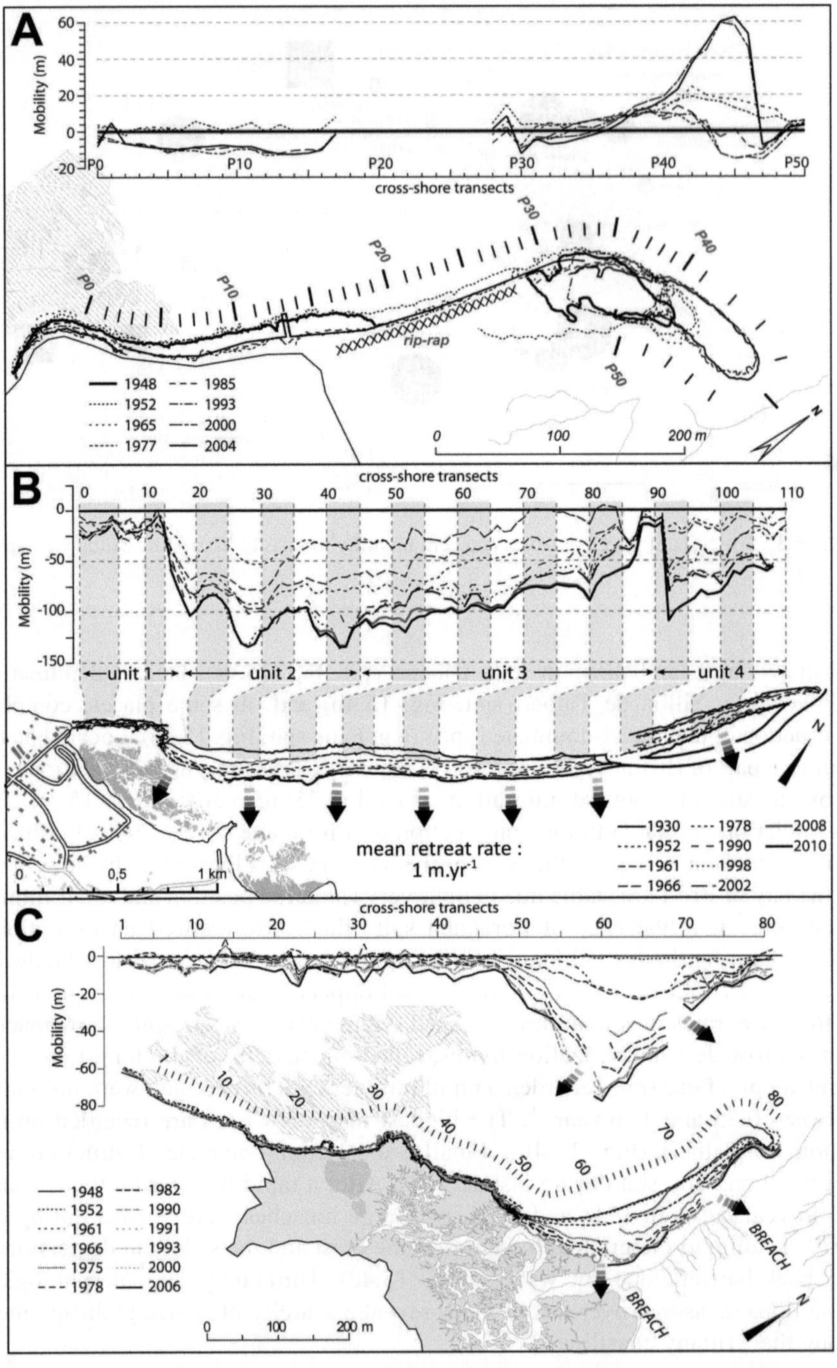

Fig. 15.4 Recent mobility of some gravel spits in Brittany, (**a**) Cannibalisation of Dibenn spit (Stéphan 2011c), (**b**) Retreat by rapid rollover of Sillon de Talbert spit (Stéphan et al. 2012), (**c**) Dislocation of Faou spit (Stéphan 2011c)

15.4 The Question of Sediment Supply from Periglacial Cliff Erosion and Sediment Budget

The recent mobility of the gravel spits highlights the lack of significant sediment input. Because the unconsolidated cliffs formed by periglacial deposits (head) are nowadays the main sediment sources of coarse sediment to the coast, the recent volumes provided by their erosion were assessed closed to three study areas. Cliff erosion rates were determined near Betahon and Dibenn spits and in the bay of Caro (Bay of Brest) by analysing changes of the backscar (cliff-top) plotted from aerial photographs. Volumes were calculated as Shuisky and Schwartz (1983) and have been estimated at 570 $m^3.year^{-1}$ in the bay of Betahon. This supply explains the recent seaward migration of the spit. However, the bay of Betahon appears to be an exception at regional scale. Generally in Brittany, the retreat of unconsolidated cliffs is slow. For instance, erosion of Dibenn cliffs provides a sediment volume estimated at 20 $m^3.year^{-1}$ which is not sufficient to prevent the cannibalization of the spit. In the bay of Brest, taking the example of Caro studied site, the maxima of volumes of sediment supply provided by unconsolidated cliff erosion is estimated at only 48 $m^3.year^{-1}$.

Gravel barriers are formed by fossil deposits which have within a context of sediment abundance and a decreasing of Holocene of sea-level rise around 6,000 cal.year BP. In the eastern part of the Channel, authors have shown that a large amount of the gravel accumulations were originated from sea erosion of a periglacial head deposits present on the continental shelf during the post-glacial transgression (Bray et al. 1995). This sediment source was gradually depleted during the last millenniums because of the decrease of the relative sea-level rise. On Nova Scotia coast (Canada), the authors have explained the erosion of gravel barrier and spit by the absence of major sea-level fluctuations during the past millennia (Forbes et al. 1995; Orford et al. 2002). In that context, no sediment supply on the foreshore by cliff erosion is realised. Stéphan (2011c) showed that in western Brittany (Bay of Brest), rates of relative sea-level rise have gradually declined since 2,700 cal.year.BP and are below 1 $mm.year^{-1}$ for the last millennium. This rise of sea-level is not sufficient to enable a widespread of large amount of sediments. Field observations indicate that the base of unconsolidated cliffs is nowadays rarely reached by waves attack (Stéphan 2011a). Furthermore, the presence of a large platform in front of the current cliffs promotes the dissipation of wave energy before it reaches the foot of the cliff.

15.5 Impact of Human Forcings

In Brittany, anthropogenic forcing have often been overstated to explain coastal erosion of gravel barriers. Sediment mining has played a role only in places. Pebbles were only extracted for industrial use in the Bay of Audierne during the

Second World War by the Wehrmacht Todt Organisation, to built the german defense (Atlantic Wall). In this area, sediment extraction represented a volume of around 1.10^6 m^3. For the Sillon de Talbert, no data is available to estimate the volumes extracted, but historical documents indicate that during the eighteenth and nineteenth centuries, the barrier was also used for pebble mining. However, this practice was prohibited in the early twentieth century. In the bay of Brest, sediment mining undertaken during the 1960s by oyster-farming activity has caused a reduction of 10 % and 50 % of the total sediment volumes of Mengleuz and Kersanton spits respectively.

Jetties constructed downdrift have led to a reduction of longshore sediment supply towards the Troaon, Auberlac'h (Fig. 15.2a) and Mengleuz (Fig. 15.2c) spits. This process is explaining cannibalization of Troaon spit and the erosion of Mengleuz and Auberlac'h spits. On the Sillon de Talbert, following the 5 April 1962 storm, major defense management of the spit was undertaken. In 1967, the breaches in the dune were artificially plugged. A riprap jetty was built between 1974 and 1982 in order to protect the dune section and to trap eastward sediment transit. A frontal rocky armor was also built at the crest of the spit over a distance of 1,500 m to prevent overwash and rollover processes. However, the armor perturbated the natural self-organisation of the spit and increased wave reflection causing the erosion of the lower beachface. During winter 1989–1990, despite the existence of this defense structure, series of storms caused considerable lowering of the barrier, the formation of new breaches and the retreat of the spit. In 2004, the main part of the frontal armor was removed by local authorities (Fig. 15.2g, h).

15.6 Morphodynamic Behaviours of Gravel Barriers at Mid-Term

Natural forcing, such as waves and tide, was analysed between 1953 and 2011 in the bay of Brest (Stéphan 2011c) and between 1979 and 2011 on the Sillon de Talbert spit (Stéphan et al. 2012). This analysis was based on the reconstruction of high water levels inducing crest erosion and/or overwash processes. On the Sillon de Talbert spit, the rate of landward retreat by rollover is strongly controlled by overwash frequency and magnitude. Between 1979 and 2010, Unit 1 (see on Fig. 15.4b) was affected by very few overwash episodes, which explains the stability of this spit section from the 1970s. Unit 2 was more frequently affected by overwash, in particular after the 1989–1990 winter storms which created two main breaches in this section. At the same time, the increase of overwash events during the 1990s can also be explained by the lowering of the crest. Units 3 and 4 showed more frequent overwash events between 1979 and 2010. The most severe ones were generated by 1989–1990 winter storms and the 10 March 2008 Johanna storm. During these two storm periods, landward retreat by sluicing overwash reached −15 m and −13.2 m for unit 3 and −17 m and −14.7 m for Unit 4. Several

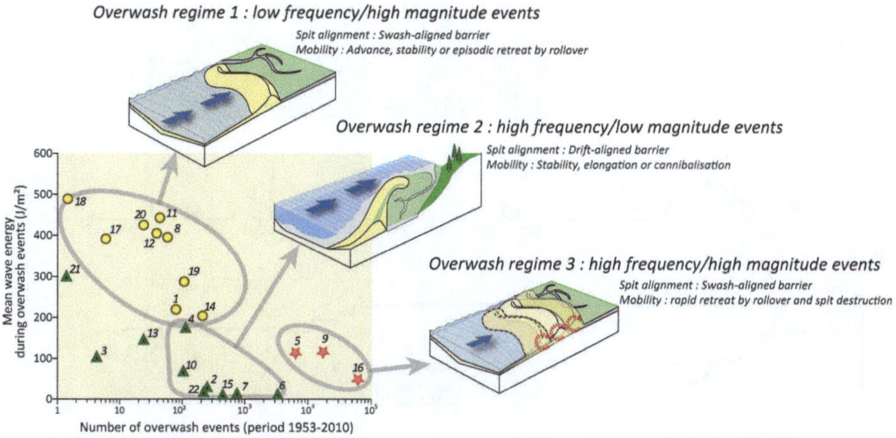

Fig. 15.5 Overwash regimes and morphodynamic behaviours according to gravel-spits alignment. Numbers on the graph refer to the list of gravel spits in the bay of Brest (see Fig. 15.1)

secondary discrete overwash events were also recorded during the spit's low mobility phases.

In the bay of Brest, spit mobility over the last decades was not so clearly linked to overwash events. Nevertheless, three overwash regimes are identified according to barrier alignment (Fig. 15.5). A first overwash regime (R1), characterized by low frequency and high magnitude of overwash events, is driving swash-aligned gravel spit processes causing slow landward retreat by rollover. A second regime (R2), defined by high frequency and low magnitude of overwash events, is driving drift aligned gravel spit evolution dominated by relative stability and cannibalization. A third regime (R3) concerns swash-aligned spits frequently reached by energetic overwash events causing rapid landward retreat by rollover or complete destruction of spits.

15.7 Short-Term Evolutions: Barrier Responses to Extreme Storm Events

Between 2002 and 2012, morphological evolution of five gravel barriers affected by erosion was annually surveyed using DGPS. The role played by overwash dynamic was also studied by analysing hydrodynamic conditions. Results show different pluri-annual morphogenetic phases. On Sillon de Talbert spit, the major storm event of 10 March 2008 caused a barrier average retreat up to 10 m and a crest lowering reaching -1.5 m. Volume of removed sediments due to barrier retreat were estimated to 120,000 m^3 (10 % of the total volume of the spit). The morphogenetic impact of this extreme event is explained by the conjunction of high spring tide levels and energetic wave conditions (hs = 9.5 m) inducing high surge level on

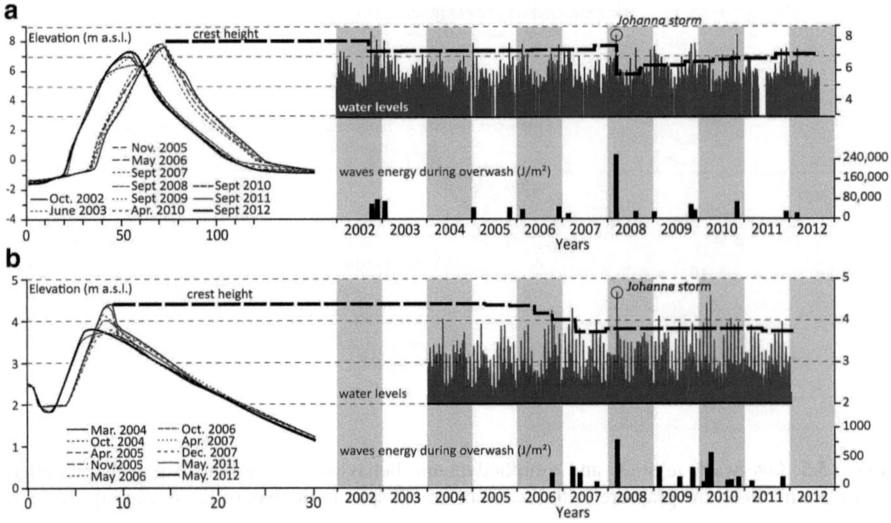

Fig. 15.6 Chronology of high water levels, waves energy during overwash events and morphological evolutions along cross-shore transects, (**a**) Sillon de Talbert spit, (**b**) Loc'h spit

the coast. The occurrence of this event is between 50 and 100 years on the coast of Brittany. This sluicing overwash phase occurred after a period of stability related to a low morphogenetic condition between 2002 and 2007 (Fig. 15.6a). After the Johanna storm event, crestal overtopping phase corresponding to the progressive spit recovery was recorded. The "ridge rebuilding activity" as defined by Orford (2011), was undertaken between March 2008 and September 2012, showing the great resilience of the barrier. During this post-storm recovery period, the average of crest elevation was up to 1 m, reaching locally 2 m. These different morphogenetic phases were also recognised on other coastal areas along the Brittany coastline (Suanez et al. 2011; Suanez and Stéphan 2011). In contrast, in the bay of Brest, the recover of spits and barriers after the 10 March 2008 storm event was not so obvious. For instance, Loc'h spit showed a tendency of erosion for the whole period of measurements, especially since the 10 March 2008 Johanna storm (Stéphan and Laforge 2013) (Fig. 15.6b).

15.8 Management Strategies of Gravel Spits

After several decades of coastal erosion management strategy mainly based on hard defense structures, the Sillon de Talbert has experienced since 2001 a new management policy provided by the '*Conservatoire du Littoral*'. The principal goal is not let the spit retreating naturally while attempting to prevent its breaching. In October 2004, a large part of the frontal armor located on the Units 2 and 3 was removed (Fig. 15.2g, h). The riprap was crushed and the material was deposited on the rear of the spit, forming three embankments in a way to slow down the landward retreat barrier. After the major morphological changes caused by the Johanna storm, human interventions were only restricted to plug breaches in the dune, and to set up new fences in a way to recover dune morphology. It denotes a radical change in terms of coastal management strategy facing erosion in France. The '*Conservatoire du Littoral*' is a national organisation in charge of the acquisition of coastal lands for there preservation. Since the 2000s, the '*Conservatoire du Littoral*' initiated a policy of acquisition of low-lying coastal lands threatened to flooding, situated on the rear of the Sillon de Talbert spit. This policy promotes the strategic withdrawal of human settlements. In the bay of Brest, a management strategy of gravel spits and back-barrier salt-marshes is establishing by local authorities (Stéphan et al. 2012) in order to protect barriers affected by erosion and restore ecological value of marsh environments. Gravel spits are also considered as geological heritage that must be protected from any human forcing using management plans.

References

Bond G, Showers W, Cheseby M et al (1997) A pervasive millennial-scale cycle in North Atlantic Holocene and glacial climates. Science 278:1257–1266

Bray MJ, Carter DJ, Hooke JJ (1995) Littoral cell definition and budgets for central southern England. J Coastal Res 11(2):381–400

Carter RWG, Orford JD (1993) The morphodynamics of coarse clastic beaches and barriers: a short- and long-term perspective. J Coast Res Spec Issue 15:158–179

Devoy RJN, Delaney C, Carter RWG et al (1996) Coastal stratigraphies as indicators of environmental changes upon European Atlantic coasts in the Late Holocene. J Coast Res 12:564–588

Forbes DL, Orford JD, Carter RWG et al (1995) Morphodynamic evolution, self-organisation, and instability of coarse-clastic barriers on paraglacial coast. Mar Geol 126(1–4):63–85

Giot PR (1998) La dune ancienne de la baie d'Audierne. Norois 45:487–494

Guilcher A, Vallantin P, Angrand JP et al (1957) Les cordons littoraux de la rade de Brest. Cahiers Océanographiques 1:21–54

Guilcher A, Bodéré JC, Hallégouët B (1990) Coastal evolution in western, southwestern and northern Brittany as a regional test of impact of sea level rise. J Coast Res Spec Issue 9:67–90

Jennings R, Shulmeister J (2002) A field based classification scheme for gravel beaches. Mar Geol 186(3–4):211–228

Lamb HH, Frydendahl K (2005) Historic storms of the North Sea, British Isles and Northwest Europe. Cambridge University Press, Cambridge

Morzadec-Kerfourn MT (1995) Coastline changes in the Armorican Massif (France) during the Holocene. J Coast Res Spec Issue 17:197–203

Orford JD (2011) Gravel-dominated coastal barrier reorganisation variability as a function of coastal susceptibility and barrier resilience. Coast Sediment 11:1257–1270

Orford JD, Forbes DL, Jennings SC (2002) Organisational controls, typologies and time scales of paraglacial gravel-dominated coastal systems. Geomorphology 48:51–85

Shuisky YD, Schwartz ML (1983) Basic principles of sediment budget study in the coastal zone. Shore Beach 51(1):34–40

Stéphan P (2011a) Colmatage sédimentaire des marais maritimes et variations relatives du niveau marin au cours des 6000 dernières années en rade de Brest (Finistère). Norois 220:9–37

Stéphan P (2011b) Les flèches de galets de Bretagne: évolution passée, présente et future. L'Harmattan, Paris

Stéphan P (2011c) Quelques données nouvelles sur la mobilité récente et le bilan sédimentaire des flèches de galets de Bretagne. Géomorphologie: relief, processus, environnement 2011(2):205–232

Stéphan P, Laforge M (2013) Mise au point sur l'évolution géomorphologique et le devenir des flèches de galets du Loc'h de Landévennec (Bretagne, France). Géomorphologie: relief, processus, environnement 2013(2):81–98

Stéphan P, Suanez S, Fichaut B (2012) Long-term morphodynamic evolution of the Sillon de Talbert gravel barrier spit, Brittany, France. Shore Beach 80(1):19–36

Suanez S, Stéphan P (2011) Effects of natural and human forcing on mesoscale shoreline dynamics of Saint-Michel-en-Grève Bay (Brittany, France). Shore Beach 79(2):19–38

Suanez S, Fichaut B, Magne R et al (2011) Changements morphologiques et budget sédimentaire des formes fuyantes en queue de comète de l'archipel de Molène (Bretagne, France). Géomorphologie: relief, processus, environnement 2:187–204

Chapter 16
Morphological Characterization and Evolution of Tahadart Littoral Spit, Atlantic Coast of Morocco

M. Taaouati, G. Anfuso, and D. Nachite

Abstract Tahadart littoral spit on the Atlantic coast of Morocco is a mesotidal environment composed of 6 km long sand spit with flat and wide beaches, foredunes and vegetated dune ridges and well developed salt marshes. Tahadart's littoral zone is affected by east and west winds and is exposed to waves approaching from the NE with an associated SW-directed littoral drift. Detailed 3D surveys reveal the spit as a flat, dissipative beach with small changes that take place homogeneously along the foreshore according to the parallel retreat model. Such a dissipative morphodynamic state was confirmed by means of classical parameters such as the Surf Scaling, Surf Similarity and the Dean number. The littoral evolution of the area was determined through analysis of the position of the dune foot evolution in 1981 and 1997 aerial photographs as well as the 2007 SPOT image by means of GIS tools. Coastal evolution was revealed by examining changes between the earliest (1981) and the latest (2007) shoreline positions. Accretion clearly prevailed over erosion and was particularly so in the northern part of the littoral (95 m, 3.7 m/year), at the river mouth (90 m, 3.5 m/year) and at southern part (the free edge) of the sand spit, with seaward and southward displacement of shoreline values increasing from north (12 m) to south (c. 180 m, 6.7 m/year). Behaviour observed at the free spit's edge matched well a model of continued beach growth/progradation and allowed three different areas with specific beach growth mechanisms and rates to be determined.

M. Taaouati (✉)
Département de Géologie, Faculté des Sciences, Université Abdelmalek Essaâdi, BP 2121, 93030 Tetouan, Morocco
e-mail: mtaaouati@gmail.com

G. Anfuso
Departamento de Ciencias de la Tierra, Facultad de Ciencias del Mar y Ambientales, Universidad de Cádiz, Polígono Río San Pedro s/n, 11510 Puerto Real, Cádiz, Spain

D. Nachite
Natural Resources Department, Polydisciplinary Faculty of Larache, Abdelmalek Essaâdi University, BP 745, 92004 Larache, Morocco

G. Randazzo et al. (eds.), *Sand and Gravel Spits*, Coastal Research Library 12,
DOI 10.1007/978-3-319-13716-2_16

16.1 Introduction

Morocco, located between Europe and Africa (Fig. 16.1), the Atlantic Ocean and the Mediterranean Sea, has around 2,934 km of coastline that extends along the Atlantic coast, from Cape Spartel (in the North), to Lagouira (in the South), and 512 km of Mediterranean coast, from Cape Spartel (in the West), to Saidia (in the East).

The nature of the coastline varies hugely from one region to another in relation to geological and tectonic factors and coastal dynamics. Rocky shores are relatively extended on the Mediterranean coast, while sandy beaches, coastal dunes, rocky platforms, marshes, lagoons and estuaries are well represented alongside the Atlantic coast. In this sense, the littoral of Morocco hosts many environments, with a particular ecological and socio-economic interest, such as lagoons (Nador, Khnifiss Moulay Bousselham, Oualidia and Sidi Moussa), estuaries (Tahadart, Loukkos Oum Errabia, Bouregreg, Sebou, Moulouya etc.), salt marshes and bays (M'diq, Agadir, Dakhla and Cintra). See Castro et al. (2006) and Nachite et al. (2008) for a fuller description.

Such important environmental heritage is quite often threatened by socio-economic activities. The majority of large-scale industrial and economic activities of Morocco are positioned at the coast, particularly the most powerful industrial complexes such as the oil refining and petrochemical activities at Mohammedia, phosphate industries at Safi and Jorf Lasfa and steel industries at Nador. In this sense, the Kenitra-Safi axis, concentrates 60 % of industrial units and 80 % of national industrial jobs (Nakhli 2010).

The littoral zone has also attracted a lot of tourist investments over the past 30 years and today it concentrates around 70 % of classified beds capacity.

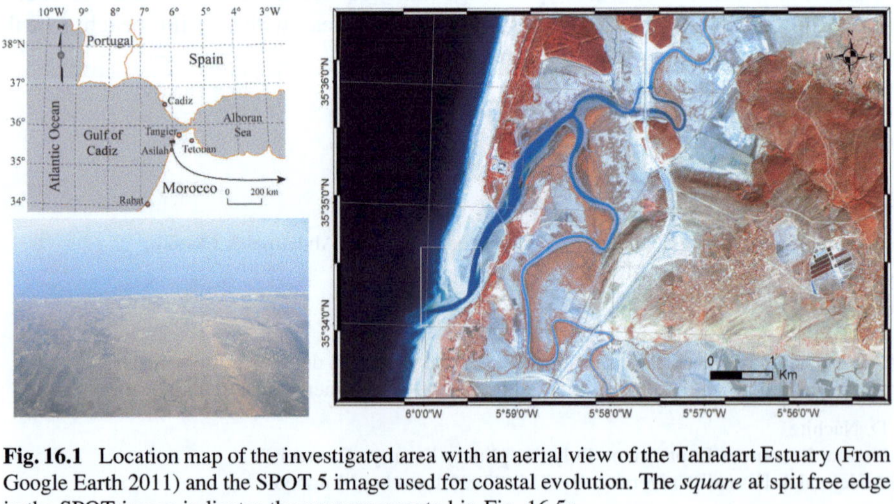

Fig. 16.1 Location map of the investigated area with an aerial view of the Tahadart Estuary (From Google Earth 2011) and the SPOT 5 image used for coastal evolution. The *square* at spit free edge in the SPOT image indicates the area represented in Fig. 16.5a

Four major tourist areas can be distinguished: Agadir and its region, El Jadida-Casablanca-Mohammedia, Rabat-Temara-Skhira Bouznika and Tanger-Tétouan. New important development projects encompassing marinas and golf camps are planned in Tangier area within the Plan Blue 2020 (Nachite 2009; CCIST 2013).

This chapter deals with morphodynamic characterisation and evolution of a littoral spit on the Atlantic littoral of Northern Morocco, a long sandy coast constituting the southern part of the Cadiz Gulf which incorporates as part of its littoral arch Southern Portugal and the Atlantic coast of Spain and Morocco (Fig. 16.1).

16.2 Study Area

The coastal sector is located in the Tahadart area, between Tangier and Asilah towns, on the Northern Atlantic coast of Morocco (Fig. 16.1). The sector is composed of a 6 km long spit that encloses an estuary with well-developed salt marshes (Nachite et al. 2008). It is the most important and extended marshland of the Atlantic coast of Morocco, with other important areas of similar type being Loukkos estuary and Moulay Bouselham and Khnifiss lagoons.

The NNE-SSW oriented littoral, consists of a sandy spit with a field of well developed vegetated dunes in the northern and central parts and foredunes in its southward edge. Along the area, beaches show a smooth profile and are backed by well developed and vegetated sand ridges (Jaaidi et al. 1993). The estuary shows salt marshes with tidal creeks (Fig. 16.1), well developed slikke and shorre areas, and it represents an area of great biological and ecological interest, e.g. SIBE L11 of the National Plan of Protected Areas of Morocco (Nachite et al. 2008, 2010).

The climatic characteristics of the area shows temperatures and precipitations levels typical of an occidental Mediterranean area, with a short, cold and rainy winter (from November to April) and a dry and warm summer period (from May to September). Precipitations show relatively strong values, ranging from 655.8 to 765.3 mm (Karrouk 1990), that are irregularly distributed throughout the year with maximum values in December. Main annual temperature is 18.2 °C, with maximum monthly values recorded in August (27.3 °C) and minimum in January (14.3 °C).

Predominant winds, named "Chergui", blow from the East (27 % occurrence), i.e. from land, are especially abundant in spring and summer and reach maximum velocities of 130 km/h.

Secondary winds (Rharbi) blow from the West (16 %), e.g., the Atlantic Ocean, are wet and prevail in winter and autumn.

Tidal range is 2.7 m (mesotidal) that has a semidiurnal periodicity (ONE 2002). The hydrodynamic conditions are principally controlled by storms approaching from occidental quadrants (Jaaidi and Cirac 1987).

The littoral area evolved during last decades because of natural processes and especially human interventions since the progressive urbanization of the estuarine and coastal areas (Nachite et al. 2008). Two rural settlements, e.g. Qouasse Briech and Boukhalef, are present in the area with a total population of 22,000 inhabitants, main activities being traditional agriculture, salt harvesting, traditional fishing and shellfish recollection (ONE 2002).

16.3 Methodology

16.3.1 Morphological Characteristics of the Coastal Area

Sediments samples were taken from salt marsh (slikke, schorre and tidal creeks) and beach-dune environments and sieved in laboratory with a Ro-tap machine using a nest of sieves at 0.5-phi intervals. Granulometric parameters were calculated according to Folk and Ward (1957) and visual observations concerning grain morphoscopy were carried out by means of a microscope.

Beach and dunes topographic surveys were carried out in July 2005 and January and March 2006. A total station (GTS-Topcon 225) was used for obtaining detailed 3-D surveys extending from the dry beach to a depth equivalent to mean spring tide low water level. Beach width and beach face slope were calculated later. The adjacent dune field located in the southern part of the spit was also surveyed.

Various parameters and formulae commonly used in coastal geomorphologic studies were employed to characterise beach morphodynamic state (Short and Jackson 2013; Jackson et al. 2005). Wave breaking type was calculated using the Surf Similarity index (Battjes 1974):

$$\xi = \tan \beta / (H_b/L_0)^{0.5} \qquad (16.1)$$

where β is the intertidal beach slope, H_b is the significant breaking wave height, and L_0 represents the deep water wave length. The index differentiates from surging ($\xi > 2$) to plunging ($0.4 < \xi < 2$) and spilling breakers ($\xi < 0.4$) (Fredsoe and Deigaard 1992).

In order to determine the morphodynamic state of the beach, Guza and Inman (1975) proposed the following Surf Scaling parameter:

$$\varepsilon = \sigma^2 H_b / 2g \tan^2\beta \qquad (16.2)$$

where σ is the wave frequency in radians, g is the acceleration of gravity and $\tan \beta$ is the beach slope. The formula allows us to distinguish between reflective ($\varepsilon < 2.5$), intermediate ($2.5 < \varepsilon < 30$) and dissipative ($\varepsilon > 30$) surf zones (Guza and Inman 1975). Significant breaking wave height (H_b) was calculated following Komar and Gaughan (1972) equation by considering the mean value recorded offshore during the month before the beach survey.

The Dean number was also calculated (Dalrymple and Thompson 1976):

$$\Omega = H/W_sT \tag{16.3}$$

where H and T represent, respectively, the height and time interval of the wave in deep water and W_s is the settling velocity:

$$W_s = 273(D_{50})^{1.1} \tag{16.4}$$

where D_{50} is the grain size (the median) in metres. Since D_{50} showed a minimum seasonal variability range, a mean representative value was used (Taaouati 2012). This parameter differentiates between reflective beaches with bars and beach cusps ($\Omega < 1$), intermediate beaches with bars and cusps ($1 < \Omega < 6$) and dissipative beaches ($\Omega > 6$). The Dean number, employed in conjunction with the relative tidal range (RTR) obtained by the relationship between tidal range and wave height (Davis and Hayes 1984), was used to classify all investigated beaches according to the model proposed by Masselink and Short (1993).

16.3.2 Coastal Evolution

A satellite image and aerial photographs from different years and timescales were used to reconstruct and quantify shoreline evolution over a medium-term trend period (26 years), (Crowell et al. 1993).

These documents were scanned, geo-referenced, and computer-rectified to eliminate effects of scale and distortion (Crowell et al. 1991; Moore 2000). Ground Control Points (GCP_s) for photo registration were obtained from the geo-referenced satellite SPOT5 image of 2007, and all information was presented in UTM/WGS84. A polynomial transformation was applied in the registration process because of the smooth topography of the study area (Chuvieco 2000) and GCP_s were located in unequivocal places (Thieler and Danforth 1994).

The error associated with the distortion of the photographs was solved and controlled in the geo-referenced documents with visual estimations by comparing the registered photo with the base map, and through the root mean square error (RMSE), giving the photographs a geometrical error of approximately ±5 m.

The precision and accuracy of aerial photogrammetric measurements depends also on the difficulties of locating the shoreline position. Typically, the shoreline is taken as the water line, especially in microtidal environments (Boak and Tarner 2005), or as the seaward vegetation limit, dune foot, or cliff top, in mesotidal environments (Crowell et al. 1993; Boak and Tarner 2005).

16.4 Results and Discussion

16.4.1 Wave Climate

Wave data were obtained from the prediction point SIMAR 1056043 (6.00° W, 35.75° N), developed by *Puertos del Estado* (Ministry of Development, Spain) using the wave model WAM. Waves essentially approach from WNW and secondary from W (Fig. 16.2). Significant wave height is between 0.5 and 1.0 m (40 % of records) and between 1.0 and 1.5 m (20 %). Higher waves are about 4 m (c. 2 % of records) and approach from the West. Maximum significant wave height with 10 years return period is 7.8 m. Wave period ranges from 3 to 16 s, being 3–4 s most common values, e.g. c. 90 % of cases. Wave climate shows an important seasonal trend with the main storms observed during the winter period, especially December and January. Wave height and period as well as storm distribution were very similar to observations carried out at Cadiz area, in the northern part of the Gulf of Cadiz, by Rangel and Anfuso (2011, 2013) which related storm distribution with negative phases of AO and NAO indices.

Coastal orientation (NNE-SSW) and predominant wave approaching direction (WNW) give rise to a main littoral drift that is clearly southward directed. This is reflected by the orientation of all sand spits along the Atlantic littoral zone of Morocco (Taaouati 2012). An opposite secondary drift is also observed and associated with wave fronts approaching from the west.

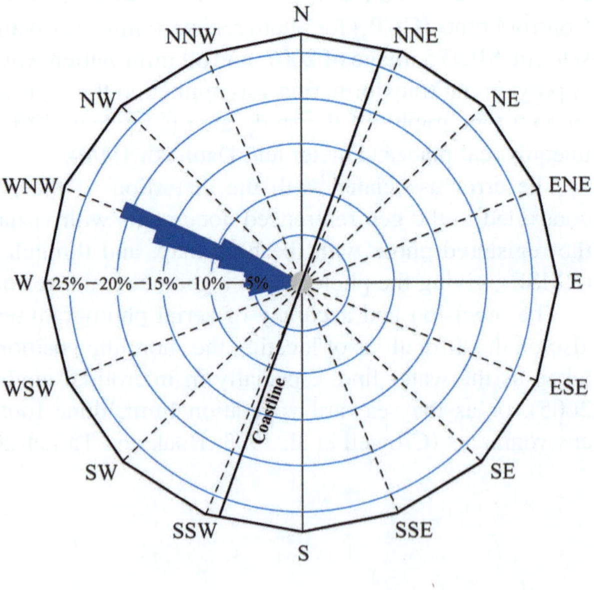

Fig. 16.2 Wave rose corresponding to the prediction point SIMAR 1056043

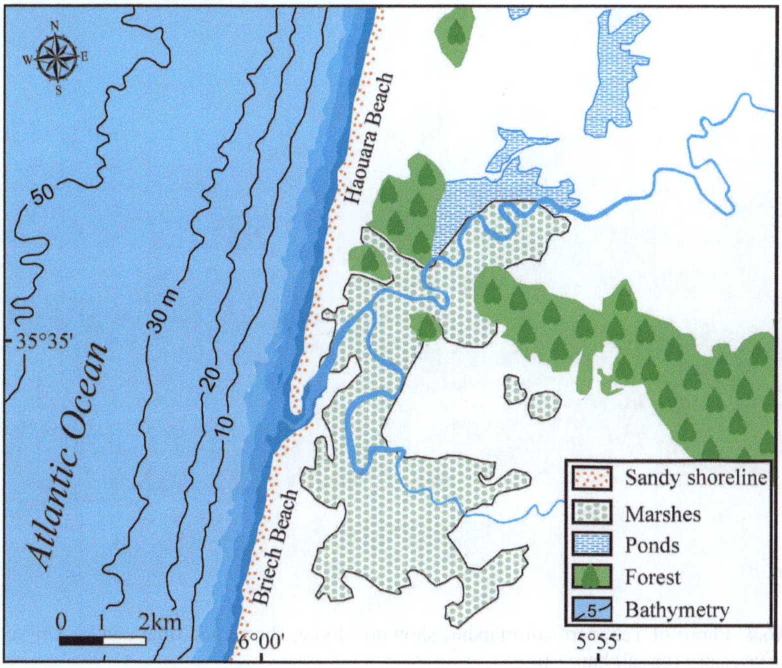

Fig. 16.3 Main morphological characteristics of the Tahadart spit and estuary. Cartography was based on the topographic map of 1963 and updated using the satellite Spot image of 2007

16.4.2 Salt Marsh Characteristics

Tahadart spit encloses an estuary with developed salt marshes including slikke and schorre areas and a complex network of tidal creeks (Nachite et al. 2008, 2010). The estuary is developed along the margins of Meharhar and El Hachef rivers that merge at about 1.2 km from the coastline (Figs. 16.1 and 16.3).

The main estuarine elements are (Fig. 16.4): (i) the main channel, continuously submerged and composed of clastic clay with low percentages of sand (<25 %). Sedimentary supplies to the littoral are almost null and composed by fine suspended particles. Vegetation is almost absent and represented by *Zoostera* and algae; (ii) Slikke, the unvegetated area flooded at every high tide, shows a slope ranging from 5 %, in the upper, to 10 %, in the lower part. Ganulometric analysis showed the presence of clay (15 %) and sand (85 %) composed by fine rounded grains of quartz; (iii) Schorre, upper, vegetated part covered only at spring tides, is separated from slikke by an escarpment of a few centimetres. It is composed of medium (29 %) and fine (70 %) sand and clay (1 %). Microscopy observations revealed the presence of bioclastic sediments and minute vegetation debris. Vegetation is represented by *Spartina maritima*, *Sarcocornia perennis*, *S. fruticosa* and *Salicornia europea*.

Fig. 16.4 Photo of Tahadart salt marshes showing slikke, shorre and tidal creeks. Un-vegetated flat surfaces are for salt harvesting

Within terms of land use evolution of the estuary area, a progressive reduction of marshes and ponds and an increase of cultivated and urban areas were progressively observed within the past few decades. In 1958, salt marshes and ponds presented an extension of 3,200 ha and occupied 37.3 % of the area. They were followed by forests (1,200 ha, 14 %), cultivated lands, beaches, dunes and urbanised areas (Nachite et al. 2008). In 1992, about 500 ha of marshes and ponds were lost, meanwhile, an increase of urban occupation linked to the expansion of existing villages and contraction of industrial emplacements was recorded. General re-vegetation of coastal areas took place with forests replaced with acacias and other inland areas with *Quercus suber*. Recorded trends in land occupation continued in successive decades but land use changes were actually very small because in 1996 the area was declared a site of biological and ecological interest (SIBE L11 of the National Plan of Protected Areas of Morocco).

16.4.3 Sedimentological and Morphodynamic Spit Characteristics

Location, morphodynamic and shape of spits are controlled by several different geological factors including alongshore drift, tidal and fluvial currents, glacial and fluvial sediment supplies, underlying geological framework, wave refraction and

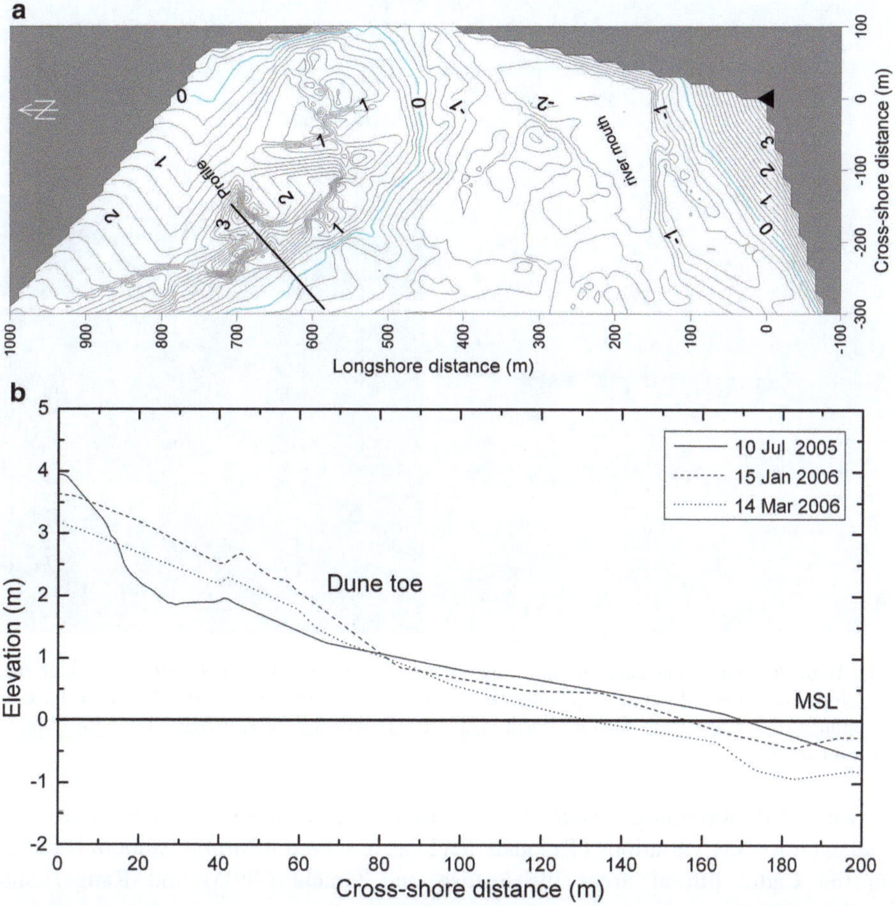

Fig. 16.5 Results of topographic surveys: (**a**) 3D model of the southernmost spit portion (see Fig. 16.1 for location) and (**b**) beach profiles of the investigated area

angle of wave approach and sea level rise (May and Tanner 1973; Hine 1979; Park and Wells 2005). Further, alongshore sedimentological differences give rise to different morphodynamic states even along the same littoral spit (Carter 1988; Calliari 1994).

Tahadart spit sediments are composed of fine, coarse to medium, quartz rich sand with a mean grain size value (D_{50}) of 0.24 mm with a homogeneous alongshore and cross-shore grain size distribution.

Beach morphology showed a smooth and wide foreshore with minor alongshore variations (Fig. 16.5a). Such homogeneous characteristics are associated with unidirectional wave approach and related drift, and the absence of controlling geomorphologic features such as rocky shore platforms (not observed at the investigation site) but common at Asilah (Taaouati 2012).

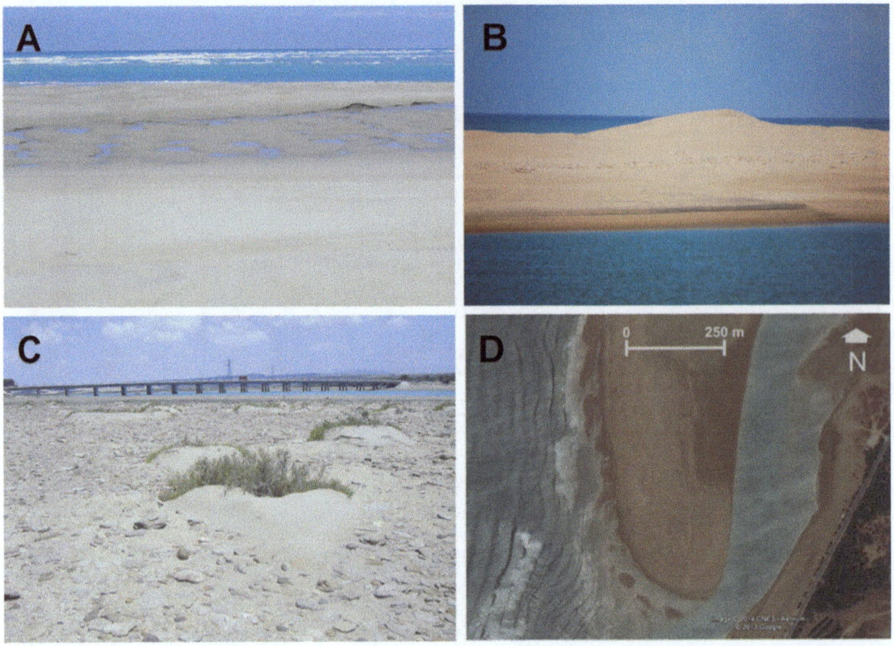

Fig. 16.6 Photographs at Tahadart spit showing a swash bar and associated landward trough (**a**), developed foredunes at southern spit edge (**b**) and a wide deflation surface extended on the back of foredunes (**c**). Image from Google Earth (7/30/2013) showing the spit free edge with oblique swash bars (**d**)

Recorded morphological changes were usually slow, resulting in negligible daily and even weekly variations (Taaouati 2012) as observed in similar smooth beaches in the Cadiz littoral areas by Anfuso and Gracia (2005) and Rangel and Anfuso (2011).

Beach profiles showed vertical changes of few decimetres, with the most important variations being recorded at the lower foreshore areas due to welding and migration of swash bars (Fig. 16.5b). Such morphologies were close to the ones observed by Taaouati et al. (2011) and to bars observed in flat beaches of Cadiz area by Anfuso and Gracia (2005) and Rangel and Anfuso (2011) and in Northern Ireland by Jackson et al. (2007). They were characterised by a wide and smooth crest with not much evidence of an avalanche face (Fig. 16.6a).

Regarding beach morphodynamic state, the beach was visually close to the "Dissipative" beach state of Wright and Short (1984) and presented a flat foreshore slope (tan $\beta = 0.013$) that did not record relevant variations during studied period (Fig. 16.5b). Morphological changes essentially took place under the "parallel retreat" model of Nordstrom and Jackson (1992) which produced small morphological and seasonal slope changes usually homogeneous along the foreshore, and very typical of dissipative beach states (Fig. 16.5b), as also observed in similar types of beaches by Hughes and Cowell (1987) and Shih and Komar (1994) and, in the Gulf of Cadiz, by Anfuso and Gracia (2005) and Rangel and Anfuso (2011).

Such dissipative beach state behaviour characterised by spilling breaks was confirmed through the use of various morphodynamic parameters, i.e. the Dean number, Surf Similarity and Surf Scaling parameters. The Surf Similarity index recorded low values corresponding to spilling conditions ($\xi = 0.15$, Fredsoe and Deigaard 1992). Regarding the Surf Scaling parameter, the beach clearly showed dissipative values ($\varepsilon = 177$) according to the limits of Guza and Inman (1975). On considering the Dean number, the mean obtained value ($\Omega = 7.2$) is typical of dissipative beach states.

In addition to previously described classic morphodynamic parameters, the RTR and Dean number were also used to classify beach according to Masselink and Short (1993): all data collected corresponded to the "dissipative barred beach state", this way confirming field observations.

In the northern part of the spit, the dune field is comprised of a continuous ridge, vegetated by Mediterranean bushes and trees, essentially *Juniperus oxycedrus*, *Pistacia lentiscus* and *Pinus Pinea*. The effects of *Chergui* and *Rharbi* winds are very important in the Southern part of the spit giving rise to 3–4 m high and rapidly migrating foredunes (Fig. 16.6b) with a wide deflation surface landward of the dune field. This is characterised by the presence of flat, centimetric pebbles and small sand accumulations on lee side of pioneer plants (Fig. 16.6c).

16.4.4 Medium Term Evolution and Shoreline Change Rates

This sand spit is part of a c. 35 km long uniform sector enclosed between Cape Spartel (at North) and Asilah (at South, Figs. 16.1 and 16.3). Coastal evolution was calculated by taking into account movements of the dune toe position in 1981 and 1997 aerial photographs and in 2007 satellite image (Fig. 16.7). The coastline along sectors from transect 1 to 120 and from 174 to 250 was identified with a seaward limit of dune vegetation. The coastline including transects 121–173 which corresponds to the southern part of the spit, was identified by the position of dune foot – because dunes were not vegetated. Mapped coastlines showed a linear and regular trend in plane confirming this way the homogeneity of the investigated area. Temporal and spatial coastal evolution was examined using shoreline changes at each transect over the whole period (Fig. 16.8a). The first method (Fig. 16.8a), describes the movement of the shoreline between the earliest (1981) and the most recent (2007) shoreline, although this movement may be not the maximum shoreline displacement recorded. The second method (Fig. 16.8a) deals with variability at each transect taking into account the maximum spatial recorded displacement of shoreline without mind about considered time span (1981–1997, 1981–2007 and 1997–2007).

Concerning the first method, accretion was recorded along (i) the northern part of the littoral, in a sector of about 600 m between transects 1 and 28 that showed maximum accretion values of 95 m (3.7 m/year); (ii) at the southern side of the river mouth, along a c. 300 m long sector between transects 181–192 with a maximum

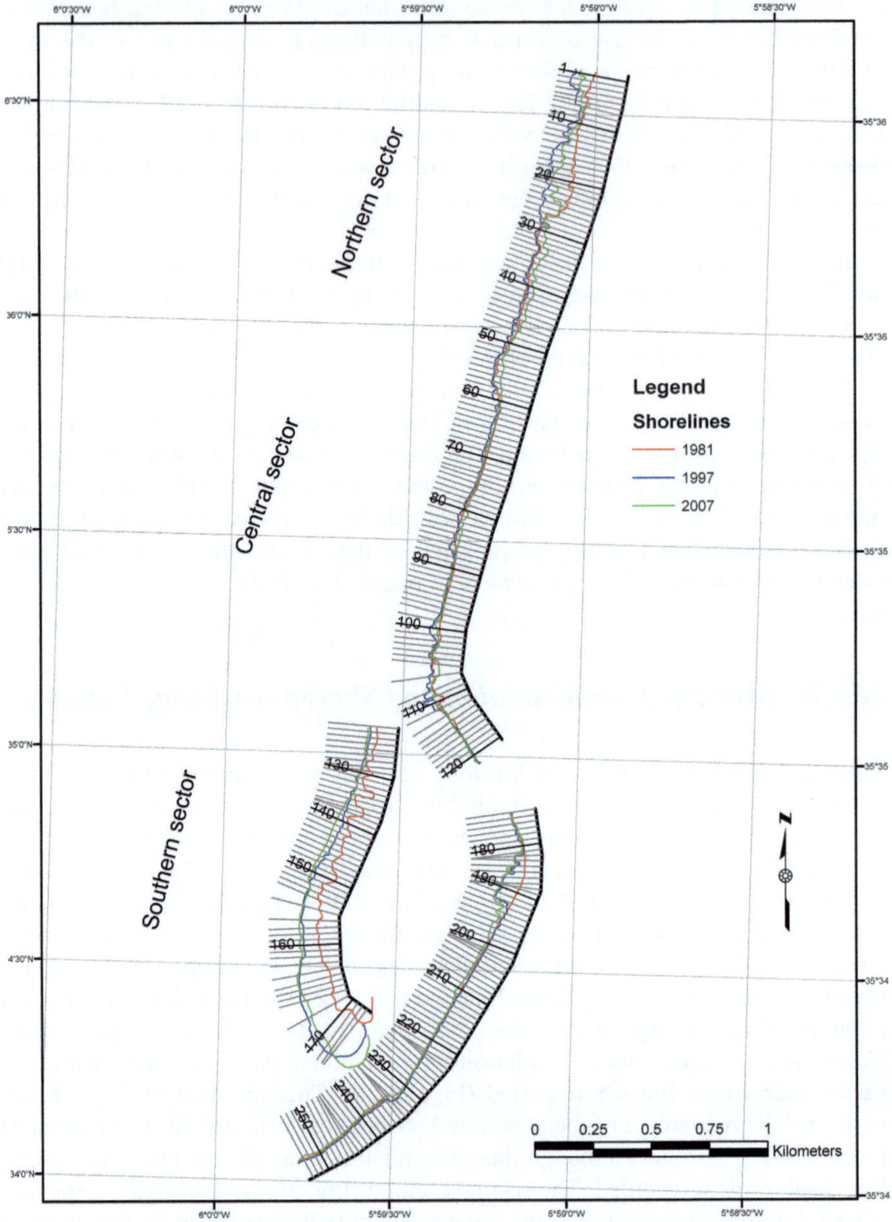

Fig. 16.7 Coastal evolution along the investigated area with location of transects

shoreline increase of 90 m (3.5 m/year); and (iii) at southern part – the free edge – of the sand spit, along a c. 1 km long sector (transects 121–173) that recorded both seaward and southward displacements of shoreline with values increasing from north (12 m) to south (c. 180 m, 6.7 m/year, Figs. 16.7 and 16.8).

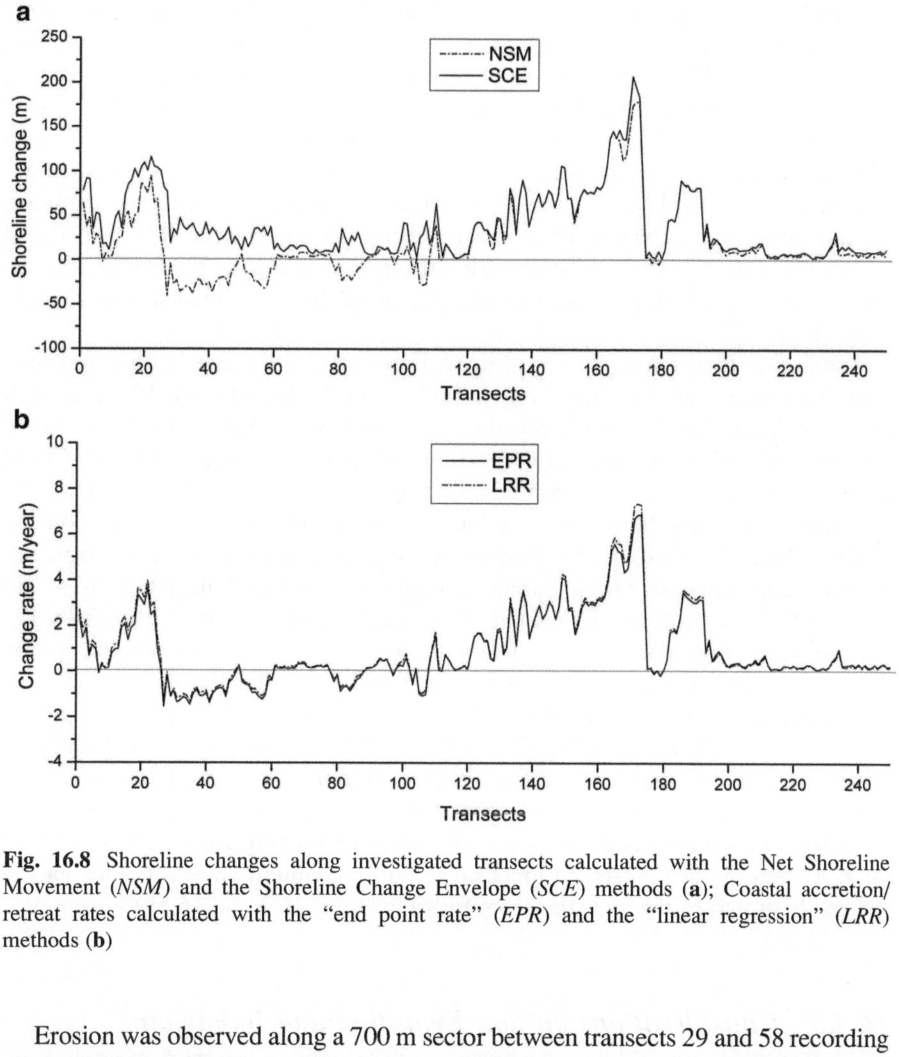

Fig. 16.8 Shoreline changes along investigated transects calculated with the Net Shoreline Movement (*NSM*) and the Shoreline Change Envelope (*SCE*) methods (**a**); Coastal accretion/retreat rates calculated with the "end point rate" (*EPR*) and the "linear regression" (*LRR*) methods (**b**)

Erosion was observed along a 700 m sector between transects 29 and 58 recording retreat values of 30–40 m, and at other small sectors, with values lower than 25 m (Figs. 16.7 and 16.8).

The second method allowed identification of the most variable sectors as well as any homogeneity in their evolutionary pattern. Analysis of Fig. 16.8 is evidence of a very good correspondence between the two methods employed, indicating that any displacement recorded in the 1981–2007 period (NSM method) generally corresponded with maximum variation at each transect (SCE method). The 1981–2007 period was one that recorded the greatest variations with respect to 1981–1997 and 1997–2007 periods.

Specifically, along transects 1–29 (Fig. 16.8a), the NSM and SCE lines presented similar trends but did not coincided because the SCE line showed greater accretion than the NSM line. This was because maximum accretion was not recorded in the longer investigated time span (1981–2007) but in the 1981–1997 period since the 1997–2007 time span showed erosion. Opposite behaviour along transects 30–50, 52–63 and 105–110 was linked to general erosion recorded during the 1981–2007 period (NSM line) and small accretion during 1981–1997 period (SCE lines). The NSM and SCE lines coincided perfectly from transect 110 to 254 indicating: (i) great prevalence of accretion along transects 125–175 and 180–194 during the 1981–2007 period and small changes in other time intervals, (ii) almost absolute absence of shoreline changes from transects 110 to 120 and 195 to 254, due to this coastal sector being strongly protected by the free edge of the spit and influenced only by river mouth migration that was small or non-existent (Fig. 16.7).

With regards to retreat/accretion rates, these were calculated according to the "end point rate" and the "linear regression" methods (Dolan et al. 1991; Crowell et al. 1997, Fig. 16.8b). The first is the most common method and determines the amount of shoreline movement by selecting two shorelines, usually the earliest and the most recent ones. The main disadvantages of this method are that all the potential information between the two-used shorelines is ignored and the position of one or both shorelines can be aberrant, consequently generating a large error. The second method consists of calculating the regression for the whole data set (Crowell et al. 1997). A problem associated with this method is that, when shorelines positions are clustered, the same data will have more influence on the regression than others (Dolan et al. 1991; Jiménez and Sánchez-Arcilla 1993).

In this study, results obtained with the two methods (Fig. 16.8b) presented a very similar trend with a mean difference value of 0.08 m/year. Such small recorded differences are due to the limited number of used shorelines, the small clustering of the data and the steady coastal trend. Specifically, erosion rates – especially recorded along transects 28–61 – presented values ranging from 0.25 to 1.55 m/year. Accretion rates presented great range with maximum values (6.9 m/year) in the southern part of the spit (transects 130–174, Figs. 16.7 and 16.8b).

16.4.5 Considerations on Spit Evolution and Behaviour

In terms of the Tahadart spit evolution, two main periods have to be considered. During the 1981–1997 period, spit accretion took place at both the northern and southern sectors with no changes occurring in the central one (Fig. 16.7). Such a trend follows the general development pattern of a spit with increasing sediment supply (Carter 1988). According to Taaouati (2012), coastal sedimentation locally observed at Haouara Beach in the northern area of Tahadart spit (Fig. 16.3) was linked to nearshore supplies and to implantation of acacia trees carried out in the 1970s to slow down landward dune migration impacting the adjacent littoral road (Nachite et al. 2008).

During the 1997–2007 period, the littoral presented – even if less evidently – the prevalence of accretionary processes. Specifically, erosion prevailed in the northern part, at the foot of the spit, and accretion or no changes at the southern part, e.g. the free edge of the spit. This trend indicated a declining sediment supply (Carter 1988). Such a sedimentary deficit was probably enhanced by important sand mining started in 1988 of the beach foreshore and backshore at an area north of Tahadart spit in order to reduce excessive sedimentation and associated infilling problems to the littoral road. About 500 trucks on a daily basis transported c. 5,500 m^3 of sand representing approximately 15–20 million cubic metres extracted over 15 years, from 1988 to 2003.

In relation to evolution during the 1981–2007 period, the majority of the littoral showed accretion (95.2 % of transects), confirming results obtained by Taaouati (2012) working on the coastal evolution of the Tangier-Asilah sector during the 1958–2007 period. The author investigated evolution of seaward limit of dune vegetation at 344 transects and recorded a prevalence of accretion (77.3 % of transects) over erosion (22.7 %) processes. This work also showed that there are few human actuations in the littoral area and that sedimentary supplies from rivers are small while important supplies arrive from the nearshore favouring beach accretion (Abdellaoui and Ozer 2005). At a regional level, prevalence of accretion processes linked to important sedimentary supplies from the nearshore was observed in the southern part of Cadiz littoral (Anfuso and Gracia 2005).

Specific behaviour and evolution relating to the southern part of the spit (transects 130–174), demonstrates an increase in beach growth towards the tip of the recurved spit linked to the shoreline orientation change and associated increase of breaker angle, as observed in the theoretical model proposed by Hine (1979). Such factors produce a decrease of longshore transport and associated deposition of transported sediments which give rise to widespread swash bars formation and welding and/or berm ridges development (Carter 1988). Examination of aerial photographs and field observations at Tahadart spit indicated that swash bars develop by onshore movement of nearshore bars that approach almost normally to the coastline. Near the active spit where the beach is curving away from the ocean into the inlet, the northern parts of swash bars weld to the shore first thus leaving the southern ends open (Fig. 16.6d). According to Stewart and Davidson-Arnott (1988), down drift direction of oblique sand bars indicates an environment having a large net volume of longshore sediment transport (in this case SW directed) with a low volume in the opposite direction.

The result is an idealised barrier-spit complex linked to continued beach growth/progradation that takes place in a different manner along the spit: it is possible to distinguish three different areas with specific beach growth mechanisms (Hine 1979).

The first area, nearest the tip of the recurved spit (transects 171–174, Fig. 16.7), represents the location of greatest accretion (mean value 179 m) because it is the place where most swash bars are formed and weld to the shoreline. At this point, sand shoals are also formed at the river mouth entrance and are clearly wave-dominated features, parallel to the coastline that control wave refraction processes (Fig. 16.6d, Carter 1988; Park and Wells 2005). The second area is located further up the recurved spit (transects 155–169, Fig. 16.7) and records less significant accretion values (mean 101 m) because the swash bars here are not as broad or extensive but can migrate up to and weld onto the beach face (Hine 1979). The third area includes the straighter portions of the beach (transects 121–154, Fig. 16.7) where longshore transport is dominant and only the growth of berms formed at neap tides occurs with small beach accretion (mean value: 56 m).

16.5 Conclusions

Tahadart spit is located in a 35 km long homogeneous coastal sector between Tangier and Asilah, on the Atlantic coast of Morocco and encloses a large estuary with well-developed salt marshes. Detailed 3D surveys were carried out with a total station to investigate morphodynamic behaviour and state of ocean beach spit. The beach showed small and constant changes along the foreshore according to the parallel retreat mechanism associated with a clear dissipative beach state. Such a morphodynamic behaviour was confirmed by applying classic parameters as the Surf Scaling, the Surf Similarity and the Dean number.

The coastal evolution of the site between 1981 and 1997 from aerial photographs and a 2007 SPOT 5 satellite image were processed by means of GIS tools to obtain medium-term spit behavioural trends. The coastline was identified with the seaward limit of dune vegetation and dune foot and both indicators showed a linear and regular trend.

Coastal evolution was examined by taking into account shoreline changes considering differences between the earliest (1981) and the latest (2007) shoreline. Erosion/accretion rates were determined by means of the end point rate and the linear regression rate methods which presented a very good correspondence. Erosion affected a 700 m long sector. Accretion clearly prevailed over erosion and was especially recorded in the northernmost part of the littoral (95 m, 3.7 m/year), at a short sector close to the river mouth (90 m, 3.5 m/year), and at the southern part – the free edge – of the sand spit, along c. 1 km long sector. In the latter seaward and southward shoreline displacements were recorded with maximum values of c. 180 m (6.7 m/year). Such a trend follows the general development pattern of a growing spit. The results matched perfectly predictions of an existing model concerning the evolution of the recurved spit sector, the location of the decrease in longshore transport, and where associated deposition of transported sediments takes place. According to the model, it was possible to distinguish three different areas: (i) the nearest to the tip of recurved spit presenting the greatest accretion (179 m) due to the proximity where most of the swash bars were generated and welded to the shoreline; (ii) the area located further up the recurved spit, which recorded less significant accretion (101 m) and (iii) the straighter portion of the beach where longshore transport was dominant and small beach accretion took place (56 m).

Acknowledgments This work is a contribution to the Andalusia Research PAI group RNM-328. The SPOT-5 HRS/HRG image was purchased under a price reduction granted by the ISIS program.

References

Abdellaoui JE, Ozer A (2005) Apport des images satellitaires à haute resolution spatiale pour l'étude des cordons d'avant-côte sur le littoral atlantique de Tanger (Maroc). Télédétection 5:81–94

Anfuso G, Gracia FJ (2005) Morphodynamic characteristics and short-term evolution of a coastal sector in SW Spain: implications for coastal erosion management. J Coast Res 21(6):1139–1153

Battjes JA (1974) Surf similarity. In: Proceedings of 14th international conference on coastal engineering ASCE, Copenhagen, pp 466–480

Boak EH, Tarner IL (2005) Shoreline definition and detection: a review. J Coast Res 21:668–703

Calliari L (1994) Cross-shore and longshore sediment size distribution on southern Currituck spit, North Carolina: implications for beach differentiation. J Coast Res 10(2):360–373

Carter RWG (1988) Coastal environments. Academic, London, 617 pp

Castro M, Arroyo G, Bekkali R, Nachite D, Anfuso G (2006) Características ambientales del entorno de la laguna de Smir. AUE-UCA, 40 pp

CCIST (2013) Chambre de Commerce, d'Industrie et de Services de la Wilaya de Tanger. http://www.ccist.gov.ma

Chuvieco E (2000) Fundamentos de Teledetección Espacial. Ediciones Rialp, Madrid, 567 pp

Crowell M, Leatherman SP, Buckley M (1991) Historical shoreline change: error analysis and mapping accuracy. J Coast Res 7(3):839–852

Crowell M, Leatherman SP, Buckley M (1993) Shore-line change rate analysis: long term versus short term data. Shore Beach 61(2):13–20

Crowell M, Douglas B, Leatherman SP (1997) On forecasting future shoreline positions: a test of algorithms. J Coast Res 13(4):1245–1255

Dalrymple RA, Thompson WW (1976) Study of equilibrium beach profiles. In: Proceedings of 15th conference on coastal engineering ASCE, Honolulu, pp 1277–1296

Davis RA Jr, Hayes MO (1984) What is a wave-dominated coast? Mar Geol 60:313–329

Dolan R, Fester MS, Holme SJ (1991) Temporal analysis of shoreline recession and accretion. J Coast Res 7(3):723–744

Folk RL, Ward WC (1957) Brazos River bar. A study in the significance of grain size parameters. J Sed Petrol 27:3–26

Fredsoe J, Deigaard R (1992) Mechanics of coastal sediment transport, vol 3, Advanced series on ocean engineering. World Scientific, Singapore, 366 pp

Guza RT, Inman DL (1975) Edge waves and beach cusps. J Geophys Res 80(21):2997–3012

Hine A (1979) Mechanisms of berm development and resulting beach growth along a barrier spit complex. Sedimentology 26:333–351

Hughes MG, Crowell PJ (1987) Adjustments of reflective beaches to waves. J Coast Res 3:153–167

Jaaidi EB, Cirac P (1987) La couverture sédimentaire meuble du plateau continental Atlantique Marocain entre Larache et Agadir. Bull Inst Géol. Bassin d'Aquitaine, Bordeaux, 1987, N° 42, pp 33–51

Jaaidi EB, Ahmamou M, Zougary R, Chatre B, El Moutchou B, Malek F, Naim K (1993) Le littoral Méditerranéen entre Tétouan et Ceuta et Atlantique entre Tanger et Asilah (MAROC). Impact des aménagements portuaires sur la dynamique côtière: cas des ports de M'diq, Restinga- Smir, Tanger et Asilah. Publication du Comité National de Géographie du Maroc, pp 21–33

Jackson D, Cooper JAG, del Rio L (2005) Geological control of beach morphodynamic state. Mar Geol 216(4):297–314

Jackson D, Anfuso G, Lynch K (2007) Swash bar dynamics on a high-energy mesotidal beach. J Coast Res Spec Issue 50:738–745

Jiménez J, Sánchez-Arcilla A (1993) Medium-term coastal response at the Ebro delta, Spain. Mar Geol 114:105–118

Karrouk MS (1990) Aperçus sur les mécanismes climatiques rifains. Rev Fac Lettre, Tétouan, "le Rif, l'espace et l'homme 4", 4° année, pp 11–36
Komar PD, Gaughan MK (1972) Airy wave theory and breaker height prediction. In: Proceedings of 13th international coastal engineering conference ASCE, Vancouver, pp 405–418
Masselink G, Short AD (1993) The effect of tide range on beach morphodynamics and morphology: a conceptual beach model. J Coast Res 9:785–800
May JP, Tanner WF (1973) The littoral power gradient and shoreline changes. In: Coates DR (ed) Coastal geomorphology. State University New York, Binghamton, p 404
Moore L (2000) Shoreline mapping techniques. J Coast Res 16(1):111–124
Nachite D (2009) El desarrollo turístico del litoral en la Región Tánger-Tetuán: una evolución hacia unos escenarios no deseables. In: Domínguez Bella S, Maate A (eds) Geología Y Geoturismo En La Orilla Sur Del Estrecho De Gibraltar. University of Cadiz, Cadiz, pp 57–78
Nachite D, Bekkali R, Macias A, Anfuso G (2008) El estuario de Tahadart: las bases para una gestión integrada de un espacio en plena transformación. AUE-UCA, 38 pp
Nachite D, Rodríguez-Lázaro J, Martín-Rubio M, Pascual A, Bekkali R (2010) Distribution et écologie des associations d'ostracodes récents de l'estuaire de Tahadart (Maroc Nord-Occidental). Revue de micropaléontologie 53:3–15
Nakhli S (2010) Pressions environnementales et nouvelles stratégies de gestion sur le littoral marocain. Rivages méditerranéens 115:31–420
Nordstrom KF, Jackson NL (1992) Two-dimensional change on sandy beaches in meso-tidal estuaries. Z Geomorphol 36(4):465–478
ONE (2002) Rapport d'Etude d'Impact sur l'Environnement de la Centrale à Cycles Combinées de Tahaddart, vol II. Rapport Final, Rabat, p 186
Park JY, Wells JT (2005) Longshore transport at Cape Lookout, North Carolina: shoal evolution and the regional sediment budget. J Coast Res 21(1):77–92
Rangel N, Anfuso G (2011) Coastal storm characterization and morphological impacts on sandy coasts. Earth Surf Process Land 36:1997–2010
Rangel N, Anfuso G (2013) Winter wave climate, storms and regional cycles: the SW Spanish Atlantic coast. Int J Climatol 33:2142–2156
Shih S, Komar P (1994) Sediments, beach morphology and sea cliff erosion within an Oregon coast littoral cell. J Coast Res 10(1):144–157
Short AD, Jackson DWT (2013) Beach morphodynamics. In: Shroder JF (ed) Treatise on geomorphology. Elsevier/Academic, Amsterdam/San Diego, pp 106–129. ISBN 9780123747396
Stewart CJ, Davidson-Arnott RGD (1988) Morphology, formation and migration of longshore sandway; Long Point, Lake Eire, Canada. Mar Geol 81(1–4):63–77
Taaouati M (2012) Morphodynamique des plages et évolution du trait de côte sur le littoral atlantique du Tangérois (Maroc Nord Occidental): approches saisonnière et pluri-décennale par techniques de la Géomatique. Université Abdelmalek Essaadi (Tetuan, Morocco) – Université de Nantes (France). 271 pp (unpublished)
Taaouati M, Nachite D, Benavente J, Elmrini A (2011) Seasonal changes and morphodynamic behavior of a high-energy mesotidal beach: case study of Charf el Akab beach on the North Atlantic coast of Morocco. Environ Earth Sci 64:1225–1236
Thieler E, Danforth W (1994) Historical shoreline mapping: improving techniques and reducing positioning errors. J Coast Res 10(3):549–563
Wright LD, Short AD (1984) Morphodynamic variability of surf zones and beaches: a synthesis. Mar Geol 56:93–118

Chapter 17
Geomorphology and Internal Sedimentary Structure of a Landward Migrating Barrier Spit (Southern Sylt/German Bight): Insights from GPR Surveys

Tanja Tillmann

Abstract Barriers comprise approximately 15 % of the world's coastlines and are formed due to the combined action of wind, waves, and longshore currents. In this study ground-penetrating radar data (GPR) of different antenna frequencies and sedimentological data were combined to reveal the sedimentary structure and architecture of the southern barrier island spit of the North-Frisian island of Sylt, to define different coastal environments and to set up a barrier island stratigraphy.

Based on these data, a sedimentological model has been generated for Southern Sylt which describes the interaction between extreme events, coastal processes and sedimentary development and elucidates the major episodes of barrier island evolution. The model is concerned with the spit add-on zone where the barrier spit is attached to the central island moraine core and shows a landward migration through barrier rollover involving an interplay of barrier retreat and washover associated with accumulation of sediment in a backbarrier environment as a result of storm surges. With the exception of the uppermost dune facies the spit add-on zone reveals a transgressive coarsening upward sequence starting with sandy mud flat deposits at the bottom which grade into coarser sandy tidal flat deposits toward the top. Sandy tidal flat deposits are overlain by washover sheet and washover fan deposits. Eroded sediment was transported along the west coast of Sylt by longshore drift and was added to the southern spit-end during fair weather conditions.

T. Tillmann (✉)
Lower Saxony Institute for Historical Coastal Research (NIhK), Viktoriastr. 26/28, D-26382 Wilhelmshaven, Germany
e-mail: tanja.tillmann@nihk.de

G. Randazzo et al. (eds.), *Sand and Gravel Spits*, Coastal Research Library 12,
DOI 10.1007/978-3-319-13716-2_17

17.1 Introduction

Barrier islands are elongated sandy islands parallel to the shoreline and are separated from the mainland by tidal flats, shallow bays, lagoons or marsh systems (Nielsen et al. 1988; Prothero and Schwab 1996; Davis and FitzGerald 2004). Barrier islands and barrier spits are geologically young, highly dynamic and represent a complex coastal system that includes a number of different but closely related sedimentary depositional environments and geomorphological elements of varying origin, genesis and evolution.

Barrier spits are attached to the land mass at one end and terminate in open water at the other end (Cooper 2012; Pilkey et al. 2009). Hence, the barrier spit is younger than the land mass to which it is attached (Evans 1942). Spits generally grow in the direction of the predominant longshore drift (Uda 2005) when shoreface deposits are added to the spit-end.

The stratification of barrier islands is often investigated only by borehole data. Therefore, the processes of evolution and the internal structure of barrier islands are often poorly described. In recent years the ground-penetrating radar (GPR) method has been widely used to examine the internal structures of Holocene sediments. As opposed to coring, ground-penetrating radar provides continuous two- or three-dimensional profiles with a high degree of geometrical and architectural detail. In this study ground-penetrating radar data and sedimentological data were combined to define different coastal environments, to reveal the sedimentary structure and architecture of a barrier island, to describe the barrier island stratigraphy and to reconstruct the geological development. Based on these data, a sedimentological model of barrier spit development has been generated.

17.2 Study Area and Geological Setting

The North-Frisian Island of Sylt is part of a sandy barrier system that belongs to an island chain lining the coast of the southern North Sea/German Bight. With an area of approximately 99 km^2 and a south-north extension of nearly 40 km, Sylt is the largest of these islands and is separated from the German mainland by a 15 km-wide tidal flat area that is part of the Wadden Sea (Fig. 17.1). Sylt is subjected to a semidiurnal lower to upper mesotidal regime with a tidal range of 1.8–2.2 m (Hayes 1979). Tidal currents are strong, particularly at the northern and southern spit because of the narrow inlets between the adjacent islands and the large tidal area between the islands and the mainland (Kelletat 1992). Due to large fetch and dominant westerly winds, Sylt is characterised by a high-energy wave regime (Hundt 1957).

Sylt as a whole consists of two main landscape elements: a Pleistocene core of till and moraine deposits in the centre of the island and two long barrier spits in the northern and southern part with extended Holocene dune areas (Dietz and Heck 1952).

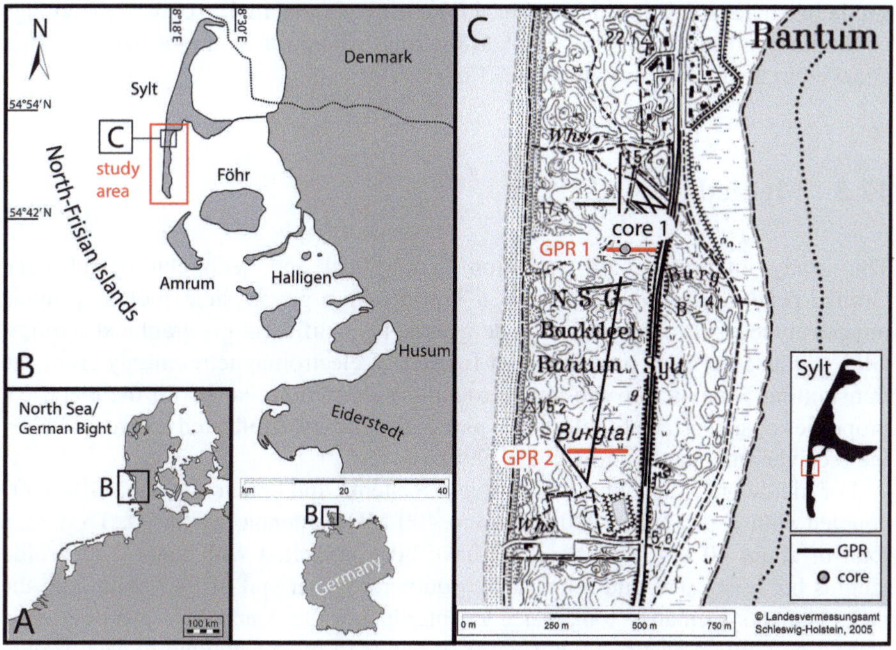

Fig. 17.1 (**a**) Location of the southern North Sea/German Bight and the position of Sylt. (**b**) Overview of the North-Frisian Islands with the position of the study area of Southern Sylt presented in (**c**). Close-up with the position of GPR lines 1 and 2 as well as the sediment core 1

With ongoing post-glacial sea-level rise during the Holocene and associated coastal erosion, the coast retreated to the east because of the unconsolidated nature of the glacial material and the abraded sediments of the central moraine core were reworked and transported by longshore drift to the north and south to build up two large sandy spit systems (Ahrendt 2007). Further sediment sources of eroded moraine deposits occur in the Amrumbankmoraine and Pisamoraine northwest of the modern coastline of Sylt (Bantelmann 1966; Ahrendt 1994). Additionally, a smaller moraine core close to the village of Rantum was suggested by Hoffmann (1974a) and Newig (1995, 2001) as a sediment source.

The moraine cores of Sylt have been settled by humans since Mesolithic times. As a result of ongoing erosion of the island's west coast during storm surges Bronze Age grave mounds and other remains of early settlement activity were revealed beneath the dune cliff next to the village of Rantum (e. g. Kersten 1967; Hinz 1968; Reichstein 2005). Since the Middle Ages that were marked by several devastating storm surges in the North Sea basin, such as the 2nd "Marcellusflut" in 1362 and the "Burchardiflut" in 1634 (compare Panten 2000; Ahrendt and Thiede 2002; Behre 2004), the village of Old-Rantum had to be given up from time to time and has been rebuilt and resettled again farther eastwards (Newig 2001).

The focus of this investigation is on the reconstruction of the geological development of the southern barrier spit of Sylt. In this study a representative area of the southern barrier spit is presented (Fig. 17.1) that is located in the south of the village of Rantum and represents the spit add-on zone where the southern barrier spit is attached to the central moraine core (Fig. 17.1).

17.3 Methods

This study integrates high-resolution geophysical and sedimentological data. Ground-penetrating radar (GPR) is a non-invasive geophysical tool to provide images of the shallow subsurface underground from different geographical settings. GPR produces a short pulse of high-frequency electromagnetic energy from the transmitting antenna into the underground. With abrupt changes in the electrical properties of subsurface lithologies, a part of the energy is reflected and recorded by the receiver antenna (Bristow et al. 1996).

For fieldwork a geophysical Survey Systems Inc. radar system, SIR-2000 coupled with a 100 MHz, 200 MHz and 400 MHz antenna, was used. To date, a total of about 40 km GPR transects have been collected with individual profile lengths between 30 m and 5 km at the southern barrier spit of Sylt. Although the measurements primarily took place in interdune valleys and flats a topographic correction with an accuracy between 0.10 and 0.15 m was applied to each profile using a differential global positioning system (either Ashtech ProMark II or Topcon GPS G3).

The software ReflexWin from Sandmeier Scientific Software was used for editing and processing the GPR data. Different standard processing steps (e. g., time-zero correction, dewow-filter, bandpass filters, background removal filter, gain function, fk-migration) were chosen to increase signal-to-noise ratio, to improve resolution, to restore real reflector geometry and to provide a more realistic image of the subsurface.

The local velocity of the radar wave was determined by the analysis of diffraction hyperbolas in each profile. Velocity analysis based on hyperbolas tends to produce approximate velocity values with errors and variance of +/− 10 % (Cassidy 2009). Consequently, a direct correlation from radar data with sediment core data was made. Values of radar wave velocity range between 0.04 and 0.09 m/ns with an average of around 0.065 m/ns in fresh water saturated sand and silt. The groundwater level was detected in a number of radar reflection profiles as a laterally continuous and horizontal reflection with high amplitude. In dune sands which are located above the groundwater level, the radar-wave velocity range was between 0.09 and 0.14 m/ns, with an average of around 0.11–0.12 m/ns. These values correspond to radar-wave velocities estimated in previous studies for saturated and dry sands as mentioned by Girardi and Davis (2010) and Neal et al. (2003) and correlate with findings in comparable settings (Lindhorst et al. 2008, 2010; Tillmann and Wunderlich 2011a, b, 2012; Tillmann et al. 2013).

GPR data interpretation is the most challenging component of GPR work and is primarily based on imaging. For that reason, it is important to understand the nature and origin of reflections and to recognise reflection patterns characteristics of specific sedimentary deposits in order to be able to interpret radar images with confidence (Van Overmeeren 1998). Radar reflection profiles were interpreted following the principles of radar stratigraphy, using the terminology suggested by Neal (2004) that is originally based on the principles of seismic stratigraphy. Sedimentary sequences, particularly those dominated by clastic sediments, appear to display a hierarchy of depositional units (Miall 1991), starting from individual laminae to whole sedimentary basin fill (Campbell 1967; Allen 1982). Radar stratigraphy interpretation technique allows a nearly objective description and comparison of the reflections patterns and configurations, as for example the shape and dip of reflections, the relationship between the reflections and the reflection continuity (Van Overmeeren 1998; Neal 2004). Different radar reflection patterns are often caused by bedding and lithological variations in clastic sediments, for example differences in grain compositions, size, shape, orientation and packing of grains (Bristow et al. 1996). Moreover, changes in grain-size parameters, porosity and sorting of the sediments are usually associated with altering water content that governs reflections in GPR data (Van Dam et al. 2002; Van Dam 2012). Bearing this in mind, radar facies analysis ideally leads to the interpretation and determination of environmental setting, depositional processes and lithofacies (Mitchum et al. 1977; Jol and Bristow 2003).

In order to be able to interpret and define depositional environments as well as to link ground-penetrating radar data with sedimentological data, vibracores were drilled at selected positions along the radar profiles (Fig. 17.1). Samples were analyzed by wet sieving (0.5 phi intervals) to determine grain size distribution. Grain size statistics were calculated using the computer program GRADISTAT (Blott and Pye 2001). All statistical grain size parameters mentioned in the text are based on the graphical method (logarithmic phi scale) by Folk and Ward (1957). Borehole data were also used to correlate the determined depth of the radargramms.

17.4 Results

17.4.1 Typical Radar Facies of a Migrating Barrier Spit

A total of eight different radar facies were identified (Fig. 17.2) and two radargramms were chosen to demonstrate main sedimentological features (Figs. 17.3 and 17.4) of the barrier spit. As a first step the identified radar facies were subdivided into four groups. The first group consists of radar facies with inclined reflection patterns, the second group encloses radar facies with horizontal and curved reflections. The third group represents irregular reflections and reflection-free parts of the GPR profiles. Radar surfaces and unconformities have been put into the last group (Fig. 17.2).

Radargramm	Interpretation	Radar facies (Rf)	Amplitude	Shape, dip, continuity, position, depth range, peculiarity	Sedimentary environment
A. Inclined reflections					
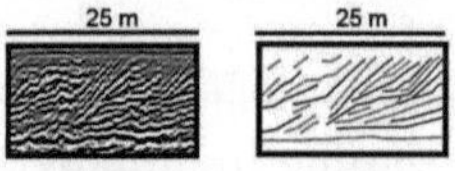		Rf-A-1	high - medium	uppermost part, above Rf-B-1, continuous - moderately continuous, partly with soil development, roots, internal bounding surfaces	dune facies, aeolian cross-stratification
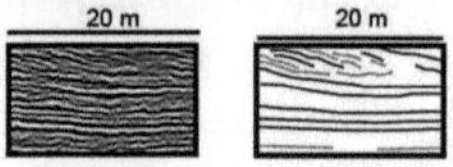		Rf-A-3a	medium - low	moderately continuous-discontinuous, tangential, horizontal layers (2°-6°) and short foresets (25°-30°) with gently land-ward dipping, mostly beneath Rf-B-1	washover foresets strata, parallel to the dominant flow direction, shaping a washover fan or a washover sheet
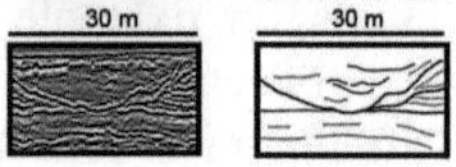		Rf-A-3b	medium	continuous, curved concave, trough - channel shaped cut-and-fill structures, mostly beneath Rf-B-1, cutting underlying reflections, occasionally accumulation of hyperbolas on the bottom	washover channel
B. Horizontal reflections					
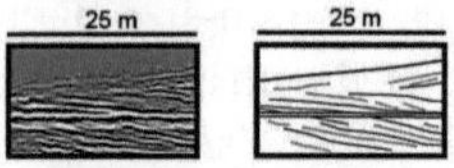		Rf-B-1	high	laterally continuous, linear, horizontal- subhorizontal, cutting other reflections	ground water table
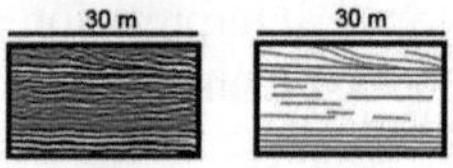		Rf-B-2a	medium- low	horizontal planar, moderat continuous - discontinuous, partly weak refelctions, parallel - subparallel, beneath Rf-B-1	backbarrier facies, sandy mud flat deposits
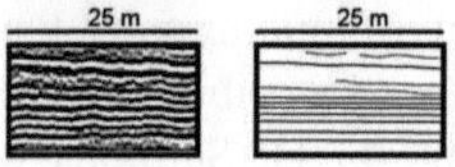		Rf-B-2b	high	horizontal planar,continuous - moderately continuous, parallel - subparallel, beneath Rf-B-1	backbarrier facies, sandy tidal flat deposits
C. Irregular reflections					
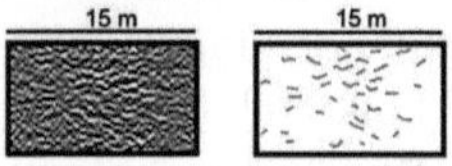		Rf-C-1	medium- very low	discontinuous, oblique chaotic, hummocky, partly weak reflections and free of reflections, occasionally hyperbolas	undefined, partly overlapping reflections, signal attenuation by salty groundwater
D. Radar surface					
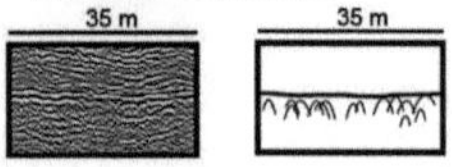		Rf-D-1	very high- high	horizontal, continuous, lowermost part, beneath Rf-B-1,accumulation of hyperbolas directly at the main reflector	Pleistocene bedrock, till deposits, surface of the Saalian moraine

Fig. 17.2 Radar facies analysis. Eight different radar facies were recognized in the study area of Southern Sylt. Based on geometrical criteria, the radar facies were divided into four categories: (**a**) inclined reflections, (**b**) horizontal reflections, (**c**) irregular reflections and (**d**) radar surfaces

17.4.1.1 Rf-A-1: Dune Facies

Radar Facies

Rf-A-1 shows a high to medium amplitude and inclining continuous to moderately continuous reflection pattern. The single reflections are parallel to subparallel orientated. Reflectors of Rf-A-1 shows a medium to high dip angle in different directions and are often bundled by numerous bounding surfaces into several radar packages.

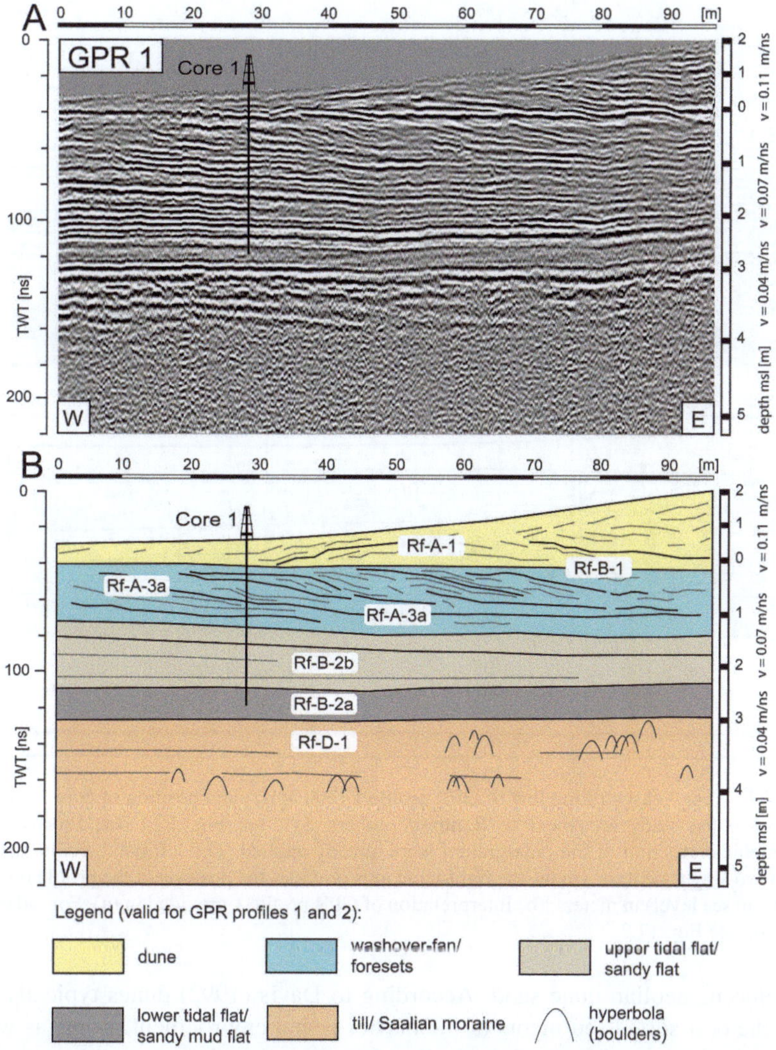

Fig. 17.3 West – east striking line of GPR profile 2 (200 MHz) and position of core 1. For exact location in the study area next to Rantum/Southern Sylt see Fig. 17.1. (**a**) Processing and topographic correction of the radargramm were already applied. The left hand axis presents the TWT (two-way traveltime) in ns, the right-hand axis provides the position of the profile relative to msl (mean sea level) in meters. (**b**) Interpretation of GPR profile 1 provided in (**a**). For radar facies definition see Fig. 17.2. Tillmann and Wunderlich (2013a)

Lithological Facies

RF-A-1 consists of well sorted medium sand which exhibits only little variation in grain size and sorting. The mean grain size, according to Folk and Ward (1957), ranges between 1.79 φ (medium sand) and 1.08 φ (medium sand) and is interpreted

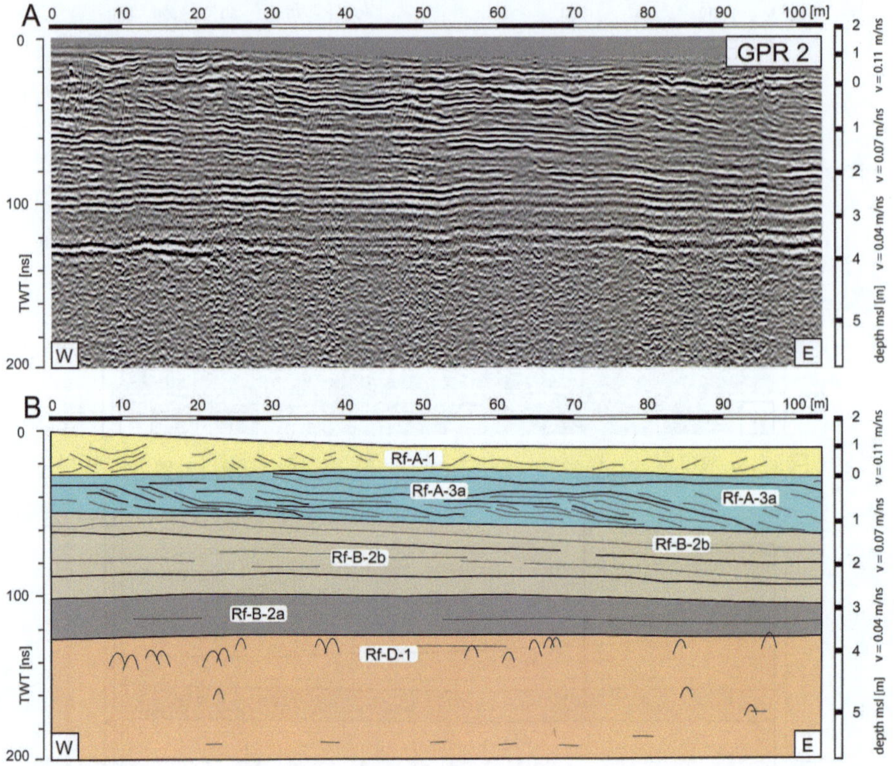

Fig. 17.4 West – east striking line of GPR profile 1 (200 MHz) and position of core 1. For exact location in the study area next to Rantum/Southern Sylt see Fig. 17.1. (**a**) Processing and topographic correction of the radargramm were already applied. The left hand axis presents the TWT (two-way traveltime) in ns, the right-hand axis provides the position of the profile relative to msl (mean sea level) in meters. (**b**) Interpretation of GPR profile 1 provided in (**a**). For radar facies definition see Fig. 17.2

as Holocene aeolian dune sand. According to Davis (1992) dunes typically comprise the best sorted sediments of the barrier island environment. Roots as well as organic horizons within the dune facies are common.

Interpretation

Corresponding to similar GPR studies measured in aeolian environments (e.g., Bristow et al. 1996, 2000a, b, 2010a, b; Harari 1996; Botha et al. 2003; Girardi and Davis 2010) the facies of Rf-A-1 is interpreted as cross-bedded aeolian strata formed by migrating dunes of various dimensions. Bristow et al. (1996) found, that small changes in capillary water reflect changes in grain size in dune sands stratification and are the main cause of the dielectric contrast detected by the radar. The cross-stratification is occasionally interrupted by paleosols associated with roots and other organic material. Organic layers and soil development within the Rf-A-1 facies may indicate several dune building stages.

Dunes are a main element on the German North Sea islands. On Sylt beach sand is reworked and blown out to build up primary foredune ridges. Prevailing westerly wind causes inland dune migration from the west coast to the eastern part of the barriers. The dune facies is found in the uppermost part of the GPR profiles above the groundwater table.

17.4.1.2 Rf-A-3a and Rf-A-3b: Washover Facies

Radar Facies

Rf-A-3a comprises moderately continuous to discontinuous reflections of medium to low amplitude. Bedding features include gently landward-dipping sigmoidal to tangential clinoforms as well as more or less horizontal planar stratification nearly always beneath the groundwater table. Rf-A-3b often cuts underlying reflections and is identified as cut-and-fill structures. Trough and channel shaped reflections of Rf-A-3b are assumed to be washover channel deposits caused by several flooding washover events which are orientated perpendicular to the dominant washover flow direction. On the bottom of Rf-A-3b, accumulations of hyperbolas occur quite frequently.

Lithological Facies

Washover deposits differ depending on the process of inundation and the source of sediment. Therefore, it is difficult to identify and separate washover deposits simply due to their textural and bedding similarities. Core samples show for the most part poorly sorted medium to coarse sand associated with lag layers of shells and shell fragments. Sedimentary series of washover deposits often enclose graded bedding sequences. Pebbles are common and are often located on the bottom of a fining upward sequence.

Interpretation

Coastal overwash is defined as the transport of sea water and sediment during storms across the island from the beachface to the backbarrier environment (Leatherman et al. 1977; Leatherman and Zaremba 1987). According to Schwartz (1982) overwash generally occurs where foredunes do not exist or the foredune relief is low as well as where high foredunes have been breached or channeled.

Landward-dipping reflections of Rf-A-3a are interpreted as washover foreset strata caused by flooding and inundation during several washover events. These washover foresets are orientated parallel to the dominant washover flow direction from west to. Foreset strata and almost plane beds are the typical sedimentary structure associated morphologically with a washover fan (Schwartz 1982) or a washover sheet (Davis 1992; Buynevich et al. 2004). Similar features in GPR data are also identified by Anthony and Møller (2002) for the Danish Holmsland Barrier and by Wang and Horwitz (2007) for hurricane caused overwash fans along the coastline of Florida (USA).

A number of diffraction hyperbolas on the bottom of these washover channels are interpreted as coarse gravels, cobbles, boulders and buried flotsam material which indicates a higher flow velocity at the beginning of the washover inundation. Afterwards, graded bedding including a fining upward sequence indicates deposition in standing water during and after washover events. Graded bedding in washover sediments was previously confirmed by Leatherman et al. (1977) on Assateague Island (USA) and can also be a result of the decreasing surge energy to the end of the storm when overwash water flows back to the sea.

17.4.1.3 Rf-B-1: Groundwater Table

RF-B-1 is a laterally continuous and sub-horizontal reflector which appears to cut other reflections. This almost linear reflector is recognizable in a number of GPR profiles and is interpreted as the groundwater table. Because of the strong contrast in dielectric properties between dry and water saturated sand the groundwater table shows continuous high amplitude.

17.4.1.4 Rf-B-2a and Rf-B-2b: Backbarrier Facies (Tidal Flat Deposits)

Radar Facies

The facies of Rf-B-2a and Rf-B-2b include planar nearly even reflections mainly below the groundwater table. Rf-B-2a shows a medium to low amplitude whereas the amplitude of Rf-B-2b is rather high. The even reflectors are continuous (Rf-B-2b) to moderately discontinuous (Rf-B-2a) and predominantly orientated in a parallel way. Rf-B-2a often causes attenuation to the underlying reflectors.

Lithological Facies

Rf-B-2b comprises well sorted, fine to medium sand and silt sediment layers. Theses sandy tidal flat deposits are well bedded containing plant material as well as scattered shell fragments. On several sites thin and finely laminated peat layers are also common within the sandy tidal flat sequence. In comparison, silt to fine sand layers of Rf-B-2b are organic rich and generally finer laminated and more frequently interbedded by clay and peat layers. Almost certainly sediments of Rf-B-2a and Rf-B-2b were accumulated in a low-energy marine setting that may was always protected by an island barrier.

Interpretation

Almost planar reflections patterns of Rf-B-2a and Rf-B2b are assumed to be tidal flat deposits of a backbarrier environment. Rf-B-2b is interpreted as well bedded sandy deposits accumulated on an upper tidal flat area. Evidently, reflections of Rf-B-2a represent sandy mud flat deposits of a lower tidal flat. In GPR data where Rf-B-2a represents the uppermost unit, for instance in GPR profiles measured in the present tidal flat area or on salt marshes, underlying reflections are strongly attenuated in order that the ground penetration decreases. In that case strong attenuation is affected by interbedded muddy sediments like peat layers and clay.

17.4.1.5 Rf-C-1: Reflection Free and Undefined

Rf-C-1 shows discontinuous irregular and oblique chaotic reflections with medium amplitude which are partly hummocky. Hyperbolas are also common. It seems that internal bedding of the sediments caused by altering layers is less clear and less defined. Rf-C-1 also comprises weak reflections as well as parts which are even completely free of reflections. The virtual absence of reflections in GPR data from coastal environments often indicates brackish or salty groundwater as well as the presence of clay or salt marsh peat near the subsurface which both cause an increased attenuation of the electromagnetic wave. Massive homogeneous lithologic units without altering bedding structure can also lead to reflection free configurations.

17.4.1.6 Rf-D-1: Surface of Saalian Moraine (Till Deposits)

Radar Facies

The radarsurface of Rf-D-1 shows an almost horizontal and continuous key reflector of high amplitude. Radar patterns of Rf-D-1 always occur in the lower part of the profiles of the northern region of the study area in the vicinity of Rantum and are dominated by numerous diffraction hyperbolas which are mainly located just beneath the key reflector. Rf-D-1 is independent of GPR profile orientation.

Interpretation

Rf-D-1 is assumed to be the boundary between overlying Holocene sediments and underlying Pleistocene bedrock of the Saalian moraine. Boulder-size stones and pebbles within the till deposits are considered to be diffraction hyperbolas producing objects. Additional horizontal reflection patterns indicate minor internal bedding within the Saalian moraine.

17.4.2 GPR Lines 1 and 2

GPR lines 1 and 2 (Figs. 17.3 and 17.4) are located in the south of the village of Rantum next to a dune valley which is called Rantum Baakdeel (Fig. 17.1). Both lines run almost perpendicular to the present coastline and were chosen to represent the typical radar facies and stratigraphic order of this spit add-on region.

The profiles can be subdivided into five main parts. The upper part of the profile which is located above the groundwater table (Rf-B-1) merely contains reflections of the aeolian dune facies (Rf-A-1) with an estimated radar wave velocity of 0.11 m/ns. Below the groundwater table (Rf-B-1) the radar wave velocity decreases to an average of 0.07 m/ns up to 0.04 m/ns in the lowermost part. Directly beneath the groundwater table reflections of Rf-A-3a occur which indicate a gentle landward dipping and are interpreted as washover foresets strata caused by flooding during several washover events. These washover foresets are orientated parallel to the dominant washover flow direction and belong morphologically to a washover fan or a washover sheet. Post-storm aeolian modification of at least the upper-most washover layers is common (Schwartz 1982). Hence, the washover unit sometimes interfingers with overlying aeolian dune deposits.

In the lower part of the radargramm reflections of the washover facies (Rf-A-3a) merge into the underlying facies of Rf-B-2b which comprises almost horizontal reflections of medium amplitude which are considered to be layers of a backbarrier sandy tidal flat. The underlying unit of Rf-B-2a contains significantly minor internal horizontal reflections compared to Rf-B-2b and is interpreted as sandy mud flat deposits of lower tidal flat environment. At approximately 3 m (GPR line 1) to 3.8 m depth (GPR line 2) below mean sea level reflections of Rf-D-1 occur that are interpreted as till deposits of the Saalian moraine core. Due to increasing attenuation, the lower boundary of facies Rf-D-1 was not observed.

17.4.3 Ground-Truthing: Core Data

A 3.5 m long core was collected 28 m from the western starting point of GPR profile 1 (Figs. 17.1 and 17.3). The topmost unit was composed of well to moderately well sorted medium sand and represents Holocene dune deposits. Just below the present groundwater table up to approximately 2.2 m below the surface poorly sorted medium to coarse sand with bivalve shells were interpreted as washover fan deposits. In a depth of 2.2–3.2 m below surface well bedded and well sorted layers of fine to medium sand and silt are common. Several layers contain shells

(*Cerastoderma edule*) as well as plant material such as roots. This unit is considered to be sandy tidal flat deposits. From a depth of 3.2 m below the surface thin and finely laminated silt and peat layers are located which correspond to sandy mud flat deposits of the lower tidal flat. Bivalves (*Macoma baltica*) and gastropods, (*Hydrobia ulvae*) are common. The core did not reach the lowermost unit defined in the radargramm of GPR Profile 1 at this position. However, coring at other sites (Dietz and Heck 1952) indicates Saalian moraine core deposits at a comparable depth. Moraine core deposits of the spit's add-on zone next to Rantum consist of sand at the top that turns into sandy moraine till with increasing depth including a Baltic and Scandinavian till spectrum of flint, quartzite, granite and gneiss boulders.

17.5 Discussion

Comparison of GPR data and sedimentological borehole data provides new detailed insights and a thorough understanding of the barrier island stratigraphy and architecture of the southern island spit of Sylt. The internal structure of the southern barrier spit of Sylt documents diverse types of sedimentary deposits and environments in a barrier island setting.

17.5.1 Barrier Rollover Model of the Trangressive Barrier Add-On Zone

The sequence of the spit add-on zone next to Rantum starts with Pleistocene deposits (Figs. 17.3, 17.4, and 17.5). From GPR and coring data as well as previous data (Dietz and Heck 1952) into consideration the top of the central moraine core of Sylt can be detected in a depth of 5–6 m below the surface (3–3.80 m below mean sea level). A gentle southward dipping of the moraine core as was previously supposed by Hoffmann (1974b) is quite likely. An unconformity marks the boundary between Pleistocene till and overlying Holocene tidal flat deposits and leads to the assumption of erosion during sea level rising that is also in a line with previous suggestions of Hoffmann (1974b).

Overlying sandy mud flat deposits of a lower tidal flat were induced by marine inundation, maintained by an ongoing gentle sea-level rise. The presence of fine grained sand and peat in a tidal flat environment suggests that the sequence was initiated during a time of a protected and low-energy backbarrier environment. With the exception of the uppermost dune facies the spit add-on zone reveals a coarsening upward sequence starting with sandy mud flat deposits at the bottom which turn into coarser sandy tidal flat deposits toward the top (Figs. 17.3, 17.4, and 17.5).

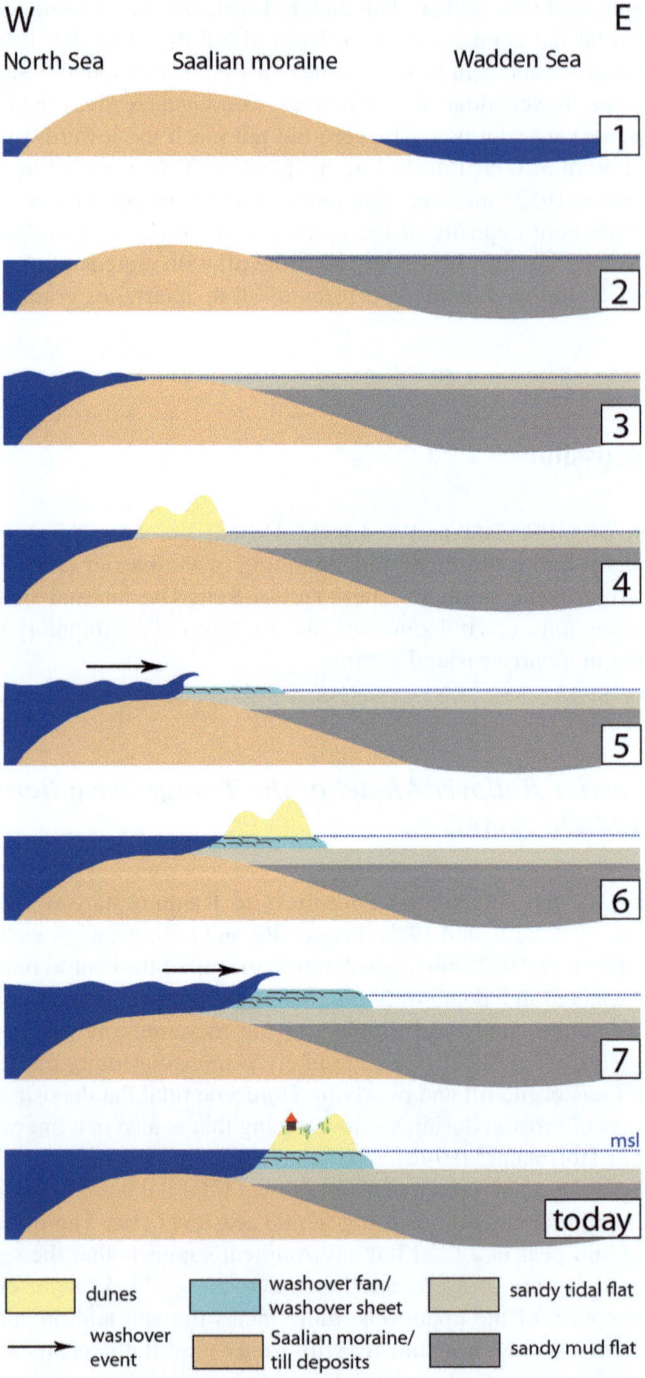

Fig. 17.5 The simplified "barrier rollover model" shows the geomorphological development of the barrier's spit add-on zone which was dominated by an interplay of storm surge erosion, washover events and barrier retreat

Sandy tidal flat deposits are overlain by washover sheet and washover fan deposits. In relation to Davis (1992) this sequence corresponds to a typical transgressive barrier sequence associated with landward migration under rising sea level. Accordingly, the barrier spit add-on zone next to Rantum indicates transgressive characteristics related to onshore west-east directed barrier migration during its initial stage of development.

The consequence of this morphological behaviour termed "barrier rollover" by Davis and Fitzgerald (2004: 151) is erosion of the front of the barrier occasionally exposing backbarrier sediments such as sandy flats and sandy mud flats in the current surf zone during ongoing retreat. That is also the reason why Bronze Age grave mounds (Reichstein 2005), sometimes appear in the surf zone of the west coast after storm surges.

The eroded sediment was transported along the west coast by longshore drift and was added to the southern spit-end (Tillmann and Wunderlich 2011a, 2013). The interplay of barrier retreat and washover accumulation in a backbarrier environment are chiefly responsible for the barrier rollover and eastward migration of the spit add-on zone to the south of Rantum.

A stepwise and simplified model of barrier rollover and stratigraphy is provided in Fig. 17.5. The first stage of the model represents the geomorphological situation at the beginning of the post-glacial sea-level rise (Fig. 17.5 picture 1).

Ongoing sea-level rise caused coastal erosion of the Saalian moraine core and the island's west coast retreated to the east because of the unconsolidated nature of the till deposits. As a second step, accumulation of fine grained sandy mud flat deposits in a backbarrier environment occurred (Fig. 17.5 picture 2). Sandy tidal mud flat deposits grade upwards into coarser sandy tidal flat deposits (Fig. 17.5 picture 3). Fine to medium grained sand was reworked to build up some primary foredunes (Fig. 17.5 picture 4).

Several storm surges caused washover events which were associated with breakthroughs of the primary foredune ridges. As a result, the eroded sediment has accumulated in the backbarrier environment as a washover fan or a washover sheet (Fig. 17.5 pictures 5–7). Storm surge erosion, washover flooding and accumulation caused barrier retreat but did not lead to the total destruction of the island. Rather, the island was rebuilt as it shifted eastward.

Today, the situation is completely different. Infrastructure, settlement and tourism demand a permanent conservation of the status quo. Hard coastal protection structures as well as recurring beach nourishments are used to preserve the modern position of the southern barrier spit. In this way, the natural barrier rollover process as well as the erosion of the west coast is being prevented. As a consequence of this lack of eroded sediment, the spit end will receive only little sediment by longshore drift in the future and will be more and more vulnerable.

17.6 Conclusion

GPR provides a successful tool for investigation concerning the shallow subsurface of barrier islands. In contrast to coring, ground-penetrating radar presents a continuous two- or three-dimensional profile with great geometrical and architectural detail. GPR data revealed different sedimentary structures and provides new insights into the sedimentary architecture of the southern barrier island spit of Sylt. Additional borehole data led to the definition of different sedimentary facies and environments as well as provided a tool for velocity and depth correlation of the radargramms. Based on these results a barrier island stratigraphy for the spit add-on zone has been generated.

GPR data as well as cores taken from the spit add-on zone of Southern Sylt indicate a transgressive barrier spit sequence that led to an eastward migration. This "barrier rollover" was mainly accomplished by an interplay of barrier retreat and washover accumulation in a backbarrier island environment. The results show that the development of the southern barrier spit of Sylt is much more complex than previously supposed, and it is not only characterized by simple spit attachment on a moraine core. In the past, high energy events like storm surges and washover processes did not cause the totally destruction of the barrier. Rather, the barrier was rebuilt and shifted eastward by a natural barrier rollover.

References

Ahrendt K (1994) Geologie und Küstenschutz am Beispiel Sylt. Berichte des Forschungs- und Technologiezentrum Westküste der Universität Kiel 4:1–135

Ahrendt K (2007) Vergangenheit und Zukunft des nordfriesischen Wattenmeeres. Coastline Rep 9:45–57

Ahrendt K, Thiede J (2002) Naturräumliche Entwicklung Sylts: Vergangenheit und Zukunft. In: Daschkeit A, Schottes P (eds) Klimafolgen für Mensch und Küste am Beispiel der Nordseeinsel Sylt. Springer, Berlin/Heidelberg, pp 69–112

Allen JRL (1982) Sedimentary structures: their character and physical basis, vol 1, Developments in sedimentology, vol 30 A. Elsevier, Amsterdam, 593 p

Anthony D, Møller I (2002) The geological architecture and development of the Holmsland Barrier and Ringkøbing Fjord area, Danish North Sea Coast. Geografisk Tidsskrift 102(1):27–36

Bantelmann A (1966) Die Landschaftsentwicklung an der schleswig-holsteinischen Westküste, dargestellt am Beispiel Nordfriesland: Eine Funktionschronik durch fünf Jahrtausende. Die Küste 14(2):5–99

Behre K-E (2004) Coastal development, sea-level change and settlement history during the later Holocene in the Clay District of Lower Saxony (Niedersachsen), northern Germany. Quat Int 112(1):37–53

Blott SJ, Pye K (2001) Gradistat: a grain size distribution and statistics package for the analyses of unconsolidated sediments. Earth Surf Proc Land 26:1237–1248

Botha GA, Bristow CS, Porat N, Duller G, Armitage SJ, Roberts HM, Clarke BM, Kota MW, Schoeman P (2003) Evidence for dune reactivation from GPR profiles on the Maputaland

coastal plain, South Africa. In: Bristow CS, Jol HM (eds) Ground penetrating radar in sediment, Geological Society, special publications 211. Geological Society, London, pp 29–46
Bristow CS, Pugh J, Goodall T (1996) Internal structure of aeolian dunes in Abu Dhabi determined using ground-penetrating radar. Sedimentology 43(6):995–1003
Bristow CS, Chroston PN, Bailey SD (2000a) The structure and development of foredunes on a locally prograding coast: insights from ground-penetrating radar surveys, Norfolk, UK. Sedimentology 47(5):923–944
Bristow CS, Bailey SD, Lancaster N (2000b) The sedimentary structure of linear sand dunes. Nature 406:56–59
Bristow CS, Augustinus P, Wallis IC, Jol HM, Rhodes EJ (2010a) Investigations of the age and migration of reversing dunes in Antarctica using GPR and OSL, with implications of GPR on Mars. Earth Planet Sci Lett 289:30–42
Bristow CS, Jol HM, Augustinus P, Wallis I (2010b) Slipfaceless whaleback dunes in a polar desert, Victoria Valley, Antarctica: insights from ground penetrating radar. Geomorphology 114:361–372
Buynevich IV, FitzGerald DM, van Heteren S (2004) Sedimentary records of intense storms in Holocene barrier sequences, Maine, USA. Mar Geol 210:135–148
Campbell CV (1967) Lamina, laminaset, bed and bedset. Sedimentology 8:7–26
Cassidy N (2009) Ground penetrating radar data processing, modelling and analysis. In: Jol HM (ed) Ground penetrating radar: theory and applications. Elsevier, Amsterdam, pp 141–176
Cooper JAG (2012) Mesoscale geomorphic change on low energy barrier islands in Chesapeake Bay, U.S.A. Geomorphology. doi:10.1016/j.geomorph.2012.06.019
Davis RA (1992) Depositional systems: an introduction to sedimentology and stratigraphy. Prentice Hall, Englewood Cliffs, 591 p
Davis RA, FitzGerald DM (2004) Beaches and coasts. Blackwell, Oxford, 419 p
Dietz C, Heck H-L (1952) Geologische Karte von Deutschland, Erläuterungen zu den Blättern Sylt-Nord und Sylt-Süd. Landesanstalt für Angewandte Geologie Kiel, 123 p
Evans OF (1942) The origin of spits, bars and related features. In: Schwartz ML (ed) Spits and bars. Dowden, Hutchinson and Ross, Stroudsburg, pp 53–72
Folk RL, Ward WC (1957) Brazos River bar (Texas): a study in the significance of grain size parameters. J Sediment Petrol 27:3–26
Girardi JD, Davis DM (2010) Parabolic dune reactivation and migration at Napeague, NY, USA: insights from aerial and GPR imagery. Geomorphology 114:530–541
Harari Z (1996) Ground-penetrating radar (GPR) for imaging stratigraphic features and groundwater in sand dunes. J Appl Geophys 36(1):43–52
Hayes MO (1979) Barrier island morphology as a function of tidal and wave regime. In: Leatherman SP (ed) Barrier islands from Gulf of St. Lawrence to the Gulf of Mexico. Academic, New York, pp 1–27
Hinz H (1968) Die Burgwälle auf den Nordfriesischen Inseln. In: Römisch-Germanisches Zentralmuseum Mainz (ed) Führer zu vor- und frühgeschichtlichen Denkmälern 9, Römisch-Germanisches Zentralmuseum Mainz, Mainz, pp 108–111
Hoffmann D (1974a) Aufbau und Alter der Marsch im Kern der Insel Sylt. Sonderdruck aus: Bericht der Römisch-Germanischen Kommission 55(2):358–378
Hoffmann D (1974b) Zum geologischen Aufbau der Hörnumer Halbinsel auf Sylt. Meyniana 23:63–68
Hundt C (1957) Die Abbruchursachen an der Nordwestküste des Ellenbogens auf Sylt. Die Küste 6(2):5–37
Jol HM, Bristow CS (2003) GPR in sediments: advice on data collection, basic processing and interpretation, a good practice guide. In: Bristow CS, Jol HM (eds) Ground penetrating radar in sediments, Geological Society, special publications, 211. Geological Society, London, pp 9–27
Kelletat D (1992) Coastal erosion and protection measures at the German North Sea Coast. J Coast Res 8(3):699–711

Kersten K (1967) Vorgeschichte der Insel Sylt. In: Hansen M, Hansen N (eds) Sylt: Geschichte und Gestalt einer Insel. Verlag Hansen & Hansen, Itzehoe, pp 11–34

Leatherman SP, Zaremba RE (1987) Overwash and aeolian processes on a U.S. Northeast Coast Barrier. Sed Geol 52:183–206

Leatherman SP, Williams AT, Fisher JS (1977) Overwash sedimentation associated with a large Northeaster. Mar Geol 24:109–121

Lindhorst S, Betzler C, Hass HC (2008) The sedimentary architecture of a Holocene barrier spit (Sylt, German Bight): Swash-bar accretion and storm erosion. Sed Geol 206:1–16

Lindhorst S, Fürstenau J, Hass HC, Betzler C (2010) Anatomy and sedimentary model of a hooked spit (Sylt, southern Northsea). Sedimentologija 57:935–955

Miall AD (1991) Hierarchies of architectural units in terrigenous clastic rocks, and their relationship to sedimentation rate. In: Miall AD, Tayler N (eds) The tree-dimensional facies architecture of terrigenous clastic sediments and its implications for hydrocarbon discovery and recovery, vol 3, Concepts in sedimentology and palaeontology. SEPM, Tulsa, pp 6–12

Mitchum RM, Vail PR, Sangree JB (1977) Stratigraphic interpretation of seismic reflection patterns in depositional sequences. In: Payton CE (ed) Seismic stratigraphy: applications to hydrocarbon exploration, AAPG Memoir 16, Payton, CE, Tulsa, Oklahoma, pp 117–123

Neal A (2004) Ground-penetrating radar and its use in sedimentology: principles, problems and progress. Earth Sci Rev 66:261–330

Neal A, Richards J, Pye K (2003) Sedimentology of coarse-clastic beach-ridge deposits, Essex, southeast England. Sed Geol 162(3–4):167–198

Newig J (1995) Zur langfristigen Gestaltänderung der Insel Sylt. Kölner Geographische Arbeiten 66:121–138

Newig J (2001) Rantum auf Sylt unter dem Einfluss von Küstenrückgang und Sandwanderung. Vechtaer Studien zur Angewandten Geographie und Regionalwissenschaft 22:17–33

Nielsen LH, Johannessen PN, Surlyk F (1988) A Late Pleistocene coarse-grained spit-platform sequence in northern Jylland, Denmark. Sedimentology 35:915–937

Panten A (2000) Die schwersten Sturmfluten an der deutschen Nordseeküste. In: Newig J, Theede H (eds) Sturmflut: Gefährdetes Land an der Nordseeküste. Ellert & Richter Verlag, Hamburg, pp 56–75

Pilkey OH, Cooper JAG, Lewis DA (2009) Global distribution and geomorphology of fetch-limited barrier islands. J Coast Res 25:818–837

Prothero DR, Schwab F (1996) Sedimentary geology: an introduction to sedimentary rocks and stratigraphy. W.H. Freeman and Company, New York, 567 p

Reichstein J (2005) Jahrtausendlange Bedrohung des Lebens auf Sylt: Ergebnisse archäologischer Siedlungsforschung. In: Jessel H (ed) Das große Sylt Buch. Ellert & Richter Verlag, Hamburg, pp 92–107

Schwartz RK (1982) Bedform and stratification characteristics of some modern small-scale washover sand bodies. Sedimentology 29:835–849

Tillmann T, Wunderlich J (2011a) Facies and development of a holocene barrier spit (Southern Sylt/German North Sea). In: Proceedings of the 6th international workshop on advanced ground penetrating radar, IWAGPR 2011, 22–24 June 2011, Aachen, Germany, pp 188–194

Tillmann T, Wunderlich J (2011b) Genese eines Strandhakens am Beispiel der Hörnum-Odde (Süd-Sylt): Untersuchungen des oberflächennahen Untergrundes durch die Kombination von geophysikalischen und sedimentologischen Methoden. Coastline Rep 17:177–190

Tillmann T, Wunderlich J (2012) Ground-penetrating radar in coastal environments: examples from the islands Sylt and Amrum. In: Vött A, Venske J-F (eds) Bremer Beiträge zur Geographie und Raumplanung, 44, pp 60–76

Tillmann T, Wunderlich J (2013) Barrier rollover and spit accretion due to the combined action of storm surge induced washover events and progradation: insights from ground-penetrating radar surveys and sedimentological data. J Coast Res Spec Issue 65:600–605 doi: 10.2112/SI65-102.1

Tillmann T, Ziehe D, Wunderlich J (2013) Holozäne Landschaftsentwicklung an der Westküste der Nordseeinsel Amrum. Quat Sci J 62(2):98–119

Uda T (2005) Bars. In: Schwartz ML (ed) Encyclopedia of coastal science, Schwartz, ML, Springer, Dordrecht, pp 909–912

Van Dam RL (2012) Landform characterization using geophysics – recent advances, applications, and emerging tools. Geomorphology 137(1):57–73

Van Dam RL, Schlager W, Dekkers MJ, Huisman JA (2002) Iron oxides as a cause of GPR reflections. Geophysics 67(2):536–546

Van Overmeeren RA (1998) Radar facies of unconsolidated sediments in The Netherlands: a radar stratigraphy interpretation method for hydrogeology. J Appl Geophys 40:1–18

Wang P, Horwitz MH (2007) Erosional and depositional characteristics of regional overwash deposits caused by multiple hurricanes. Sedimentology 54(3):545–564

Chapter 18
Morphology and the Cyclic Evolution of Danube Delta Spits

Alfred Vespremeanu-Stroe and Luminiţa Preoteasa

Abstract This chapter is about the sand spits formation in relation with the evolution and morphology of the Danube mouths and deltaic lobes, developed within the general framework of the virtually tideless Black Sea basin in a medium-wave energy environment. It deals with both barrier spits and islands, in respect to their processual development as most of the Danube delta spits derive from former barrier islands. The morphodynamics of the modern spit barriers associated to Danube mouths is discussed in relation with human induced fluvial discharge decline and long-term storminess variability. Regarding the Danube delta evolution, based on the morphology, internal structure and revised chronologies, the authors argue for the first time that all open-coast deltaic lobes had or have an evolution marked by the cyclic development of sandy barrier spits and islands in front of their downdrift units.

18.1 Introduction

A description of the modern barrier spits and islands morphodynamics, followed by an overview of sand spits development as a cyclic sequence of the wave-influenced deltaic lobes evolution ends up with the analysis of the imprints these morphological features leave in the deltaic lobes architecture to conclude upon the key morphogenetic role they play in organizing large scale coastal units such as deltas. This is the particular case of the barrier islands and sand spits developed at the mouths of the Danube's distributaries where individual spits barriers are documented to have been formed since the last post-glacial stabilization of the sea level about 5,000 years ago (Giosan et al. 2006) and which continues to form (Dan et al. 2011).

Most of the successive spits generations constructed during this interval are nowadays preserved as barrier-marsh plain within the downdrift wings of deltaic lobes; in places, morphological evidences (e.g. beach ridges, dune ridges or truncation ridges) stand as important proxies for reconstructing the morphodynamics of the Danube delta. The attempt to understand the evolution of such fragile

A. Vespremeanu-Stroe • L. Preoteasa (✉)
Faculty of Geography, University of Bucharest, Bucharest, Romania
e-mail: luminita@geo.unibuc.ro

G. Randazzo et al. (eds.), *Sand and Gravel Spits*, Coastal Research Library 12,
DOI 10.1007/978-3-319-13716-2_18

morphological features recently resulted in several conceptual models of a sand spit formed at a river mouth. Bhattacharya and Giosan (2003) based on qualitative analyses developed the theoretical model of the barrier island construction proposed by Vespremeanu (1983), Dan et al. (2011) modeled the wave-induced sediment circulation along and cross-shore an idealized ellipsoidal spit formed at a river mouth and assessed the major role played by overwash in Sacalin spit evolution while a recent quantitative approach (Preoteasa et al. submitted) documents the cyclic evolution of the spit barriers development based on geochronological and sedimentological data interpretation.

18.2 Danube Delta: Regional Settings

The Black Sea is one of the largest semi-enclosed seas in the world (0.46×10^6 km^2) delineated by a 4,350 km long shoreline with a highly diverse morphology related to different geological settings and wind/wave climates with a marked contrast between northwestern region (low-lying coast and macroenergetic wind regime) and southeastern region (rocky coasts with low-wind/wave energy). The most favorable conditions for barrier spits formation are primarily encountered in the northwest Black Sea basin where some of the largest European rivers converge to the sea: Danube with a medium water discharge of about 200 km^3/year and the East European rivers (Dnieper, Southern Bug and Dniester) amounting to 66 km^3/year, which altogether cumulate 80 % of the fresh water discharge into the Black Sea (Bulgakov and Yurkova 1999). Consequently, the largest width of the Black Sea continental shelf is located in front of the Danube delta (over 200 km wide). A characteristic of the sand spits developed in this region is the wide spectrum of shapes, ages and orientations, particularly in the Dnieper – Bug – Dniester river mouth area and the regular pattern of barrier island and spits formation in front of the downdrift part of the Danube's distributaries mouths.

Danube is 2,870 km long and the watershed extends over 817,000 km^2. Its delta develops in the north-western part of the Black Sea where the river flows into the sea through three main branches from north to south (Fig. 18.1): (1) Chilia, which transports approximately 58 % of the water and sediment discharge, (2) Sulina, the major waterway, 19 %, and (3) Sf. Gheorghe, 23 %, (Bondar and Panin 2001).

The wave climate along the Danube delta is medium-energy with a mean offshore significant wave height of 1.43 m, increasing to 2–5 m during regular storm events; the maximum wave height exceeds 5 m during severe storms. The mean offshore wave period is about 5.5 s reflecting the short and medium fetches (<250 km) specific to prevalent northern and eastern winds. The waves arrive obliquely to the shoreline (e.g. the mean angle formed between the shoreline and the wave crest is of 0–60°) yielding a strong longshore currents which in most sectors transport $0.65–1 \times 10^6$ m^3/year of sediments (Vespremeanu-Stroe 2004, 2007; Dan et al. 2007; Fig. 18.2a, b). In front of the Musura, Bistroe and Sacalin spits the dominant northeasterly waves approach shoreline at acute

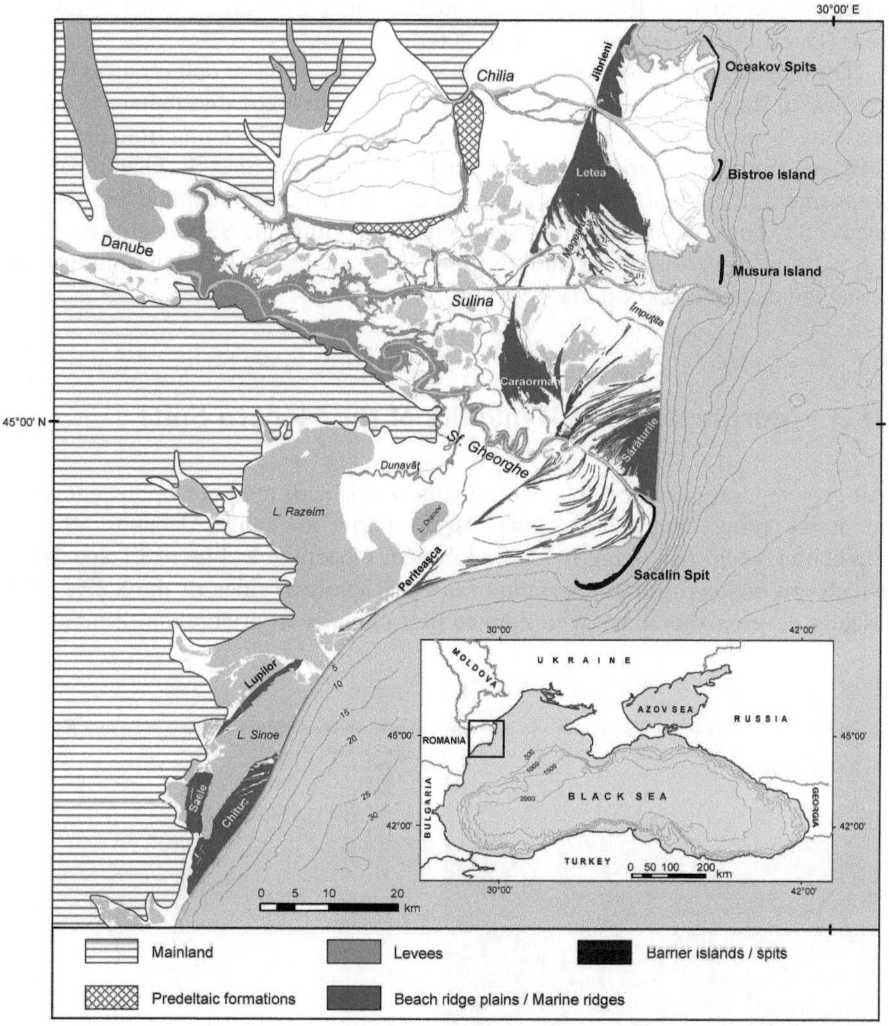

Fig. 18.1 Danube delta geography. The *inset* indicates the location of the Danube delta in the NW Black Sea

angle producing a net longshore sediment transport (LST) with maximal values of $0.8–1 \times 10^6$ m^3/year resulted from three times more sediment drifted southward than in a reverse direction (Vespremeanu-Stroe 2004). The monthly multiannual analysis of the LST shows a highly skewed distribution with ca. 60 % occurring during winter, and 8 % during summer.

The Black Sea has a microtidal regime with amplitude of 0.12 m (Bondar and Panin 2001) which has a neglectful morphologic influence. Black Sea level is recorded at Sulina gauge since 1858. Data analysis shows a relative sea level rise rate of 2.47 mm/year (Fig. 18.2c); otherwise, the eustatic oscillations are smaller if taken into account the local deltaic subsidence which measures about 1.5–1.8 mm/year (Panin 1999; Vespremeanu et al. 2004).

The biggest and oldest modern barrier spit from the Danube delta coast is Sacalin spit which started to form at the mouth of the Sf. Gheorghe distributary as part of the actively prograding deltaic lobe, with the sediment load coming from the littoral drift and from the riverine input. Recent estimates show that St. Gheorghe solid discharge is about 0.80×10^6 m^3/year (Panin and Jipa 2002), comparable with the local LST. Thus, the mean annual sediment input within the Sacalin area is of 1.5–1.8×10^6 m^3/year, the spit extending within the distal part of the biggest littoral cell along the deltaic coast (Sulina – Sacalin).

18.3 Morphology and Dynamics of the Modern Spits

All contemporary Danube delta barrier spits and islands are built-up in connection to active prograding deltaic lobes (from north to south: Oceakov, Bistroe, Stambulul Vechi belonging to Chilia lobe, Sf. Gheorghe 2; Figs. 18.1 and 18.2a) which are major features protruding seaward. The two sides of the active lobes usually develop shorelines with marked discontinuity in orientation, generating a

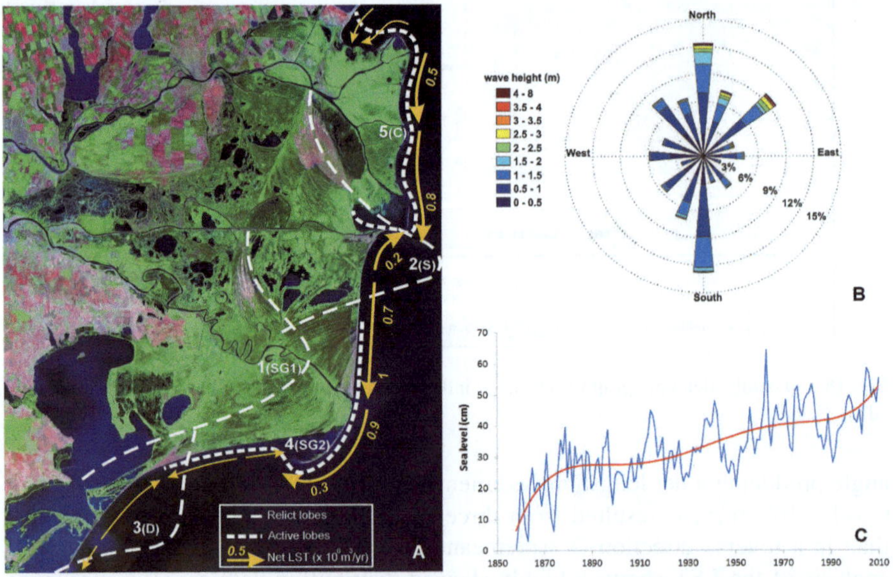

Fig. 18.2 (**a**) Delineation of the active and relict open-coast deltaic lobes: *SG1* Sfântu Gheorghe 1, *S* Sulina, *D* Dunavăţ, *SG2* Sfântu Gheorghe 2, *C* Chilia; the numbers express their chronological succession (Modified after Panin 1983 and Vespremeanu-Stroe et al. 2013). Net LST along the present coastline (values from Vespremeanu-Stroe 2004 and Dan et al. 2007); (**b**) Directional frequency distribution of wave height (H_o) at 6 km offshore on Danube delta coast (1991–2008); (**c**) Sea level oscillations at Sulina gauge (1858–2010)

negative gradient in longshore sediment transport and consequently the appearance and elongation of the sandy spits in front of the downdrift lobe (e.g. Damietta – Nile delta, New Brazos).

Due to the high shoreline mobility, Danube delta barrier spits and islands are very low, with mean heights of 0.5–0.8 m which allow frequent (annual) overwash processes and significant backwards migration (up to 60 m per event) during severe storms. The back roll-over migration preserves the general cross-shore morphology with low and moist beaches and frequent salt crusts that inhibit the aeolian transport activity. Excepting the Sacalin spit, on the others spits and islands the sand dune are missing with very rare exceptions when small incipient foredune develop during the (severe) storm-free periods (ex. 2006–2011) on the relatively stable sectors. In the back-beach, the storm waves create successive generations of wash-over fans, with most of them northeast-southwest orientated, preserving the dominant storm wave direction. All barrier spits and islands are frequently breached with inlets development depending on the storm climate and Danube discharge which are also the main factors of the local sea-level oscillations. As a direct consequence, during the extraordinary high Danube discharges of 2005–2006 interval the total number of inlets increased threefold (14 vs. 5).

The present secondary delta of the northern Danube distributary (Chilia) which collects ≈60 % of water and sediment discharge (Bondar and Panin 2001) has a bird-foot morphology developed at a rapid pace during the last 200 years with mean rates up to 2.2 km^2/year in the 1830–1922 period (Brânduş and Canciu 2011) after which decreased with more than 50 % (1 km^2/year in the 1922–2013 interval) concomitant with the appearance in the twentieth mid-century of the first wave-build features: sandy beaches and spit barriers. During last decades (1980 – present), the marine features became prevalent marking the gradual transition from a river-dominated to a wave-influenced lobe morphology. The three main branches/mouths of Chilia secondary delta (Oceakov, Bistroe and Stambulul Vechi) formed barrier islands and spits which nowadays are about to close the adjacent arm mouths (Akundinovo, Vostok, Musura) through downdrift elongation and to simplify the hydrographic network. Oceakov, the northernmost branch of the Chilia arm (Danube), advances towards northeast, fronts almost perpendicular the storm waves and it builds the only symmetrical lobe of the Danube. In fact, during the last 200 years the main channel changed the flow direction from NNE to ENE under the stress of the dominant northern waves. On the two sides of the lobe, the waves approach the coast at acute angle and generate instabilities in coastline behavior (Ashton and Giosan 2011). As a consequence, both lobe flanks are fronted by flying spits, the northern one being more fragile and wave-exposed, suffering from frequent breaching during storms while the southern spit is less mobile and wider.

Bistroe arm mouth built up its first barrier island at the end of 1990s followed by a fast southward elongation until 2006 when (it) fronted the next downdrift river mouth (Akundinovo). Dredging works started in 2004 interrupted the alongshore sedimentary bypass over the Bistroe mouth which reflected in a sudden decrease in elongation and acceleration of the backwards migration (from 18 to 32 m/year, prior and after 2006). Even though, the significantly reduced elongation process of the barrier, supplied by central sector cannibalization, laterally stresses the Akundinovo mouth and determines its southward deflection with the potential risk of closure.

Stambulul Vechi and Sfântu Gheorghe arm mouths present similarities in terms of orientation (southeast), water and sediment discharge (20–25 % of Danube), incident wave angle and LST magnitude ($0.8–1 \times 10^6$ m^3/year) which presently impose a common general aspect of their lobes, with a sandy, north-south drift-aligned updrift coast and an irregular muddy/organic downdrift coast, east-west developed in fetch-limited conditions: lagoons and the shadow-zone of Sacalin spit. Stambulul Vechi is much younger than Sfântu Gheorghe, and its evolution seems to become similar with the later, changing from a fluvial-dominated lobe to an asymmetric wave-influenced lobe whose downdrift wing evolves by a cyclic development of barrier islands and spits.

Both arms mouth build shallow subaqueous platforms (with depths lower than 2 m) at the shelter of the seaward-positioned updrift coast, whose outer edge starts to emerge after major floods and transforms into a new barrier island which evolves by constant elongation and landward migration until it attaches the northern tip to the bayhead secondary delta becoming a sandy barrier spit. In this context, the present barriers are at different stages: Musura is a young feature while Sacalin already became a mature spit.

After emergence, the new island reinforces the longshore sediment circulation which moves the main depocenter (or creates a new one) to the southern tip, favoring the rapid downdrift extension of both the island and the subaqueous platform. It is the case of Musura Island which started to emerge in the late 1980s as a series of drift-aligned islets and then merged and extended southward with a mean rate of 176 m/year (1996–2013); presently, the southern end migrates with 150 m/year (2010–2013) reaching to just 200 m near the Sulina arm jetties (Fig. 18.3a). Between Sulina jetties and the island, a flow channel of the Musura arm is still active, but will probably be soon overwhelmed by longshore drift-sourced sediments transforming the barrier into a spit connected to the Sulina jetties, whereas the Musura waters flow will switch northward continuing to divide the barrier by several mobile inlets.

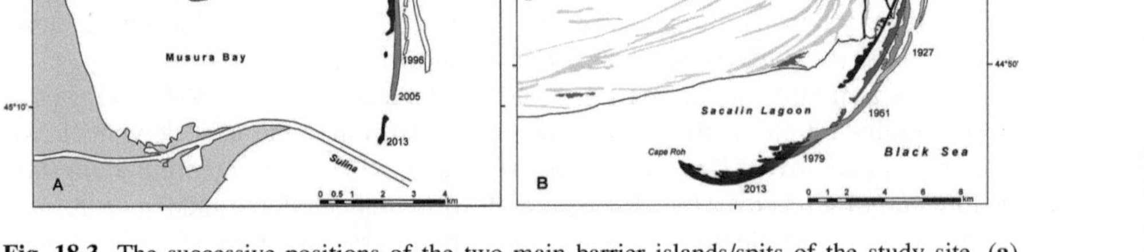

Fig. 18.3 The successive positions of the two main barrier islands/spits of the study site. (**a**) Musura barrier island (1988–2013) and (**b**) Sacalin spit barrier (1898–2013)

18.3.1 Sacalin Spit

Sacalin barrier spit formed at the southernmost Danube mouth (Sfântu Gheorghe arm), representing the youngest downdrift feature of the Sfântu Gheorghe deltaic lobe (1500 BP – present). Sacalin Island emerged at the end of nineteenth century following a major flood interval (1895–1897) and developed by constant southward elongation (Ionescu-Dobrogeanu 1909). Eighty years after the emergence, Sacalin evolved into a sandy barrier spit (1985), with the northern tip attached to the secondary delta, following a gradual backward translation (lateral migration) produced by overwash during storms. The high shoreline mobility of the barrier is mainly driven by coastal storms and associated processes: longshore and cross-shore sediment transport, breaching and overtopping, wash-over fan building. The barrier island/spit was frequently breached in the central part (narrow inlets) and episodically it experiences large elongation and retreat rates, up to 300 m/year and respectively 60 m/year.

The barrier morphology and dynamics changed in time depending on the state (island or spit), the cross-shore positions or plan-shape and length which resulted in different exposures to wave climate. During first stage (until the end of 1920s) the island was very exposed, occupying a 1.7 km seaward position than the present, and usually split in two or more islets. The wave – fluvial current interactions imposed a complex morphodynamics of the barrier with repeated advances and retreats and new localized emersions of the platform outer bar crests which lagged the welding of the shoals and islets into a firm barrier island. Consequently, the downdrift elongation was moderate (120–150 m/year;) whereas the backwards migration started later, around 1910, at very slow rates (<5 m/year) due to a high water discharge specific to 1910–1920 wet period and until the Danube flow behind the island significantly decreased in the favour of the northern channel. Following a rapidly evolving intermediary-stage, from 1930s to the early 1960s (Fig. 18.3b), when relatively high rates occurred for both the cross-shore and alongshore migration (27 m/year, respectively 189 m/year, with ca. 30 % bigger than the mean rates of 20.5 m/year, respectively of 143 m/year), the frequency of the severe storms significantly increased during the late period of the island stage (1960s and 1970s). Storm reactivation overlapped on the strong negative North Atlantic Oscillation phase and promoted the highest retreating rates of the central Sacalin (37 m/year). As a result, the barrier was split in five to six islets divided by large inlets (100–400 m widths). During the island stage, the central part of the barrier remained narrow and low, susceptible to breaching and overtopping, while both the island ends stayed wide and curved by diffraction, with the northern one less mobile and usually accommodating free-vegetation sand dune; the southern end is highly dynamic with a net tendency to become wider by the juxtaposition of the previous tips and the initiation of temporary beach ridge plains (persisting several decades).

Once transformed into a spit, Sacalin morphology evolved from a symmetrical arch, which defined the main aspect of the former island, to an asymmetric increasingly downdrift curved barrier (Fig. 18.3b). The central part is almost linear with an aspect of N 125°, perpendicular on the river flow direction, and records the highest shoreline mobility with frequent inlets. Nowadays, the Sacalin spit barrier measures about 21 km length and is composed of very well sorted fine siliciclastic material and a small fraction of residual organic matter drifted from inner delta. Spit width increases from north to south in accordance with the back-barrier accommodation space being limited by the bayhead delta in the north, forming large washover fans in the center due to shallow depths (<2 m) of the lagoon and reaching highest widths in the southern part where the former distal ends of barriers are preserved as subparallel ridges and create small beach ridge plains of 300–600 m width. The northern quarter of the barrier rolls backward over a reedmarsh bed owing to the Sf. Gheorghe bayhead delta while the rest of the barrier rolls over a shallow lagoon. Retreating rates are mainly dependent on the accommodation space and barrier height which controls the overwash. During the last decade (2000–2013) characterized by low storm frequency and high Danube discharge, the northern sector where the spit joined the mainland become quasi-stable (in fact a succession of slightly accumulative and eroding sectors), the central sector remained erosive with retreating rates of 12–20 m/year smaller than before, while the 7-km long southern sector where the island arcuates the most continued the accretion regime established in the late 1980s. The erosive hotspot of the barrier is placed just downdrift from the spit detachment point from delta plain and migrates southward, parallel with the advance of the bayhead delta front.

18.4 Cyclic Spit Development and the Significance for Deltaic Lobes Evolution

Spit barriers are assumed to play a key role in the development of Danube delta, as most of the hypotheses regarding its evolution, despite the differences, have commonly assumed the appearance during the mid-Holocene of an initial-spit which blocked the Danube gulf and favored the rapid extension of a bay-head delta (Antipa 1917; de Martonne 1931; Zenkovich 1956; Coteţ 1960; Diaconu and Mihailov 1963; Panin 1974, 2003; Giosan et al. 2006). The age and evolution of the initial-spit are still in debate as the first C14 datings (7500–11700 BP; Panin 1983, 2003) seem to be affected by reworked samples (i.e. single-valve shells), whereas the only so far reported OSL age indicate an younger age of 5200 BP (Giosan et al. 2006) which still differs from the interpretation the recent geo-archaeological findings at Taraschina Neolithic settlement (e.g. 6400 BP the minimum age of the spit, cf. Carozza et al. 2012).

Thereafter, a succession of five deltaic lobes developed seaward from the initial-spit during the last six millennia composes the so-called "marine delta": Old Sf. Gheorghe (SG1), Sulina, Dunavăţ, Modern Sf. Gheorghe (SG2) and Chilia (listed in chronological order). Excepting the oldest lobe (Sf. Gheorghe 1) which downdrift side is completely subsided today and cannot be directly explored and the newest lobe (Chilia) which recently built its first generation of sandy spits, the others had or still have an evolution marked by the cyclic development of sand spits in front of their downdrift part.

The morphology and the internal structure of the five deltaic lobes indicate a common morphodynamic model specific to the asymmetric wave-influenced river mouth (Bhattacharya and Giosan 2003). They commonly develop an accretionary updrift compartment creating a firm beach ridge plain and a downdrift unit composed by a subparallel series of sandy ridges encased in delta plain (reed marsh or lagoons). These ridges originate as barrier islands or spits on the river mouth bar and then migrates landward until they attach to the delta plain, usually to a bayhead delta developed in the backbarrier lagoon. Once a new cycle starts, with the emergence of another barrier island or spit, the former barrier remains sequestered within the delta plain as an isolated sandy ridge or as a ridgeset composed by several spits ramified from the initial spit barrier.

18.4.1 Sulina and Dunavăţ Lobes

Sulina lobe formed during 3400–2000 BP (Giosan et al. 2006). Its rapid expansion, characterized by a mean progradation rate of 22.4 m/year (double than Sf. Gheorghe lobe ≈10 m/year), recorded a gradual acceleration up to 30 m/year during the later centuries (2700–2000 BP) most probable due to the intensified human-induced land use change within Danube watershed during late Iron Age and Roman Empire (Giosan et al. 2012; Maselli and Tricardi 2013). The updrift wing is composed by a large beach ridge plain with two distinct units (low/high in western/eastern part) corresponding to different progradation rates which enabled or hampered the foredune development (Fig. 18.2a). During most of its lifespan Sulina lobe had three main channels – from north to south: Magearu, Sulina and Împuţita – excepting the final stage (2500–2000 BP) when the northern distributary (Magearu) was abandoned concomitant with the initiation of Letea dunefield (Preoteasa et al. 2009). Barrier islands and spits developed on the downdrift parts of the Magearu and Împuţita secondary distributaries but never in front of the river-dominated delta of the main Sulina channel. The faster the river mouth advances seaward, the lower the barrier spit frequency and bigger the lagoons developed within the downdrift lobeside. Magearu which advanced at slower rates (≈9 m/year) left the morphological imprint of 11 cycles of barrier island/spit during ca. 600 years contributing to the formation of a rhythmic barrier plain, whereas the higher rates of Împuţita distributary (≈21 m/year) imposed a fast growth of the downdrift lobe with big lagoons enclosed by island/spit barriers built during only four cycles which occurred during 2800–2000 BP interval.

Southward of Sf. Gheorghe deltaic lobes (1 and 2) extends the young southern Danube delta which is distinct in terms of altimetry, morphology and chronology. This unit built during the last 2,000 years under the influence of rapidly changing distributary courses, ancient deltaic lobes sediment reworking and active neotectonics (subsidence) which generate large lakes delineated by beach ridge plains and narrow barriers which are either remnants of the ancient subsided deltaic lobes or lacustrine barriers. Once with the sudden flow decrease on Sulina branch the former southern branch of Danube (Sf. Gheorghe) was reinforced and it started to develop a new lobe (Dunavăţ lobe) during 2000–1400 BP following a southward track (Dunavăţ channel; Preoteasa et al. 2013; Fig. 18.2a). The central and outer parts of the Dunavăţ lobe are affected by intense neotectonic movements which led to their recent drowning and the formation of tectonic lakes. The only remnants of the former barrier spits are currently visible at the western side of the lobe as Lupilor barrier. It shares the same evolutionary pattern with Crasnicol and Periteaşca barriers which similarly formed on the downdrift wing of the Sf. Gheorghe 2 contemporary lobe. All these fan-like barriers converge at one or several downdrift points, expressing an evolutionary signature specific to the downdrift compartment of the asymmetric wave-dominated deltaic lobes of the Danube delta. In fact, all the sub-parallel sandy ridges that originated as barrier islands and spits or beach barriers are entrapped within large volumes of fluvial sediments (mostly muds and fine sands) imposing the rapid downdrift extension of the deltaic lobe and the building up of a barrier-marsh plain (Preoteasa et al. submitted).

18.4.2 Sf. Gheorghe 2 Lobe

The modern Sf. Gheorghe lobe is the only in the Danube delta where spits formed continuously during the last two millenia offering the opportunity to examine their initiation and morphodynamics and to assess the role they play in the large scale coastal development. Sf. Gheorghe lobe is the last feature of the old southern Danube branch (Panin 2003). The presence of many fossil beach ridges sets resembling former spits (Fig. 18.4) suggests that modern Sf. Gheorghe lobe has an evolution controlled by the successive formation of sand spits. It started to form 1,500 years ago when Sf. Gheorghe branch suddenly turned its main flow to the south-east in change of Dunavăţ channel; moreover, its mean sedimentation rate (7–9 Mt/year) is similar to that of the abandoned Dunavăţ lobe which together with their chronological succession is an evidence of the lobe switching of the Sf. Gheorghe southern Danube arm (Preoteasa et al. submitted). Recent investigations revealed the succession of ten evolutionary cycles deployed as a four-phase morphodynamic model (Figs. 18.4 and 18.5), common for each cycle: (i) building-up of an extensive subaqueous deltaic platform in front and downdrift of the river mouth, (ii) the emergence of the platform outer edge as a new barrier island, (iii) conversion into a sandy spit through landward roll-over of the island and attachment to the bayhead delta of the updrift end of the barrier, (iv) embedding of the spit barrier within delta plain as a beach ridge (Preoteasa et al. submitted).

During the first phase of intense subaqueous accumulation, a positive feedback sets once the platform extends offshore; the waves energy reaching the coast is now better dissipated enabling a more efficient entrapment of sediments on the platform (Giosan 2007). As a result, the platform becomes strongly seaward offset and

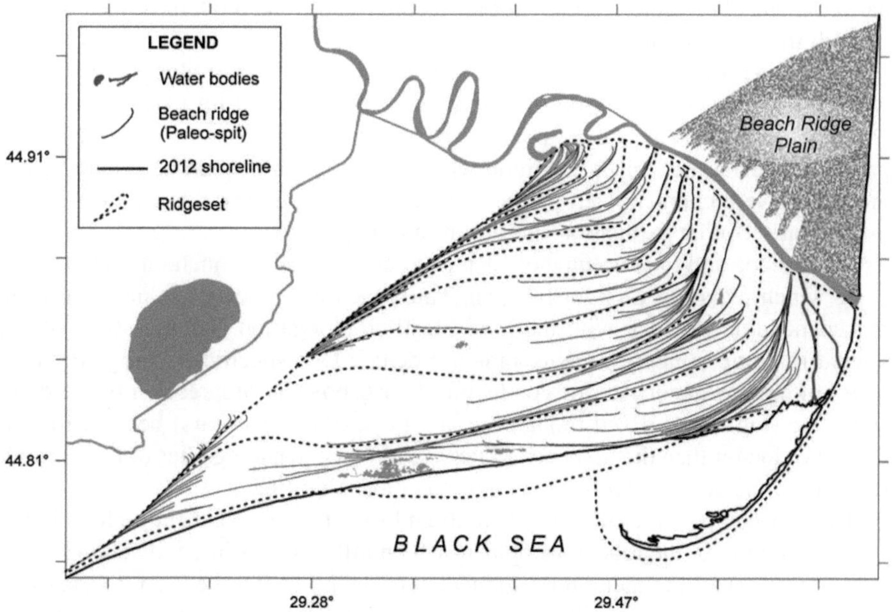

Fig. 18.4 Line drawing of the beach ridges (former barrier spits) grouped in ten ridgesets composing the downdrift part of the Sf. Gheorghe 2 deltaic lobe

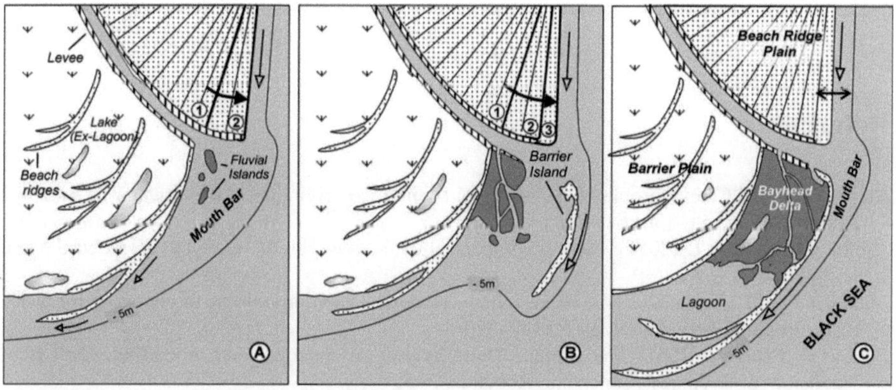

Fig. 18.5 Theoretical model of spit barrier formation through time and its contribution to an asymmetric wave-dominated deltaic lobe development. (**a**) Subaqueous platform growth and subsequent appearance of middle bars, (**b**) the emergence of a barrier island followed by rapid elongation and backwards migration, (**c**) spit barrier merging with the mainland and the construction of secondary spits at a downdrift location (Modified from Preoteasa et al. submitted)

progressively wave-exposed enabling the accommodation of a bar at the outer edge. The emergence of the bar and construction of a distinct barrier island is not an instantaneous event. It usually happens after a significant flood and takes several years necessary for the storm waves to rework the sediments and raise the bar height over the mean sea level in a few sectors to finally merge alongshore into a firm barrier island. The same gradual process records for all the modern islands born in front of the Danube mouths.

Once emerged, the barrier island evolves through constant elongation (southward) and landward migration until it attaches the northern tip to the delta plain (bayhead delta) and becomes a sandy barrier spit. The spit development cycles produced during the last millennium, when Danube solid discharge progressively increased, were characterized by the construction of complex ridgesets composed by several secondary spits branched from initial spit-barrier.

The surface and total volume of each ridgeset increase exponentially with every cycle which reclaims a longer lifespan. Thus, each new cycle lasts more than the precedent in accordance with the lobe coastline elongation. The coastline of the downdrift compartment of the asymmetric deltaic lobe specifically progrades as a series of similar quadrilaterals conducting to a geometric progression of the delta front size (Fig. 18.4). Even though the latest cycles ($\approx$350 years) became four to five times longer than the first ones (50–80 years) the advancing rates of river mouth remained constant of 10 m/year (Preoteasa et al. submitted).

The constant renewal of barrier islands and spits proves to be a key-phase in the cyclic evolution of the lobe development which influence sediment dynamics since they appear: (i) an exponential strengthening of the LST after the barrier formation, (ii) frequent initiation of sediment bypass over the river mouth, and (iii) sheltering of the downdrift coast which promotes extensive lagoon initiation, which either follow rapid siltation via secondary bayhead delta or a slow degradation into several lakes through bioaccumulation.

References

Antipa G (1917) Câteva probleme Ştiinţifice şi economice privitoare la Delta Dunării. An Acad Rom Mem Sect St 2(36):61–135

Ashton AD, Giosan L (2011) Wave-angle control of delta evolution. Geophys Res Lett 38(13): L13405

Bhattacharya JP, Giosan L (2003) Wave-influenced deltas: geomorphologic implications for facies reconstruction. Sedimentology 50:187–210

Bondar C, Panin N (2001) The Danube Delta hydrologic database and modelling. Geo- Eco-Marina, 5–6. Bucharest-Constanţa, Romania, 5–52

Brânduş C, Canciu C (2011) On the geomorphologic evolution of the Chilia secondary delta. Present Environ Sustain Dev 5(2):101–110

Bulgakov NP, Yurkova IY (1999) Contemporary state of the investigations of the influence of the discharge of rivers on the hydrologic structure of the Black Sea. Phys Oceanogr 10(6):541–554

Carozza JM, Micu C, Mihail F, Carozza L (2012) Landscape change and archaeological settlements in the lower Danube valley and delta from early Neolithic to Chalcolithic time: a review. Quat Int 261:21–31

Coteţ P (1960) Evoluţia morfohidrografică a Deltei Dunării (O sinteză a studiilor deja existente şi o nouă interpretare). Probleme de geografie VII:97–108

Dan S, Stive MJF, Walstra DJR, Sabatier F (2007) Sediment budget of the Danube delta coastal zone. In: Proceedings of *coastal sediments 2007*. ASCE, New Orleans, USA, pp 207–220

Dan S, Walstra DJR, Stive MJF, Panin N (2011) Processes controlling the development of a river mouth spit. Mar Geol 280:116–129

de Martonne E (1931) Europe centrale, vol II. A. Colin, Paris, pp 786–787

Diaconu C, Mihailov VN (1963) Procesele formarii deltei in etapa actuala. In: Diaconu C, Nichiforov ID (eds) Zona de vărsare a Dunării. Editura Tehnică, Bucharest, pp 192–222

Giosan L (2007) Morphodynamic feedbacks on deltaic coasts: lessons from the wave-dominated Danube delta. In: Proceedings of *coastal sediments 2007*. ASCE, New Orleans, USA, pp 828–841

Giosan L, Donnelly JP, Constantinescu Ş, Filip F, Ovejanu I, Vespremeanu - Stroe A, Vespremeanu E, Duller GAT (2006) Young Danube delta documents stable Black Sea level since the middle Holocene: morphodynamic, paleogeographic, and archaeological implications. Geology 34(9):757–760

Giosan L, Coolen MJL, Kaplan JO, Constantinescu S, Filip F, Filipova-Marinova M, Kettner AJ, Thom N (2012) Early anthropogenic transformation of the Danube – Black Sea system. Sci Rep 2:582. doi:10.1038/srep00582

Ionescu-Dobrogeanu I (1909) Formarea deltei Dunării. Bul Soc Rom Regale de Geografie XXX:35–48

Maselli V, Tricardi F (2013) Man-made deltas. Sci Rep 3(19):26. doi:10.1038/srep01926

Panin N (1974) Evolutia Deltei Dunãrii în timpul Holocenului. Studii tehnice şi economice – Institutul Geologic 5:107–121

Panin N (1983) Black Sea coast line changes in the last 10.000 years a new attempt after ancients. Dacia XXVII:25–36

Panin N (1999) Global changes, sea level rise and the Danube Delta: risks and responses. Geo-Eco-Marina 4:19–29. Bucharest, Romania

Panin N (2003) The Danube Delta geomorphology and Holocene evolution: a synthesis. Géomorphologie: Relief, Processus, Environnement 4:247–262

Panin N, Jipa D (2002) Danube River sediment input and its interaction with the north-western Black Sea. Estuar Coast Shelf Sci 54:551–562

Preoteasa L, Roberts HM, Duller GAT, Vespremeanu-Stroe A (2009) Late-Holocene coastal dune system evolution in the Danube Delta, NW Black Sea Basin. J Coast Res 56:347–351

Preoteasa L, Vespremeanu-Stroe A, Hanganu D, Katona O, Timar-Gabor A (2013) In: Conley DC, Masselink G, Russell PE, O'Hare TJ (eds) Coastal changes from open coast to present lagoon system in Histria region (Danube delta). Proceedings 12th international coastal symposium, Plymouth, England. J Coast Res (Spec Issue 65):564–569

Preoteasa L, Vespremeanu-Stroe A, Tătui F, Zăinescu F, Gabor-Timar A, Cârdan I, (submitted to *Geomorphology*) The evolution of the Sf. Gheorghe (Danube) asymmetric deltaic lobe in association with the cyclic development of a river-mouth bar and present adaptations to human-induced sediment depletion

Vespremeanu E (1983) Geomorphological evolution of Sfântu Gheorghe arm mouth (Danube delta, North-West of the Black Sea) in the last 200 years. Revue Roumaine de Geologie, Geographie et Geophysique 27:61–68

Vespremeanu E, Vespremeanu-Stroe A, Constantinescu SŞ (2004) The Black Sea level oscillations in the last 150 years. Analele Universităţii Bucureşti – seria Geografie LIII:69–76

Vespremeanu-Stroe A (2004) Transportul de sedimente în lungul ţărmului şi regimul valurilor pe coasta Deltei Dunării. Studii şi Cercetări de Oceanografie Costieră 1:67–82

Vespremeanu-Stroe A (2007) Ţărmul Deltei Dunării, Edit. Universitară, Bucureşti, 210 p

Vespremeanu-Stroe A (2013) The evolution of the Danube river mouths. The National Symposium of Geomorphology, Suceava, June 2013 (oral communication)

Vespremeanu-Stroe A, Preoteasa L, Hanganu D, Brown T, Bîrzescu I, Toms P, Timar-Gabor A (2013) The impact of the Late Holocene coastal changes on the rise and decay of the ancient city of Histria (Southern Danube delta). Quat Int 293:245–256

Zenkovich VP (1956) Zagadka Dunaiskoi Delty. Priroda 45:86–90

Index

G. Randazzo et al. (eds.), *Sand and Gravel Spits*, Coastal Research Library 12,
DOI 10.1007/978-3-319-13716-2